McGRAW-HILL TELEVISION,
RADIO, AND AUDIO SERVICING COURSE

(Books are listed in the recommended order of study.)

MARKUS · Television and Radio Repairing
MARCUS AND LEVY · Practical Radio Servicing
MARCUS AND LEVY · Profitable Radio Troubleshooting
ANTHONY · Profitable Television Troubleshooting
ECKLUND · Repairing Home Audio Systems
MARKUS · How to Make More Money in Your TV Servicing Business

BOOKS BY JOHN MARKUS

Electronics and Nucleonics Dictionary *(with Nelson M. Cooke)*
Electronics for Engineers *(with Vin Zeluff)*
Electronics Manual for Radio Engineers *(with Vin Zeluff)*
Electronics for Communication Engineers *(with Vin Zeluff)*
Handbook of Industrial Electronic Circuits *(with Vin Zeluff)*
Handbook of Industrial Electronic Control Circuits *(with Vin Zeluff)*
Handbook of Electronic Control Circuits
How to Get Ahead in the Television and Radio Servicing Business
How to Make More Money in Your Television Servicing Business
Television and Radio Repairing

TELEVISION AND RADIO REPAIRING

Television and Radio
REPAIRING

JOHN MARKUS
Technical Information Research Staff, McGraw-Hill Publishing Co.
Senior Member, Institute of Radio Engineers
Formerly Feature Editor, Electronics

SECOND EDITION

McGRAW-HILL BOOK COMPANY, INC.
New York   Toronto   London   1961

TELEVISION AND RADIO REPAIRING

Copyright © 1961 by the McGraw-Hill Book Company, **Inc.**
All Rights Reserved.

Copyright, 1953, by the McGraw-Hill Book Company, Inc.
All rights reserved. Printed in the United States of America.
This book, or parts thereof, may not be
reproduced in any form without permission of the publishers.
Library of Congress Catalog Card Number: 60–53221

VI

40453

Preface

By telling how to test, repair, and replace television and radio parts, this book trains a beginner to handle over 75 per cent of the repair jobs that come to the average television and radio service shop. Such rapid progress toward a professional career is possible in just one book only because the emphasis is on the simple, practical procedures that are used most often on repair jobs. Learning the easiest things first gives a feeling of confidence and a sense of real accomplishment.

This book starts from scratch. It assumes the reader has had no previous experience in television or in radio. It only assumes that he knows how to read, has average intelligence, and can follow simple, step-by-step instructions. Each tool and part is introduced as if seen for the first time. The discussion of each type of receiver begins with a simple get-acquainted description. The goal in writing has been to hold interest by giving how-to-do information that can be applied to actual receivers right from the start.

Words are short, for easy reading. Sentences are short, for easy understanding. Paragraphs are short, so important ideas are clear. Bold paragraph headings are used frequently to tell what is coming next and to help find a desired topic.

Chapter organization is arranged for easy and logical steps in learning. The first four chapters tell how to get started in servicing, what tools to buy, how to make a service call, how to fix simple troubles without removing the chassis, and how to remove the chassis of a television or radio receiver. The next two chapters tell how to test parts with a multimeter. Four following chapters cover the testing of all types of tubes, including picture tubes.

Each technique and new idea is presented when needed, not before. The reader always sees the practical use for what he is learning. The two chapters on soldering thus come after the tube-testing chapters, since tubes are checked before soldered-in parts are inspected. The reader learns how to use soldering guns on ordinary circuits, then receives instructions for using pencil-type soldering irons on printed circuits.

Ten chapters tell how to test and replace parts that are normally soldered into receiver circuits, including transistors, crystal diodes, resistors, capacitors, controls, switches, coils, transformers, and loudspeakers. Other highly practical chapters cover such important new topics as clock-radio timers, remote controls, 12-volt hybrid auto radios, and cabinet repairs. The final chapter gives step-by-step instructions for installing radio and television antennas.

It is possible to start fixing sets on a business basis after studying only 10 chapters of this book. About three out of four sets go bad because of tubes, and tube troubles have been covered by the end of the tenth chapter. For sets having troubles other than tubes, the beginner is told how to farm out the work to shops that specialize in fixing sets for servicemen or dealers on a wholesale basis. On these sets, the beginner gains experience and can usually break even profitwise. Each succeeding chapter after the tenth reduces the percentage of sets on which help is needed, thereby boosting earning-while-learning profits.

By the time the last chapter has been mastered, the percentage of sets requiring outside help has been cut down to well under 25 per cent. These remaining sets are the ones that require a knowledge of circuit operation and technical troubleshooting. To learn to repair them, the reader can continue his studies in the advanced books of the McGraw-Hill TV, Radio and Changer Servicing Course.

Appreciation is expressed here to the many firms who cooperated in furnishing practical information and illustrations pertaining to their products. Specific credits are given under the illustrations.

John Markus

Contents

Preface	v
1. Getting Started in Servicing	1
2. Tools Needed for Servicing	23
3. How Television and Radio Sets Work	43
4. Making a Service Call	74
5. Getting Acquainted with Electricity and Magnetism	106
6. How to Use a Multimeter	122
7. Removing and Replacing Receiving Tubes	143
8. Removing and Replacing Picture Tubes	170
9. Testing Tubes without a Tube Tester	205
10. Using a Tube Tester	230
11. How to Solder	248
12. Soldering and Repairing Printed Circuits	262
13. Testing and Replacing Transistors and Crystal Diodes	284
14. Power-supply Troubles in Home, Portable, and Auto Sets	308
15. Testing and Replacing Carbon Resistors	333
16. Testing and Replacing Wirewound Resistors	358
17. Testing and Replacing Controls, Switches, and Clock-radio Timers	377
18. Testing and Replacing Capacitors	396
19. Testing, Repairing, and Replacing Coils and Transformers	437
20. Adjusting and Repairing Tuning Devices and Remote Controls	469
21. Repairing and Replacing Loudspeakers	498
22. Repairing Cabinets	511
23. Installing and Repairing Television and Radio Antennas	521
Index	559

1

Getting Started in Servicing

What Servicing Involves. The main job of a television and radio serviceman is to fix television sets, radio sets, and home audio systems. In general, this involves finding out which tube, transistor, or other part is bad by testing as in Fig. 1, then putting in a new part. The work is clean and pleasant,

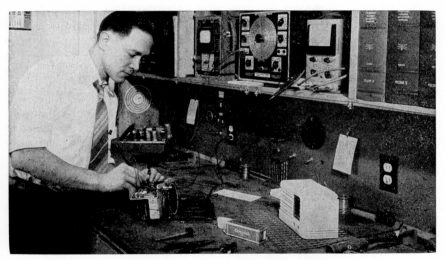

Fig. 1. Fixing a table-model radio. A few simple measurements with a multimeter locate the bad part, after which replacement with a high-quality new part is easy. (Centralab photo)

whether done as a regular full-time occupation or in spare time to make extra money.

Servicing is an ideal combination of stimulating brainwork and easy mechanical work with your hands. As you acquire experience, the brainwork becomes almost instinctive and automatic with you. For example, you hear a certain weird sound from a misbehaving radio or see a pattern of zigzag lines messing up a television picture, and your mind immediately associates it with a particular part in a particular section of the set.

Being Your Own Boss. As an expert serviceman, you will have an interesting and varied life. In a typical day, you may repair half a dozen sets, ranging in size from a tiny battery-operated portable radio to a large floor-model radio-television combination. You can alternate as you like between relaxing on a comfortable stool at your well-lighted workbench or making outside calls and deliveries. You can stand or sit at your bench, as you prefer; you can smoke at your work or not, as you prefer; you can listen to your favorite radio programs as you work, or you can enjoy silence.

On rainy days you can concentrate on benchwork, while on nice days you can be outside more—all because you are your own boss. If you feel like working in the evening instead of the morning, you can sleep till noon. There is no time clock to punch, and yet you will find yourself putting in extra hours just because you like to.

As your own boss, you can take your vacation when you feel like it. You can take a day off when and if you like. In general, however, you will find the work so interesting that your total number of hours on the job each month will be more than that of a salaried worker. Your income will then be correspondingly higher than if you were fixing sets for someone else on a salary basis. Yes, in television and radio servicing you can write your own figures on your weekly pay check.

For many men, radio is a hobby. If it is your hobby too, you can actually make a living from your hobby. Happy indeed is the man who can do this.

Earning While Learning. There is practically no other career in which you can make extra money long before you have completed training. By learning to repair and replace television and radio parts in the practical way presented in this book, you can do repair work even while you are studying.

You will be pleasantly surprised to find that television and radio servicing is not mysterious or difficult. You do not need any special talent or education to master it. You do not even need to be handy with tools; you can easily learn how to use the few simple tools required for fixing sets.

You do not have to spend a year or more struggling with dull theoretical principles; you learn right from the start how to fix troubles, starting with the simplest and easiest ones. Thus, testing and replacing tubes as in Fig. 2 is a money-making job you can do after studying just the first few chapters of this book.

Getting Started in Servicing 3

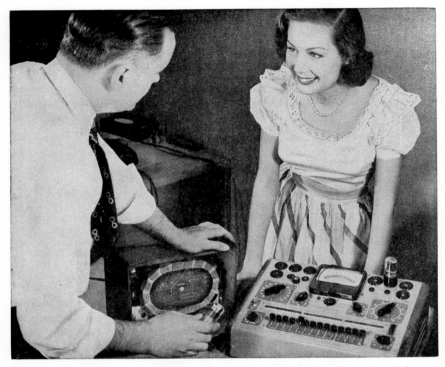

Fig. 2. Home calls provide an interesting change in the daily routine of servicing. Tubes are tested right in the home because they are the commonest cause of trouble

Cash Needed to Start. Another important advantage of servicing as a career is the fact that you can start with a cash investment of only a few hundred dollars if necessary. The income from your first repair jobs can be used to build up your stock of spare parts and get all the tools and equipment you will eventually want. Only a few tools are essential right from the start. The rest can wait until you have earned the money to pay for them.

Getting Started at Home. You do not need a shop or store in order to operate a successful television and radio servicing business. In most localities you can operate right from your own home at the start, either in spare time or on a full-time basis. Only a simple workbench like that in Fig. 3 is needed, in the basement or anywhere else that is convenient. In this way you keep overhead expenses down while building up business for your own service shop.

You can start as a spare-time serviceman, doing the work evenings and week ends while holding a regular job that provides living expenses. The more profits you put back into your business, the faster it will expand to the point where you can safely give up your regular job.

Later, when you have become established with your own shop or store, there can be additional profits beyond those you get for fixing sets. You

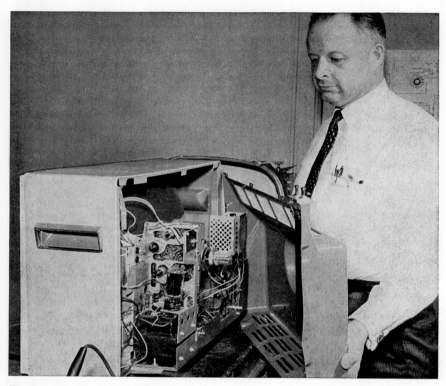

Fig. 3. A simple wood bench will get you started in profitable television servicing. The bench is the optimum height of 36 inches for working on television and radio sets. This portable television set is easy to fix because all tubes are easily reached when the back cover is taken off. (General Electric photo)

can buy radio and television receivers and accessories at wholesale discounts, for selling to your customers at a profit. You can also make arrangements with stores to install and service the sets they sell.

What Television and Radio Servicing Pays. The amount of money that you earn in servicing is entirely up to you. Your charges will vary with the rise and fall in the value of the dollar. For this reason, definite weekly or monthly salary figures for an average servicing business cannot be given. The ability to change your income in this way to keep in step with inflation is one important advantage of television and radio servicing as a career.

If your expenses for food, rent, clothing, and other cost-of-living factors go up 10 per cent during a half-year period, you simply boost your own schedule of charges the same amount. This is entirely fair and honest, because it was raises in wages of other people that made the cost of living go up. Products increase in price when the labor of making them costs more, because everything you buy involves labor. Nothing comes from the earth without work—not even diamonds.

Insurance for the Future. Once you learn how to fix television and radio sets, you have something to fall back on if your regular job is lost for any reason. This extra training makes you independent of the future and gives you peace of mind.

Depressions and loss of your regular job need not worry you, because there will always be television and radio sets to fix. There will always be a need for good men to do television repair work. Many men learn television and radio servicing primarily for protection against loss of their regular job. These men fix sets as a spare-time business while they have a good regular job. The extra income is, of course, welcome, and the work is fascinating and enjoyable relaxation.

Television, radio, and home audio are today still fast-growing. In television, particularly, the future will open still more opportunities for you to advance and gain security in the years to come, no matter what your present age or occupation may be. Furthermore, television and radio knowledge is the groundwork for the field of electronics. In the years to come, the electronic industry will have an even greater variety of opportunities for men who can fix vacuum-tube and transistor circuits.

Working for Others. Many men prefer to work regular hours for a fixed salary and have regular vacations, with no business responsibilities. Opportunities for getting the good-paying jobs broaden tremendously for those who know how to fix television and radio sets.

First, there are jobs in the larger television and radio shops where the owner cannot handle all the business or where the owner prefers to sell sets and therefore has to hire men for the servicing part of his business. A few months of work for a highly successful serviceman is a good idea if you are a bit timid about getting started in business. Choose your employer carefully so you will not be picking up the bad habits of a careless or unbusinesslike man. Many of the servicemen who are successful today started by working for others.

There are many factory jobs where television and radio training pays off. It can help the technician in the receiver-design section of a factory, who tries all kinds of circuits and parts in a new receiver design until he finds the combination that works the best. It can help the mechanic who assembles huge transmitters. It can help the technician who supervises work on assembly lines. It can help the inspector who checks finished receivers. It can help testers who make the final adjustments and check performance of sets before they are shipped out. Television and radio training also prepares you for highly responsible police radiotelephone and aircraft radio maintenance jobs.

In World War II, advanced ratings were given to many men who had previous radio training. These men played a major role in keeping military radio and radar equipment operating in the early war years before schools could be set up for training purposes. Radio, radar, and still-newer electronic equipment for guided missiles will require still more trained men for maintenance almost immediately in any future military emergency. There just will not be time to train men for this purpose.

Getting Started Right. The first thing you learn in this chapter is where and how to get replacement parts, no matter where you live. You learn how to establish yourself as a businessman so you will get discounts from wholesale distributors. Practical suggestions guide you in building up a stock of spare parts with a minimum investment of cash. You also learn about circuit diagrams, service data, and reference manuals that can speed up your servicing work. With all this practical business information on hand, you will get started right without wasting time or money, whenever you decide to earn money in servicing.

Fixing sets efficiently means getting the required new tubes and parts quickly, once you have found the trouble. Repairs are fastest when the needed parts are right at hand on your shelves. It is impractical to keep on hand every part that you might possibly need, however. The cost of doing this would be tremendous. There are now way over ten thousand different makes and models of sets in use. Some parts fit in only one or two of these sets. Such parts could gather dust for years or even forever, while you wait for those particular sets to come in.

Building Up a Stock of Parts. In the beginning, you can fix sets without a stock of parts, by making more frequent trips or orders for needed parts. Oftentimes you will not be able to get all the needed parts the first time, though. A burned-out tube may hide other defects. These show up when you put in the new tube. You then have to hold up the repair job until you can get more new parts. This means that you can fix sets faster if you have a good stock of replacement parts on hand.

The better solution is to start with a minimum stock of the tubes and parts that are most often needed. Get enough to fix the majority of sets. Make more frequent trips to your jobber or place more frequent orders by mail at the beginning. You will soon learn which tubes and parts you need most. Gradually, you can build up your stock and cut down the number of weekly orders for parts.

Stay away from bargain sales until your business is well started. Many bargains turn out to be worthless, so wait until you can afford to gamble with your money. Inferior, cheap parts are expensive in the long run, because they go bad in sets you have repaired and guaranteed. Then you have to fix these sets for free and put in the good-quality parts anyway.

Where to Buy Parts. First of all, get acquainted with all the places that sell replacement parts in your locality. This is an important part of getting started in television and radio servicing.

Two distinct types of firms should be considered: (1) parts jobbers, also called wholesalers and distributors; these have stores at which you can buy just about everything you will need; (2) mail-order houses, which correspond to jobbers but do business chiefly by mail from catalogs.

Both types of suppliers charge essentially the same wholesale prices to servicemen. On the average, they provide about the same services. Your choice of a supplier will therefore depend on your own locality and the particular firms serving it. To help you make this choice, additional information will now be given about each type of supplier in turn.

Buying from Parts Jobbers. For most servicemen the chief source of supply for radio tubes, parts, hardware, test equipment, and manuals is a parts jobber. Jobbers sell parts at wholesale prices that permit a fair and legitimate profit to you. The usual discount to servicemen will be 40 per cent off list price. On tubes this discount is most important because small sets like that in Fig. 4 often need only new tubes.

A business card is needed the first few times you buy parts from a jobber, to identify yourself as a serviceman who is entitled to discounts. Suggestions for wording on business cards are given later in this chapter.

Television and radio parts jobbers are located only in the larger towns. They are generally in out-of-the-way locations, so you may never have seen or noticed one of their stores. To find them, look in your classified

8 *Television and Radio Repairing*

Fig. 4. Putting an entire new set of tubes in a table-model radio will make it work again about half the time. The old tubes can then be put back one by one, to find the tube that makes the set go bad again. Most tube replacements can be made without even taking the chassis out of its cabinet

telephone directory under headings like "Radio Supplies and Parts—Wholesale and Manufacturers" or "Television Supplies and Parts—Wholesale and Manufacturers." Some of the firms may serve only manufacturers, but they will gladly refer you to others that want you as a customer.

Another way to locate jobbers is to get acquainted with a serviceman who has been in business for several years in the locality, and find out where he gets his parts.

Charge Accounts. Jobbers often permit charge accounts, payable once a month. An extra discount, usually 2 per cent, is generally allowed for cash payment and for bills paid within 10 days. A charge account has many advantages if used as a convenience. It establishes you as a professional businessman, builds up your credit rating, and lets the jobber know that you are one of his good customers. A charge account is bad, however, for those who yield to temptation and buy more than they can pay for at the end of the month. Overdue charge-account bills hurt your credit rating and thus spoil chances for borrowing money to expand your business in the future.

Ordering by Telephone. Many jobbers provide daily delivery service on phoned-in orders from shops in their own city. While you work during the day, you can build up your list of wants and phone it in, as in Fig. 5, thus saving the time required for a trip to the jobber.

Getting Started in Servicing

In a few localities, aggressive jobbers make the round of service shops with trucks containing stocks of parts or come around daily taking orders for delivery the next day.

Some jobbers also provide semiweekly or weekly delivery service to shops in suburban communities and nearby towns. Even once-a-week deliveries can be adequate if you keep a good supply of tubes and parts on hand. The long-distance toll charge for placing an order in time for one of these deliveries is spread over many jobs and hence amounts to only a few pennies per job. This is usually less than it would cost you in time and mileage to drive to a jobber in the heavy traffic of a city. Thus, a small-town servicing business can operate just as efficiently as one in a big city, once it is on a full-time basis so that orders for parts are large enough to rate delivery service.

In many Western states, where towns are too far apart even for delivery service, jobbers use express or parcel post to provide practically overnight service on phoned-in orders. Some jobbers even allow you to reverse the charges on long-distance calls when ordering parts.

Any parts supply store may have rush hours. Business gets so heavy at these times that you always have to wait. Clerks will gladly tell you when their rush hours come, so you can schedule your phone calls and visits earlier or later to save time.

Fig. 5. Save time by ordering needed replacement parts by telephone whenever possible, for delivery by truck, express, or parcel post

Buying from Receiver Distributors. In the larger towns are firms known as receiver distributors. Their main business is selling television and radio sets to stores in wholesale quantities. Some of these distributors also carry spare parts for the various makes of sets handled, along with a more or less complete stock of standard replacement parts and some test equipment.

You will have to go to the various receiver distributors in your locality to get special parts for the sets they handle. Therefore, learn who these distributors are and where they are located, so you can get special parts quickly when needed. Your jobber can usually give you their addresses.

You will find that stocks of general replacement parts are smaller at receiver distributors than at jobbers, and perhaps a bit higher in price. Also, some distributors may give poorer counter service than jobbers.

On most parts, distributors give servicemen the standard 40 per cent discount. A few items, such as replacement cabinets or special television parts, may have smaller discounts or net prices, however.

Manufacturers of sets rarely sell parts directly to servicemen. Writing to manufacturers for special parts is usually a waste of time, except for some of the smaller manufacturers who do not have distributor setups for handling parts.

Ordering Special Parts. Examples of parts you would get from receiver distributors are new printed tuning dials, new cabinets, and special transformers. Standard parts, such as resistors and capacitors, are easier to get from parts jobbers.

When ordering a part from a receiver distributor by phone, letter, or personal visit, be sure to give the number of the part along with the make, model number, and serial number of the set in which it is used. The part number is usually stamped on the part. If not, look it up in the service data for that set. If prices are given in the service manuals of receiver manufacturers, they are usually list prices from which servicemen deduct 40 per cent to determine their cost price.

The model number of the set is either on the back of the chassis or on a label glued inside the cabinet. The serial number is usually on the back of the chassis. If in doubt, give all numbers and letters found in these locations, and include a list of the tubes used in the set.

Manufacturers and their distributors rarely keep special parts in stock more than 5 years for lower-priced table-model sets. On larger and more expensive sets they may extend the time to 10 years. Therefore, on sets over 10 years old you will rarely be able to get special parts. Here the best thing to do is advise the customer to trade in his set on a new model.

Getting Started in Servicing 11

Buying from Mail-order Houses. Many servicemen in smaller towns and remote localities get all their needed replacement parts and supplies from mail-order television and radio supply firms. These firms put out complete catalogs describing the parts that they handle. The catalogs are sent free on request, and are well worth having on hand for reference. One use of such catalogs is for finding your cost prices on parts when you have to give an estimate on the cost of a repair job.

Quite often, mail-order firms buy surplus parts from receiver manufacturers and sell these at even lower than normal wholesale prices. Once you become acquainted with reliable well-advertised brand names, you can safely take advantage of an occasional money-saving special in the mail-order catalogs.

Prices given in mail-order catalogs are net prices. Postage or express charges are extra for most items.

Value of a Checking Account. Payment by check is accepted practice today in business. A check tends to indicate that you are established in business and entitled to wholesale discounts. A money order or cash enclosed with an order for parts may mean that you will be considered as a beginner or amateur by some firms, and not entitled to discounts. A check has the added advantage of serving as proof of business expense for income-tax purposes.

You may have to pay service charges on your checking account at first, because many banks require quite high minimum balances for free checks. Consider these charges as essential business expenses. They will generally be lower than money-order fees.

A personal checking account will do at the start, if you clearly identify business checks and business deposits on the record sheets or stubs. You will have to keep separate balances, one for business cash on hand and the other for personal cash on hand. This is admittedly complicated bookkeeping, so get a separate checking account for your business as soon as you can.

What to Charge for Parts. On tubes and parts, a serviceman always charges list price or higher to his customers to cover the time and expense involved in ordering or getting these parts and to cover a fair part of the overhead expenses associated with any business. This means that a tube which you buy at a net price of 60 cents would be sold at its list price of $1 or more. When you know the established list price of a part, always charge that price.

When you know only your cost price, multiply it by 1.7 to get a list

12   Television and Radio Repairing

price that gives you just about a 40 per cent profit based on list price. Thus, if you pay $4 for a part, multiply $4 by 1.7 to get $6.80 as the list price that you charge the customer. You then make $2.80. Table 1 gives other examples, for convenience in making up your bills.

Table 1. Recommended Charges for Parts When Only Cost Is Known

Your Cost	Your Charge	Your Cost	Your Charge	Your Cost	Your Charge	Your Cost	Your Charge
0.10	0.25	1.40	2.40	2.70	4.60	4.00	6.80
0.20	0.50	1.50	2.55	2.80	4.80	4.10	7.00
0.30	0.65	1.60	2.75	2.90	4.95	4.20	7.15
0.40	0.80	1.70	2.90	3.00	5.10	4.30	7.35
0.50	1.00	1.80	3.05	3.10	5.30	4.40	7.50
0.60	1.20	1.90	3.25	3.20	5.45	4.50	7.65
0.70	1.35	2.00	3.40	3.30	5.65	4.60	7.85
0.80	1.50	2.10	3.60	3.40	5.80	4.70	8.00
0.90	1.60	2.20	3.75	3.50	5.95	4.80	8.20
1.00	1.70	2.30	3.95	3.60	6.15	4.90	8.35
1.10	1.90	2.40	4.10	3.70	6.30	5.00	8.50
1.20	2.05	2.50	4.25	3.80	6.50		
1.30	2.25	2.60	4.45	3.90	6.65		

For costs between $5 and $50, multiply cost and charge values in Table 1 by 10. On more expensive parts such as picture tubes you may be satisfied with less than 40 per cent profit, particularly when doing a job for a retired person trying to live on a too-small pension.

On small parts costing way less than a dollar, a profit of 40 per cent is not enough to cover the time you spend ordering and getting the part. Here you should charge at least twice your cost just to break even. Thus, a pilot lamp costing 6½ cents is normally listed at 15 cents on customer bills. A small part costing, say, 11 cents, would be billed at 25 or 30 cents. This is standard business practice with successful servicemen.

Most-needed Replacement Parts. To fix television and radio sets, the parts you need most of all are tubes. Over a thousand different types of tubes are in use today, costing an average of over $1.50 apiece even at wholesale. It is thus pretty important to build up your tube stock carefully. Your parts jobber can tell you which tubes are replaced most often in your locality.

Suggestions for building up stocks of other replacement parts are given in the chapters dealing with these parts.

Keeping Your Stock Up to Date. The value of a stock of parts drops rapidly unless the stock is kept up to date. You need a system for reminding you to order a new part each time you take one from stock. An easy and foolproof reminder is a pad of scratch paper mounted on a bench or shelf with wood screws so it cannot get lost. Jot down the type number or value and rating of each part as you take it from stock. Get into the habit of doing this right from the start. Each time you go to the parts jobber or order parts by phone or mail, tear off the top sheet and use it as the basis for your order list. Use judgment, however; do not reorder older tubes or parts that you may never need again.

New tubes and some replacement parts come in cartons having labeled covers. When you use one of these parts, tear off the top of the carton and save it in a box near the pad as a reminder for reordering.

Small parts are generally much cheaper when bought in quantities. These small parts do not need to be listed on your pad when used, because their cartons will always show when your stock is getting low. Keep your stocks of new parts stored safely and neatly on shelves or in drawers.

Getting Receiver Service Data. It is always easier to fix a set if you have the service data for it. Such data include the circuit diagram and other helpful information. Some of the larger radio and television sets are almost impossible to fix unless you have the service data for them.

Service data for a particular set can usually be obtained from the manufacturer, but writing for it and waiting for the reply takes time. Collections of circuit diagrams and service data that are published regularly in loose-leaf form for servicemen are much more convenient. The different sources for service data will now be taken up individually, to help you decide which to get.

Manufacturers' Service Data. When you have a receiver that is not much over five years old and when the manufacturer is still in business, the chances are pretty good that you can obtain the service data for the set by writing for it. If you do not have printed business stationery yet, clip your business card to the letter. This helps to get a quick reply. Some manufacturers make a small charge for a service manual, so they may send you a bill with the manual.

It is always better to send neatly typewritten letters than handwritten letters. Be sure your complete name and address are clearly typed under your signature, or at least printed clearly by hand with pen and ink. Use good white bond paper if you do not yet have letterheads. Never use a pencil for business correspondence. Never use postcards for business correspondence.

When writing, be sure to give the make, model number, and serial number of the set. As a double check, include the list of tubes in it. Your letter can be short and simple, as follows:

Put date here

*Put name and address
of manufacturer here*

Attention: Service Department

Gentlemen:

Please send me service data on your _____, model _____, serial number _____, which has the following tubes: _____

Thank you for your cooperation.

Very truly yours,

Put your name here
Put your business name here
Put your business address here

In the space after the model number, describe the set briefly, as auto radio, portable television set, audio amplifier, table radio, clock radio, record changer, tape recorder, etc. The list of tubes can be omitted when the set has more than about six tubes. There are fewer models of large sets and correspondingly less chance of sending you the wrong diagram.

On many sets you will find only a trade name, with no clues to the name and address of the manufacturer. Others may have the manufacturer's name but no address. Here your jobber may be able to give you the manufacturer's name, or you can find it in indexes for published volumes or sets of service diagrams. In the receiver manufacturing business, some firms go out of business each year, and new ones take their places. The best place to get circuit data on these orphan sets is from publishers of receiver diagrams.

Although you may prefer to write for individual diagrams at the start as needed, you cannot operate a profitable business in this way. Writing letters is a waste of valuable working time. You have to wait days or even weeks for the reply, and even then you may not get the needed data.

Getting Started in Servicing **15**

Buying Service Manuals. Much faster service on circuit diagrams can be obtained by buying sets of service manuals from your jobber or from a mail-order firm. These are available in the form of Sams Photofact Folder sets.

Service data can be obtained in this way for practically any television or radio receiver, amplifier, record changer, or tape recorder built after the end of World War II in 1946, even though the manufacturer of the set went out of business long ago.

In television servicing, manuals are absolutely essential. Without the circuit diagram for a set right at hand, hours and even days can be wasted tracing out circuits on the chassis while troubleshooting.

Since receivers can last 10 years or more, you will eventually want to have a fairly complete collection of service manuals going back at least 5 years and preferably a full 10 years. This you can build up gradually from the profits of your business. The cost of manuals and reference books is a legitimate part of your overhead expense. Your fees for service must therefore be high enough to cover this along with your charges for labor and parts.

Photofact Folders. Each Photofact Folder set contains complete service information on a number of sets, specially prepared in a uniform style for maximum value to servicemen. New sets of folders are issued several times a month to cover new equipment as it comes on the market. By ordering a Photofact Folder set each time you need a diagram that is not in the folders already on hand, you can build up a library of service data on the installment basis without putting out a large sum of money at any one time. An index is available that quickly tells you which folder contains the desired diagram.

Photofact Folders can be kept in loose-leaf binders, each holding about 10 sets of folders, or in file cabinet drawers each holding 60 sets, as in Fig. 6. You can obtain the binders separately and put individual folder sets in them as you get them. The folders in file cabinets may be purchased on a time-payment plan.

Your jobber may also have specialized service manuals on such subjects as record changers, auto radios, tape recorders, and audio amplifiers. You will not ordinarily need these, however, if you plan to acquire a complete set of Photofact Folders.

Rider's Manuals. Although publication of the Rider's manuals shown in Fig. 7 has been discontinued, you will see these older manuals in some service shops. They present servicing information exactly as it was released

Fig. 6. Examples of Photofact Folders stored in a file cabinet alongside a bench. These give the essential data needed for fixing sets fast. (Howard W. Sams photo)

Fig. 7. Using a Rider's Manual for reference while hunting for trouble in the chassis of a console radio. Other manuals are within easy reach on the shelf. Rider's Manuals were discontinued in 1961. (John F. Rider photo)

by the manufacturer, without redrawing of the diagrams. A separate cumulative index was available, for determining which manual contains a particular make and model of television or radio receiver. The quality of the information in Rider's manuals varied greatly, because some manufacturers prepared much better service diagrams and instructions than others. If you can get used Rider's manuals for a few dollars each, they will give you coverage of older sets at minimum cost.

Manufacturers' Volumes of Service Data. Some manufacturers of television and radio sets publish their own volumes of service manuals, usually as yearly editions. These often contain more detailed information than is given in Sams or Rider's manuals. Manuals covering only one make of set are ideal for those who specialize in that make of set.

A few manufacturers offer their service manuals on a yearly subscription basis. By subscribing in advance, you receive each manual automatically by mail as soon as it is printed. Manufacturers' distributors can give information on this service. A typical subscription charge for television manuals is $5 a year.

Tube Manuals and Charts. The "RCA Receiving Tube Manual" gives the base-connection diagram for each type of receiving tube made, along with useful technical data. Other radio-tube manufacturers also have tube manuals, as well as charts and other useful reference books. These are generally available from jobbers at nominal prices. Always glance over the booklet counter to see what is new, when you go to your jobber for replacement parts.

Catalogs. Start now to build up a collection of catalogs of parts manufacturers and supply firms. Each catalog covers different lines of parts. You may be able to find in one catalog a badly needed part that is not listed in the others. Put up a shelf in your shop to hold these catalogs, your study and reference books, and the service manuals you will be getting.

Catalogs can be obtained by writing directly to the manufacturers. Parts jobbers usually have racks filled with catalogs and literature from which you can help yourself. A charge is made for some of the larger booklets, but they are usually worth many times this charge.

Free Publications. Some manufacturers of replacement parts send out little magazines to servicemen regularly each month or every other month, free of charge. Some of these so-called house organs are also available at jobbers. You will often find in them helpful articles covering latest developments in television and radio.

Your jobber can tell you which manufacturers publish such house organs. Send a short letter to each manufacturer, identifying yourself as a serviceman and asking to be placed on their mailing list to receive their publication regularly.

Tube manufacturers often offer service aids of various kinds free with purchase of quantities of tubes.

Magazines for Servicemen. To keep abreast of new developments, subscribe to at least one of the magazines published for servicemen, or buy individual copies regularly from your jobber. Eventually you may want to get all of them.

Getting Acquainted with Fellow Servicemen. One other way of learning practical information fast is by talking to other servicemen. Better yet, join the nearest organization of servicemen and attend their meetings regularly. You will find that, even though they may be competitors, they are as a rule friendly and congenial.

You cannot expect to get much free advice from servicemen at their shops. At meetings and gatherings the men talk more, and are usually eager to answer questions or to discuss their problems and successes. You can learn a lot in this way about local conditions affecting servicemen, including the names of customers who are unreasonable or who seldom pay their bills.

Show willingness to serve on committees and to do other work for the servicemen's organization. Ask for jobs like sending out notices of meetings or handling correspondence. Do a good job and you will be in with the gang in no time. Your own self-confidence will be bolstered tremendously once you are an officer of a servicemen's organization.

Fixing Tough Receivers. If you fix sets while still studying, there will be quite a few in the beginning that are too tough to handle. You are not expected to fix every set right from the start. If you realize this, you will not be embarrassed and will not waste a lot of valuable time on the tough sets before you are ready for them.

The sensible way is to try out only the test procedures that you have already learned, and then allow yourself not over three guesses as to suspicious parts. After this, turn the set over to a dealer or experienced serviceman with whom you have previously made arrangements for handling tough sets during your training period.

Dealer Service. Some service shops in the larger towns specialize in doing repairs on a wholesale cash-and-carry basis, which is known as dealer service. You bring the set to the shop and call for it when repaired. The serviceman charges you only for parts and labor, since he does not have to waste time making calls and talking to customers. Your own bill to the customer is much larger, to cover the time spent in dealing with the customer, travel time, overhead, and possible new trouble during your 30- or 90-day guarantee period.

Many distributors for receiver manufacturers also have shops in which they fix tough sets for servicemen on this same cash-and-carry basis.

If you take advantage of dealer service, you can profitably start fixing sets right from the beginning while you are learning television and radio servicing. Make inquiries at the nearest distributor, to find out which shops will fix sets for you.

By using dealer service, you make real money right at the start on the sets that you can fix, and break even or show a small profit on the sets the dealer fixes. You learn a lot by trying to fix all the sets yourself first, since the dealer tells you what the trouble was. You preserve your reputation as a serviceman, since you do not tell your customer who actually fixed the set.

Never leave wires disconnected when taking sets to a dealer for repair. If you unsolder wires while making tests, put them all back exactly as they were, if you want to keep the charges down. Servicemen hate to get in a set that has been bungled by an amateur. It takes a lot of time to trace the wiring piece by piece and get it all back. They have to do all this before they can even start troubleshooting. If a set comes in to you with disconnected wires and you have to pass it on to a dealer, be sure to explain the situation to preserve your own reputation.

Use dealer service any time that you run into a tough set—even after you have finished your training and are well established in business. You will be money ahead if you do, because your time is valuable. It is far better to break even or take a loss on an occasional job than to waste hours of time for which you cannot fairly charge your customer.

Importance of a Business Card. The first step in establishing yourself as a professional serviceman is getting a business card. This you can hand to clerks at a jobber or distributor to get wholesale prices on parts.

Business cards can also be inexpensive advertising if widely used. Many stores in small towns display such cards under glass counters or on bulletin boards as a courtesy to you and as a service to their own customers. Libraries, post offices, and other places may have bulletin boards on which business cards are permitted. Be sure, however, to make the rounds regularly and put up clean cards. A dirty card is poor advertising.

Your biggest and most profitable use for business cards is as a reminder

20 Television and Radio Repairing

in every set that you repair. Knowing the card is there, the customer can easily look up your phone number on it when some other set in the house goes bad or when that same set needs attention again. Attach a card neatly in a permanent position inside the cabinet or on the back cover of the set with Scotch tape.

Design of a Good Business Card. Since so much depends on the impression made by your business card during the critical beginning months of your business, do not skimp on its cost. Stay away from cheap cards. Locate a good printer who will design the entire card, using the most effective styles and sizes of type for your particular purpose.

For television, radio, and home audio servicing, a business card should have the following information:

1. Your name.
2. The name of your business, if you decide to use one.
3. Your address.
4. Your telephone number.
5. A line telling what your business is.

Keep your card simple and dignified, as in the examples shown in Fig. 8. Stay away from anything unusual. Your card should express a feeling of

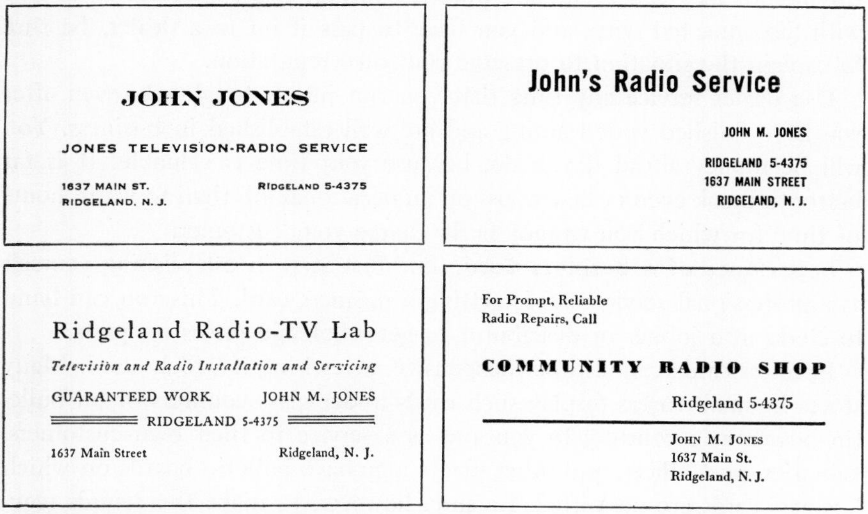

```
            JOHN JONES

     JONES TELEVISION-RADIO SERVICE

   1637 MAIN ST.         RIDGELAND 5-4375
   RIDGELAND, N. J.
```

```
         John's Radio Service

                              JOHN M. JONES

                              RIDGELAND 5-4375
                              1637 MAIN STREET
                              RIDGELAND, N. J.
```

```
      Ridgeland Radio-TV Lab
    Television and Radio Installation and Servicing
    GUARANTEED WORK      JOHN M. JONES
    ======= RIDGELAND 5-4375 =======
    1637 Main Street        Ridgeland, N. J.
```

```
   For Prompt, Reliable
   Radio Repairs, Call

    COMMUNITY RADIO SHOP

             Ridgeland 5-4375

                        JOHN M. JONES
                        1637 Main St.
                        Ridgeland, N. J.
```

Fig. 8. Ideas for your business card. A good printer can make up one of these with your own name and address, for use in establishing yourself as a professional serviceman so you can get wholesale prices on tubes and replacement parts

competence and reliability. Use high-quality plain white cards and a rich dark-colored ink. Dark green, dark maroon, dark blue, or dark brown are good colors. Black ink is also acceptable.

Many printers offer free use of illustrations for cards, but you will be better off without these. The illustrations are generally out of date anyway and do more harm than good.

Certain manufacturers of radio tubes offer to print cards for servicemen at very attractive prices. These cards always contain an emblem and sometimes other advertising of the manufacturer. You can save money at the start by using such cards.

Getting Business Know-how. When you decide to start earning money from servicing, spend a few weeks studying the business side of servicing. Get this business training and then come back to your study of practical servicing. You then will not ruin your future by making the business mistakes so many other beginners fall into. With a professional attitude right from the start, you will gain and hold the respect of customers and fellow servicemen alike. You will never have to live down such gossip as, "Oh, he'll fix your set for nothing—he's just learning."

Building Self-confidence. When you decide to fix receivers for a fee— now, next month, or next year—act as if you definitely know how. Act as if you cannot possibly fail to fix any set. Your customers will not know how much time you spend on their sets at the beginning, nor will they care. Likewise, your customers will not know or care whether you have the set fixed by a dealer. All this is behind the scenes and should be kept there for business reasons.

Remember also that a little knowledge can be dangerous. Do not be too positive in telling a customer what is wrong, because even the best servicemen make wrong estimates of troubles sometimes. Play safe—merely suggest possible troubles and say that further tests in your shop will be necessary to confirm your suspicions.

A set that cannot be fixed in 30 minutes or less in the home should be taken to your shop. Many servicemen do no more than check tubes in the home. If tubes are good, the set is taken to the shop for troubleshooting. It is extremely difficult to work on a chassis on the floor even if you have all the necessary spare parts, service data, and test equipment right alongside you in the home.

Automobile Requirements for Servicing. At the start, you can get along nicely with your own car when making home calls. The trunk compartment of the average car is large enough to hold almost any television

22 Television and Radio Repairing

Fig. 9. Some established service businesses prefer small foreign-made trucks like this, having low floors for easy loading and unloading of cartons and cabinets. Ladders for television-antenna work are easily mounted on the roof. (Sylvania photo)

chassis complete with picture tube. Your toolbox and other equipment can be placed in the trunk or on the floor inside the car.

Later, when your business is big enough to justify the cost, you may want to get a station wagon or small truck to be used only for business purposes. This can be painted professionally to advertise your business, as illustrated in Fig. 9.

2

Tools Needed for Servicing

Choosing Tools. Only a few tools are needed to get started in servicing. These are a soldering gun, long-nose pliers, diagonal cutters, combination pliers, three screwdrivers, a knife, a set of socket wrenches, a flashlight, and a toolbox, all shown in Fig. 1.

Suggestions for choosing and using each of these basic tools will be given first. Additional tools for speeding up repair work at the shop bench will then be described. Suggestions will also be given for building a good

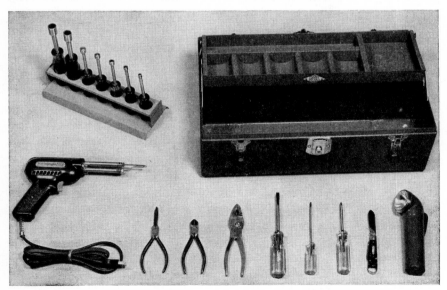

Fig. 1. Tools needed for getting started in servicing and making home calls. Top row—set of spin-type socket wrenches with stand; toolbox. Bottom row—Weller soldering gun; long-nose pliers; diagonal cutters; combination pliers; $\frac{1}{4}$- by 4-inch screwdriver; $\frac{1}{8}$- by 4-inch screwdriver; No. 2 Phillips screwdriver; two-blade pocketknife; flashlight

workbench. Separate chapters later tell you how to choose a multimeter and a tube tester to complete your basic equipment for servicing.

It pays in the long run to buy high-quality tools. They may cost more, but they will last a lot longer, do better work, and give you a feeling of pride.

Soldering Gun. Since practically all connections in television and radio sets are made with soldered joints, a soldering tool comes first on your tool list. For this, a soldering gun is strongly recommended in place of an ordinary soldering iron. A soldering gun is ready to use a few seconds after being turned on, whereas ordinary soldering irons take several minutes to heat up. A soldering gun has a long, thin tip that is ideal for working in a crowded television chassis. Ordinary soldering irons are large and clumsy in comparison and may burn or damage adjacent parts when used in crowded locations.

Additional information on the selection of a soldering tool is given in the chapter on soldering, along with complete instructions for making soldered joints.

Long-nose Pliers. Next to a soldering tool, you will need most often a pair of long-nose pliers. These are used for pulling on a wire when unsoldering a joint, for prying a loop of wire apart when unsoldering, for threading wire through a hole in a soldering lug, for bending hooks or loops in wires, for squeezing loops together to get tight joints, and for crushing insulation on wires so it can be removed easily.

Diagonal Cutters. For cutting hookup wire of all kinds in all locations, diagonal cutters are used. These are also known as side-cutting pliers. The most useful type has flat jaws between the handles for squeezing insulation on wires.

Diagonal cutters can also be used for snipping off loose ends of insulation and rubber on wires, for cutting dial cord, for cutting thin soft metal such as soldering lugs, and for cutting insulating tape.

Diagonal cutters for television and radio work range from 5 to 7 inches in length. Choose the size that fits most conveniently in your own hand.

Cheap diagonal cutters get dull quickly, and for this reason are a waste of money. Spend at least $2 right at the start to get a good pair.

Combination Pliers. These are the commonest pliers of all, so you probably have them already. They are used for holding, bending, and squeezing, for breaking off old connections and old mounting brackets, and for loosening and tightening nuts. The name comes from the two-position pivot that gives a combination of two different sizes of jaws. Slipping the pivot to the other hole permits opening the jaws wider for gripping larger objects.

If buying a new pair of combination pliers for service work, get one of the thin, lightweight types. Sometimes they are called slip-joint pliers or gas pliers.

When using combination pliers for loosening or tightening nuts, grip the pliers tightly, with the jaws squarely on the nuts. This prevents the pliers from slipping and rounding off the corners of the nut.

Screwdrivers. Screwdrivers have just one purpose—to loosen or tighten screws and slotted-head bolts. In emergencies, however, they can be used as a substitute for everything from a crowbar to a chisel. Of course, such misuse is only for real emergencies when the correct tool is not at hand.

You can get along nicely at the start with three screwdrivers. Get a medium-size standard screwdriver with a $\frac{1}{4}$- by 4-inch blade for all-round use. This will be used chiefly for removing chassis-mounting screws, tightening the bolts used for mounting new parts, and removing wood screws.

A small standard screwdriver with a $\frac{1}{8}$- by 4-inch blade is needed for loosening small screws such as are used for fastening a cartridge in a phonograph pickup.

A No. 2 Phillips screwdriver with a 4-inch shaft takes care of practically all sizes of the special Phillips screws that are encountered in television and radio sets. These screws have four-cornered depressions in the heads, to speed up inserting the screwdriver and to minimize chances of having the screwdriver jump out of the screw when tightening or loosening.

Get the best quality in screwdrivers right from the start, since even these cost less than a dollar apiece. The best types have shatterproof plastic handles that fit comfortably in the hand. Plastic handles also serve as insulation, permitting safe use in high-voltage circuits.

Knife. You will need a good two-blade pocketknife for scraping wires, cleaning soldering lugs, and cutting insulation off wires. Choose a strong knife that has two good blades. Keep the smaller blade razor-sharp all the time for cutting insulation. Use the larger blade for rougher work, such as scraping.

Socket Wrenches. A set of spin-type socket wrenches is needed for removing and replacing chassis-mounting bolts, speaker-mounting nuts, and other nuts and bolts used for mounting new and replacement parts in a television or radio set. These wrenches have handles like screwdrivers. Each has a different size of hexagonal socket at the tip. Sometimes they are called hex nut drivers.

Get a high-quality set of these wrenches at the start because you will use them a lot. They should have hollow shafts that fit over the projecting ends of long bolts to reach the nuts. Cheap wrenches are a waste of money because they wear out quickly, do not fit accurately enough, and do not usually have hollow shafts. Plan to spend at least $5 for a set of seven wrenches.

Spin-type wrenches are handled much like screwdrivers, and are far better and faster than pliers. The socket cannot slip off once it is held over a nut or a bolthead. When a nut is loosened, you can spin it off quickly by rolling the shaft of the wrench between your fingers.

Typical sizes in a set of spin-type wrenches are $3/16$, $7/32$, $1/4$, $9/32$, $5/16$, $11/32$, and $3/8$ inch. The $1/4$- and $5/16$-inch sizes will probably be the most used, so paint colored bands or dots on the handles of these two sizes if they are not made in different colors already.

Flashlight. A small flashlight is badly needed on home calls to light up dark corners inside the cabinet when replacing tubes, when loosening or tightening mounting bolts, or when looking over the chassis for defects like broken wires. An ordinary two-cell flashlight will give you the most value per set of batteries, but almost any flashlight will do. If you have one already, by all means use it.

Choosing a Toolbox. The only tools that you ordinarily need to take along on service calls to homes are those just described. For carrying these tools and an assortment of small parts and supplies, get a good-looking toolbox right from the start. Choose this box carefully because it is the first piece of equipment that your customer sees when you enter the house. Get a fairly large toolbox, attractively finished in a bright baked-on enamel. It should have an automatic lift-up tray that makes the bottom of the box accessible immediately when you lift the cover.

The better models of fishing-tackle boxes are ideal for your purpose. You can remove some of the tray partitions to make a few large partitions for tools. All-aluminum boxes are particularly attractive.

A large box is desirable so that it will hold at least half a dozen new tubes in cartons. An 18-inch-long box is a good size. On radio calls you can often obtain from the customer beforehand the make and model number of the set and can take along a set of new tubes for it. In about 7 out of 10 radio sets new tubes will clear up the trouble, allowing you to do the whole job in one call.

Toolbox Supplies. The big temptation is to pack a toolbox full of things that are seldom if ever needed on service calls in homes. Your customer's

Tools Needed for Servicing 27

home is not a workshop. The receiver chassis should always be taken to the shop when special tools are needed.

In addition to the basic tools and soldering gun shown in Fig. 1, your toolbox should have a few feet of rosin-core solder, a few feet of insulated

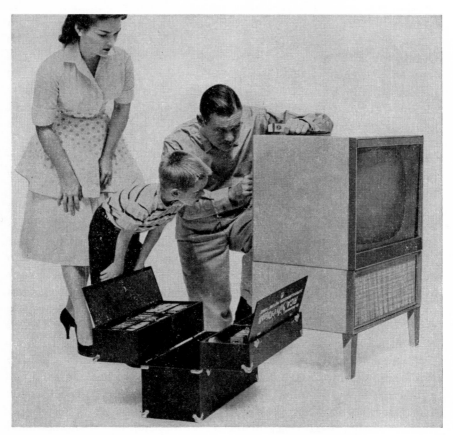

Fig. 2. Handy tube caddy can hold up to 262 tubes, all easily accessible when caddy is opened as shown. (RCA photo)

hookup wire, a roll of ordinary Scotch tape and a roll of No. 33 Scotch electrical tape, a bottle or tube of service cement, one box of assorted pilot lamps, one box of assorted television and radio fuses, a line cord with plug, and one or two extra line-cord plugs.

A tube of Lubriplate or equivalent lubricant takes care of many record-changer and tuning-mechanism jobs in homes. For noisy controls, you can get from your jobber a screw-on plunger-type control cleaner that applies lubricant under pressure through the shaft bearing of a volume or tone control without any need for removing the chassis from the cabinet. Detailed instructions for using this tool are given in the chapter on controls.

Inexpensive picture-tube brighteners can profitably be carried in your toolbox. Try one when a picture-tube replacement is indicated but the customer cannot afford the job right away. If the brightener brings up picture brightness satisfactorily, explain to the customer that she can buy it from you to keep the old tube working a few months longer. You can then offer to give a full refund on the brightener when you put in the new picture tube at some later date. This refund offer makes it practically certain that you will get the highly profitable picture-tube replacement job. The brightener can then be used on some other set. Instructions for installing brighteners are given in the picture-tube chapter.

A separate tube caddy like that in Fig. 2 is needed for the stock of new tubes that you bring into a customer's home. Suggested tube types for this stock are given in the chapters dealing with tubes.

Cheater Cord. Practically all television sets are made in such a way that the receiver end of the line cord comes off with the back cover. This is a safety precaution for users, so they cannot get a shock if they touch the high-voltage terminal inside the set. For servicing, however, you have to be able to operate the set with the back cover removed. Television-receiver line cords made for this purpose are called cheater cords. Get two from your jobber when you start working on television sets. Keep one on your bench and put the other in your toolbox for home calls, as it is needed even for replacing television tubes.

Neatness Pays. Always keep a clean drop cloth in your toolbox, on top of the supplies in the lower compartment, so you can make a good impression by carefully unfolding the clean cloth and spreading it out on the floor or rug before taking out a single tool.

Always fasten on the back of the set with Scotch tape a business card that gives your name, address, and telephone number. This pays off in bringing you future service calls, particularly in localities where service shops or servicemen have similar names.

When cleaning up in the customer's home, put all scraps of excess solder, small bits of wire, and fragments of insulation in a compartment of the toolbox. Carefulness in cleaning up will make a good impression. Be sure to remove the scraps when you get back to your shop, for otherwise your next customer will get a poor impression when studying the contents of your toolbox.

Tools Needed for Servicing   29

Care of Tools. Plan the arrangement of your toolbox carefully so that you have a definite place for each tool. Keep the box alongside you when working, as shown in Fig. 3. Put each tool back when you are finished with it. This minimizes chances of leaving tools in homes or losing them, especially when working outdoors on television antennas. Screwdrivers

Fig. 3. Toolbox is kept within easy reach while replacing 16-inch round picture tube in older set on home call. A drop cloth should be placed on the floor or rug before setting down the toolbox and tools. Keep children away from an exposed picture tube and a hot soldering gun. Wear safety goggles when handling a picture tube

are particularly likely to roll under furniture where they are overlooked.

Several of the small partitions in the toolbox tray should normally be left empty. Use these for bolts, knobs, and other hardware taken off a set when you remove a chassis for shop work.

Do not toss tools into your toolbox. You might miss and dent a valuable piece of furniture, making your customer unhappy. The clatter will also be annoying. Your customer does not understand the things you do to the receiver, and hence he judges you on the basis of little things like handling of tools.

Put a drop of oil on the pivot of each pair of pliers occasionally to keep them working smoothly. Good clean tools inspire good work. Dirty tools, on the other hand, seem to encourage slipping and carelessness that often damages radio tubes, parts, and even your own skin.

Workbench Tools. The tools described so far are used both on service calls and on your workbench. The additional tools you will need on your workbench at the start for replacing defective parts are three files, a hacksaw, a drill, a hammer, a center punch, a cold chisel, and a vise.

Files. Three files will take care of your minimum requirements when getting started.

A slim-taper fine-cut triangular file 8 or 10 inches long is used for filing a screwdriver blade to shape and for filing small parts. Do not use this for filing a soldering-iron tip or for filing soldered joints, as soft solder will quickly fill in the fine grooves and make the file useless.

A flat second-cut machinist's file $\frac{3}{4}$ or 1 inch wide and 8 or 10 inches long is used for rougher work, such as filing the tip of an ordinary soldering iron and filing mounting brackets of new parts when they are too big.

A second-cut round or rat-tail metal file $\frac{5}{16}$ inch in diameter and 8 or 10 inches long is handy for enlarging round holes in a chassis when you do not have the required large drill or reamer.

After a file has been used for a while, the grooves between the teeth become clogged with metal, and the teeth themselves become dull. You can clean a file by scraping out the metal from each groove in turn with a needle, but this does not sharpen the teeth themselves. It is far better to discard a clogged file in favor of a new one.

With files, as with so many of the other tools you will use, it is better to pay a few cents more to get a well-known make having a quality you can rely on.

Hacksaw. Replacement volume controls and some types of switches come with extra-long shafts that must be sawed to the correct length. For this you will need a hacksaw on your workbench right from the start. Here, however, you can get along with a cheaper tool, since you will use it only occasionally and for light work. It does not even have to be adjustable in length. Blades with 24 teeth to the inch are most useful and are easily obtained.

Always keep a few extra blades on hand, as they break easily. New hacksaw blades are cheap, so replace the blade whenever it gets dull. You will save time and do better work with a sharp blade.

To replace a hacksaw blade, unscrew the wing nut at the far end of the saw or unscrew the handle, according to the type of saw you have, until the blade slides off the holding pins. Put the new blade over these pins, with the blade teeth pointing *away from the handle*, and tighten.

Take long, slow strokes when sawing, to distribute the wear over more teeth and to make the blades last longer. Apply steady pressure on the pushing-away stroke so you can see filings fall down. Ease up the pressure on the pullback stroke.

To start a hacksaw when cutting metal, nick the work with a fine triangular file at the correct point and start the saw in the nick.

When using a hacksaw, put the work in a vise whenever possible. If the work slips when held in the hand, the saw blade may break and cut your hand.

Drill. When replacing defective parts, it is frequently necessary to make new mounting holes. An ordinary breast or hand drill that will take up to $\frac{1}{4}$-inch-diameter round-shank drills is quite adequate to start with. Both types are shown in Fig. 4.

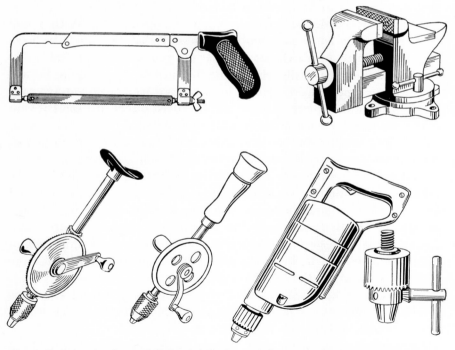

Fig. 4. Hacksaw, vise, breast drill, hand drill, electric drill, and Jacobs chuck with key, all suitable for bench use in a servicing shop

A breast drill has a curved top piece that fits against your chest for applying pressure when drilling, but usually the work must be placed on a chair or on the floor to take advantage of this. A hand drill is easier to use at service bench height, but here all pressure must be applied with the left hand while turning the drill with the right hand. Many servicemen get extra pressure on a hand drill by using their chin also, on top of the left hand.

You can get along with only five drill points at the start, because the sizes of bolts used for mounting parts are pretty well standardized. These five drill sizes are No. 4, 10, 18, 28, and 33. If ordering drills by diameter in inches, get the next larger equivalent of each, such as $7/32$ inch for No. 4, $13/64$ inch for No. 10, $11/64$ inch for No. 18, $9/64$ inch for No. 28, and $1/8$ inch for No. 33.

The cheaper carbon drills are satisfactory for a breast or hand drill. Keep at least two of each size of drill on hand, so that you can replace dull drills frequently. Since the two smaller sizes break frequently, it is even better to start with four of each of these. Carbon drills cost so little that it is cheaper to replace them than to pay for sharpening.

Sharp drills are desirable from a safety standpoint also. With a dull drill you have to apply so much extra pressure that the drill may break and damage parts in the chassis and may cut your hands as it skids.

When drilling, especially when using the smaller sizes of drills, be sure your work is rigid. All drills are highly tempered hard steel and will snap viciously when they break, producing sharp-edged pieces that can cut your hands. Never let anyone hold the work for you when drilling, because of this danger. Support the work in a vise if possible, so it cannot slip or move when you apply the required drilling pressure.

Apply enough pressure to a drill so that the cutting edges bite into the metal instead of sliding over it. If drilling on a chassis, put a block of wood under the drilling location so you can apply pressure without risk of damaging other parts. Ease up on the pressure just before the drill bites through, and turn slowly from then on. If the point of the drill jams in the last big shaving, back out the drill to get clear, then turn still more carefully to break up the shaving.

When drilling hard material such as a steel chassis, or when a large hole is required, make a small hole first. You can then change to the larger final drill and easily enlarge the hole.

Electric Drill. A $1/4$-inch electric drill is a highly desirable substitute for a breast or hand drill. A type designed for light, intermittent duty is adequate if of a reliable well-known make. A drill with a key-type Jacobs chuck like that shown in Fig. 4 is best for your use.

A good electric drill can now be obtained for around $20. It will soon pay for itself in time and energy saved if doing full-time servicing. With electric drills, use high-speed drill points rather than the carbon type.

Larger electric drills taking up to $\frac{1}{2}$-inch drills are not recommended for routine service work because they are heavier and clumsier, and generally run at slower speeds that require too much pressure when working on a chassis.

Hammer and Punch. A drill will skid on a smooth piece of metal when starting, unless you first punch a depression at the correct point. For this you need a hammer and a center punch. The hammer need not be expensive, but be sure that its head is firmly anchored to the handle. An ordinary carpenter's hammer, of medium weight such as 16 ounces, is best as it can be used also for installing antenna systems.

Spend a few pennies extra to get a really good center punch. Chrome-vanadium steel center punches are swell, and cost only about 50 cents each.

Before using a punch or chisel on a chassis, remove all the tubes. If possible, place a block of wood on the opposite side from where you plan to punch. The larger the drill point, the larger must be the starting depression and the harder you will have to hit the punch with the hammer. When the point of the punch gets dull, resharpen it on a grinder.

Drilling Out Rivets. Many larger parts are riveted to the chassis. The easiest way to remove rivets is to drill them out with an electric drill or hand drill. Use a drill slightly larger than the body of the rivet. Sometimes it is easier to start with a small drill, and then enlarge the hole. Make a starting hole for the drill point with a center punch. You can drill out either the head or the flattened end of the rivet, depending on which is easier to get at.

If the rivet begins to turn with the drill after you have drilled awhile, hold the drill at enough of an angle to stop the spinning, or try to break off the rest of the rivet with side-cutting pliers or a sharp cold chisel. A rivet usually does not start turning until the drill has almost cut through. Always remove tubes before doing any heavy pounding on the chassis.

If the rivet persists in turning yet cannot be cut off easily, move the chassis so the undrilled end of the rivet is against a bench vise or other heavy metal part. Now drive a center punch into the partly drilled hole. This tightens the rivet and prevents it from turning, allowing you to finish the drilling. If a helper is available, it may be easier to have him hold a heavy hammer against the other end of the rivet while you tighten it with the center punch.

Replace rivets with conventional nuts and bolts. Always use a lock washer under the nut. Vibration set up by the speaker or power transformer may loosen the nut if the lock washer is omitted.

Cold Chisel. A cold chisel is used chiefly for cutting off the heads of rivets. Get a high-quality chrome-vanadium steel chisel, with a ½-inch-wide cutting edge.

Heavy work with a chisel on a chassis can damage other parts and loosen soldered joints. For this reason, when enlarging a chassis opening for a power transformer or other part, drill holes first along the cutting line, as close together as possible. Light taps on a sharp cold chisel are then enough to cut out the metal between the drilled holes. Finish with a file to get a neat straight or curved edge.

Vise. A hacksaw can be used best when the object being sawed is held in a vise. One with jaws somewhere between 2 and 3 inches wide is adequate for service work. It can be the type that clamps onto the bench and is readily removed, provided your bench top projects enough to give room for the vise clamp.

Some bench tops are flush with the sides or are too thick for clamping on a vise. In this case, fasten a 1½- by 1½- by 4-inch block of hardwood to the side of the bench at a convenient location for clamping on the vise when needed, or get a vise that can be fastened with screws, as shown in Fig. 4.

Workbench Suggestions. At the start you can get along quite well with a simple bench or even a plain table. Eventually, however, you will want to build something that looks impressive and provides convenient places for all your equipment. Study the benches used by other servicemen, and make notes of the ideas and designs that appeal to you.

The construction features and dimensions given in Fig. 5 can serve as a starting point or even as the complete building plan for your own bench. The complete bench in Fig. 6 illustrates how bench dimensions and features can be changed to suit personal preferences.

The working surface of your bench should be somewhere between 36 and 40 inches high, depending on your height. This is about 10 inches higher than a carpenter's bench. The extra height is needed so you can work comfortably on sets while standing up, as well as while sitting on a high stool.

A one-man bench does not have to be more than 6 feet wide and 2 feet deep on the working surface. If you make a bigger bench, it merely becomes a storage shelf for sets and parts, and it never looks good. Build extra shelves elsewhere in your shop for storing sets.

The working surface of a bench can be a piece of linoleum, ¼-inch or thicker plywood, or ⅛- or ¼-inch Masonite Presdwood laid over or-

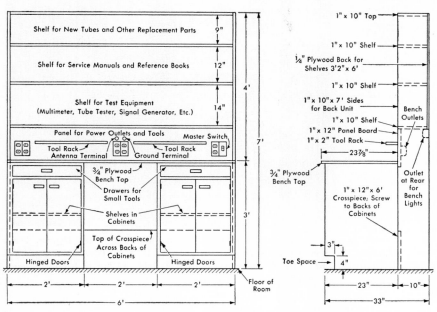

Fig. 5. Suggested construction of easy-to-build television and radio service bench. Shelves at rear are made separately and attached to bench with angle brackets and wood screws. Base units can be purchased assembled but unpainted, simplifying carpentry work. Shelf heights should be changed to fit your own equipment if necessary; thus, Rider's Television Manuals require a 16-inch-high shelf

dinary boards. Cracks and joints are undesirable in the working surface because small screws get lost in them. Most servicemen prefer Masonite for the top. Never use a metal top, because it can be a shock hazard. For the same reason, do not place decorative metal edging across the front of your bench.

Bench Supports. Leave the center portion open under the bench so you can get your knees under when sitting on the stool. Shelves or drawers can be built on each side for storage of spare parts. Drawers are best, though more difficult to make. With drawers, you can easily see everything in

36 *Television and Radio Repairing*

Fig. 6. Example of service bench having many of features described in this chapter. Bench should be located for shortest possible path to car or truck, so heavy television sets will not have to be carried too far. Note simple storage shelves at right and left of bench. (Sylvania photo)

them. With cabinets or shelves under the bench, you can only see things that are stacked in front.

Kitchen-cabinet base units with drawers are ideal for the base of a service bench. These units come completely cut and sanded, but are sometimes unassembled and unpainted. You need only a hammer, screwdriver, and glue to assemble them. Cabinet units can be obtained from some mailorder houses, department stores, or furniture stores at way less than a carpenter's charges for making them. A piece of ¾-inch plywood 2 feet wide and 6 feet long can be obtained from a lumber yard for use as a bench top. It can be fastened to the cabinet units with small angle brackets. This gives you a complete workbench base with a minimum of carpentry work.

Individual drawers in housings are also available, ready to be screwed in position under the center of the bench.

Bench Back. At the back of your bench you will want a rack for holding power outlets, tools, test equipment, service manuals, and reference books. Keeping the top of your bench 2 feet deep or less makes it easy to use test instruments without taking them off the shelf. At the beginning and for spare-time servicing, a simple shelf for test instruments is entirely adequate. A mirror can be placed under the shelf as in Fig. 7 if desired, for convenience in adjusting television receivers.

Bench Tool Rack. There should be a place for every single tool on your bench or in drawers underneath. Spring-type holders for tools can be obtained in hardware stores. Another way, used on the bench in Fig. 6, is to drill holes of various sizes in a 1- by 2-inch strip of wood and to screw this strip to the back of the bench. Screwdrivers, pliers, and many other small tools fit nicely into a rack of this type.

Bench A-C Outlets. Your bench should have plenty of electric outlets. You need one for the soldering tool, one for the tube tester, and one for the set being worked on. Additional outlets are desirable for letting sets run a few hours after they have been repaired, for the electric drill when you get one, and for a host of other purposes. Eight outlets are none too many. These can be on the back of the bench, arranged two on each side and four at the center as indicated in Fig. 5. Put another outlet box behind the bench back, for plugging in the overhead lights of the bench.

All these outlets will take a lot of punishment, so stay away from the bargain counter and dime store when buying them. Get the very best grade obtainable, at an electrical supply store, and be prepared to pay about three times the price of corresponding cheap units. Spring contacts in

Fig. 7. An inexpensive mirror at the back of your bench makes it easy to see the picture on the screen of a television set while making back-of-set adjustments. (Electrical Merchandising Week photo)

38 Television and Radio Repairing

cheap outlets lose their tension quickly and fail to grip or even touch the prongs of plugs.

Regular wall outlets with plastic mounting boxes and plastic surface plates are ideal for benches. With plastic boxes, you can connect the outlets together with nonmetallic-sheathed cable or even ordinary lamp cord instead of hard-to-handle BX.

Use a line cord and plug for the electrical system of your bench and plug it into an existing electrical outlet in the room. If bench wiring is permanently connected to the wiring system of a building, the job must be done by an electrician or at least pass official inspection if fire insurance is to remain valid.

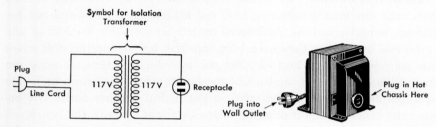

Fig. 8. Circuit and construction of isolation transformer used between power line and receiver on workbench to prevent hot-chassis shocks

Isolation Transformer. It is often necessary to make voltage measurements on a chassis that is electrically hot (connected to one side of the power line). The safe procedure here is to use an isolation transformer between the power line and the receiver, as in Figs. 8 and 9. These isolation transformers are made especially for servicing, and well worth their cost. The transformer gives out the same a-c voltage (about 115 volts) that

Fig. 9. Example of isolation transformer having three outlets, to permit testing sets at low and high line voltages. (Chicago Standard Transformer Corp. photo)

Tools Needed for Servicing 39

is applied to it, but is safe because there is no direct connection between its output terminals and the power line.

Cost of an isolation transformer ranges from $8 to $25 depending on power-handling capacity. Most universal a-c/d-c radio sets draw under 40 watts, so a 50- or 100-watt unit is large enough for small radios. Many television sets also have a hot chassis; since these sets draw considerably more power, an isolation transformer rated at 250 watts or even higher is preferable if you plan to do television servicing. Ratings in va (volt-amperes) can be considered roughly equal to ratings in watts.

Isolation transformers having additional outlets or other means for adjusting the output voltage cost a few dollars more and are well worth the price. With these, you can lower or raise the a-c voltage applied to a set to duplicate conditions in the home of a customer where the a-c line voltage is higher or lower than normal.

Master Switch for Bench. It is a good idea to have one master switch on your bench. This is connected to the line cord serving the whole bench, including bench lights, as indicated in Fig. 10. The master switch then turns off the lights along with everything else on the bench. The lights are a reminder for you to turn off the switch when you leave the bench. This arrangement eliminates the fire hazard of leaving a soldering iron and other devices plugged in all night with power on. The master switch can be a standard toggle switch on a wall plate, mounted alongside the right-hand pair of outlets.

Workbench Lighting. Provide plenty of light for your workbench. Two or preferably three 100-watt lamps in reflectors will serve nicely for overhead lighting. Aluminum clamp-on reflectors made for photographic use

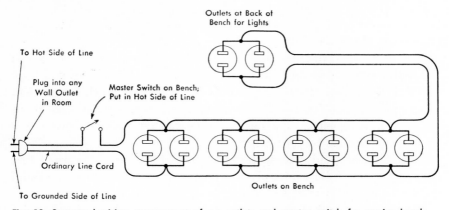

Fig. 10. Suggested wiring arrangement of a-c outlets and master switch for service bench

are ideal here because they can be adjusted to the most effective angle. They come complete with socket and line cord for around $3 each.

Photographic reflectors can be clamped to ceiling rafters in a basement or to any supporting rod mounted horizontally above the bench for this purpose. When more light is needed for close work, you can easily take one of the reflectors down temporarily. A reflector designed for No. 1 photoflood lamps will hold a 100-watt bulb perfectly.

Better still for close work on the bench is a gooseneck-type desk lamp that can be moved around and adjusted easily to give light exactly where needed on a chassis.

Fluorescent lights for a bench are expensive and at times also a nuisance. These lights can cause electrical noise interference in the radio sets on which you are working, unless well made and completely shielded and grounded. If you prefer fluorescent lighting, get high-quality two-lamp fixtures that have the proper built-in ballasts and filter capacitors to give steady light with a minimum of electrical interference.

Some servicemen feel that an electrically noisy fluorescent light helps to speed up radio repairs. The interference signal serves as a handy noise generator for judging the sensitivity of a receiver and touching up receiver adjustments. If you prefer fluorescent lighting, go right ahead and get it.

Assembling a Bench. When building a bench, bear in mind that you may some day have to move it. Assemble it in sections that fasten together with wood screws and angle brackets. The working top can be fastened to the base sections with metal brackets. Fasten these underneath the top with short wood screws, so no screwheads show on the top surface of the bench.

Make sure that each section of your bench will go through all doors. Benches represent a large investment of time and money, so preserve your investment by planning for portability.

After assembling your bench, spend at least 2 hours sanding it smooth and rounding off all sharp edges and rough corners. Get half a dozen large sheets of 1/0 garnet paper, which cuts faster and lasts much longer than ordinary sandpaper. Tear the sheets into convenient smaller sizes to fit around a sanding block. Thorough sanding can do much to give a professional appearance to your bench.

Fill in all cracks and holes with plastic wood if they are in locations that show. Overfill each hole since plastic wood shrinks a bit when drying. Sand down all filled areas after the plastic wood has dried.

Painting the Bench. Spar varnish gives an attractive finish for a service bench, and is easy to apply. For the first coat, thin out by adding 25 per cent turpentine, and sand this down lightly after it dries. This varnish is particularly good for the working surface of your bench if you are using plywood or Masonite. Another easy-to-apply finish that is good-looking on well-sanded new wood is a combination stain and wax.

Hardware and Supplies. Many different kinds of screws, bolts, washers, fasteners, springs, and other small parts are used in television and radio sets. These occasionally need replacement, but they can be obtained in convenient small assortments from your jobber as needed.

Assortments of machine screws, nuts, and lock washers are used for mounting new parts. The 6-32 and 8-32 sizes are the ones you will use most. The first number in the screw size refers to the diameter of the screw and is proportional to the diameter; thus, 8 is thicker than 6. The second number tells the number of threads per inch; most radio screws have 32 threads per inch.

Many different kinds of cements and chemicals are available from jobbers for servicing work. Here again you can order as needed. Service cement will be used most of all. Get one bottle of this for your toolbox and another for the bench. You can use it for cementing practically anything that gets loose on a receiver. Duco cement, sold in handy tubes, can be used instead.

Service cement solvent will be needed on the bench for loudspeaker repairs and for getting cement off your fingers. The chapter on loudspeakers covers its use.

A can of fine machine oil can be obtained from any gas station and is used for lubricating shafts and tuning mechanisms.

Carbon tetrachloride is used for cleaning metal parts. Be sure not to breathe its fumes, however, as they can cause severe illness.

Furniture polish and some form of scratch remover are needed for freshening up wood cabinets, as described in the chapter on cabinet repairs.

Start out with a 1-pound spool of rosin-core solder, as specified in the chapter on soldering. Unwind a few feet from this to roll up separately for your toolbox. Get the fine 50-50 rosin-core solder made especially for television and radio work.

Get a 25-foot coil of solid tinned insulated hookup wire for the bench. Take off a few feet of wire for your toolbox. You do not need much wire for repair work, so this is plenty for a start.

Workbench Antennas. For testing radio receivers, run about 25 feet of insulated wire of any type from your workbench up to the ceiling of the shop and then along the ceiling in a straight line. It is best to run the wire crosswise to pipes and other wiring, but even this is not critical. Attach a small battery clip to the end of the wire at the bench, to speed up connecting this wire to a receiver. Do not use more than about 25 feet of wire, because a longer antenna can overload some radios and cause distortion that did not exist for the customer. Most radio sets now have good built-in antennas, however, so you will seldom need to use a workbench antenna.

For testing television receivers, you will eventually need a good rooftop television antenna, with a twin-line transmission line running down to the workbench. Instructions for installing television antennas are given in a later chapter.

3

How Television and Radio Sets Work

Extra Knowledge Pays. In servicing, your interest in broadcasting begins at the receiving antenna, where the desired radio wave is picked up and fed to a television or radio set. To fix the set, you do not need to know how the program was produced, how it was changed into a radio wave, or how that radio wave got through the sky. You just need to know how to find, test, and replace the bad parts, as illustrated by Fig. 1.

To make top money in servicing, however, it helps to know a little more than how to fix sets. When a customer asks questions about television and radio, simple answers will boost your reputation and your income.

The average person knows little about what is inside a television receiver, so his questions are more likely to be about simple things outside the receiver. He may ask why television antennas have to be aimed carefully in some locations. He may ask why f-m radio is supposed to be better than ordinary radio.

Questions like these will be easy to answer if you have a conversational knowledge of the three kinds of broadcasting systems that are being used today. This chapter gives you the whole story, so you can satisfy your curious customers. They will be impressed because they can understand your answers, and will give you credit for knowing so much. They will respect you as a professional serviceman. They will recommend you to others. Best of all, they are much more likely to pay your bill, without questioning or quibbling.

Whenever servicemen discuss television or radio, the word *frequency* is heard many times. Frequency is actually one of the most important technical terms you will use in servicing. For this reason, the meaning of this

44 *Television and Radio Repairing*

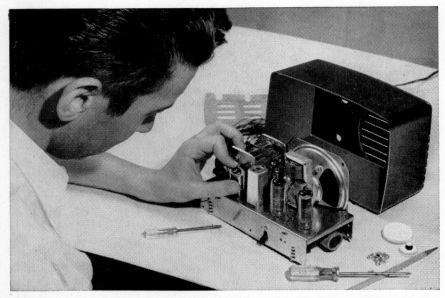

Fig. 1. This radio set was fixed by replacing one bad tube, without even thinking about how the set actually works. It pays to know more than is needed, however, because curious customers often ask semitechnical questions. Usually it is not even necessary to remove the chassis for tube replacements

word will be explained before the three kinds of broadcasting systems are taken up.

Frequencies of Sounds. The number of times per second that anything vibrates is called its *frequency*. A big bass drum might vibrate 30 times a second when hit. This means that it goes through 30 complete back-and-forth movements of its stretched skin in one second. Each of these back-and-forth movements is called *one cycle*. The drum would thus have a frequency of 30 cycles per second.

You can actually feel the compression parts of a sound wave when the drummer goes by in a parade. If you are holding an opened newspaper, you will feel the sound waves still better because the air compressions move the newspaper back and forth. Sound waves can thus make other objects vibrate.

When you strike the key for middle A on a piano, a hammer inside hits a wire string having exactly the correct length to vibrate 440 times per second. The resulting sound has a frequency of 440 cycles.

When the musician draws his bow across the thinnest string of his

How Television and Radio Sets Work

violin, a much higher frequency is heard—perhaps as high as 4,000 cycles. Each sound in this world has its own frequency. The shriller the sound, the higher is its frequency.

The human ear has definite limits on what it can hear. The lowest frequency that the normal ear can detect is about 20 cycles. The highest is somewhere around 20,000 cycles. The frequencies between these limits are known as *audio frequencies* because they can be heard.

The term audio frequency is often abbreviated as a-f. When you use abbreviations like this in conversation, just pronounce the letters.

Frequencies of Electricity. The electricity flowing through power lines to homes has a frequency also. This electricity goes through 60 complete

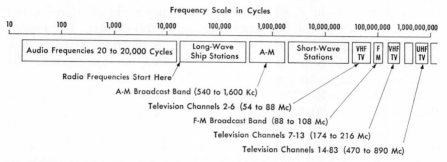

Fig. 2. Frequency chart. Frequency limits for the five bands in which home receivers operate are listed below the boxes. Remember that 1,000 cycles is 1 kc, and 1,000,000 cycles is 1 mc

cycles of change in the direction of flow each second, so its frequency is 60 cycles per second. For convenience, the expression *per second* is usually dropped.

For every frequency of a sound wave, an electrical signal can have the same frequency. Signals like this are called *audio-frequency signals* or *a-f signals*. The range of a-f signals is about 20 to 20,000 cycles per second, the same as for sound waves, as shown at the left in Fig. 2.

Electricity can easily be made to go through more than 20,000 complete cycles of change in direction per second. Electrical signals whose frequencies are higher than about 20,000 cycles per second are called *radio-frequency signals*, abbreviated as r-f signals.

The radio frequencies from 20,000 up to 540,000 cycles are not used for commercial broadcasting and cannot be picked up by ordinary radio sets. Merchant marine and military stations use these frequencies, chiefly for communication with ships at sea.

Kilocycles and Megacycles. It is awkward to write strings of zeros after numbers when dealing with radio frequencies. For this reason, two larger units are used in place of cycles.

Kilocycles means *thousands of cycles*. You will use *kilocycles* most in your work on a-m receivers. Kilocycles are abbreviated as kc. One kilocycle is equal to 1,000 cycles.

Megacycle means *millions of cycles*. You will use *megacycles* chiefly when working on f-m and television receivers. Megacycles are abbreviated as mc. One megacycle is equal to 1,000,000 cycles. Also, one megacycle is equal to 1,000 kilocycles.

Broadcasting Systems. The three distinct types of broadcasting systems in use today for bringing entertainment to homes by way of the air are a-m (amplitude-modulation) radio, f-m (frequency-modulation) radio, and television. Each type will be taken up in turn in this chapter, after a quick look at the frequencies used by the stations in the three systems. The receivers used with each system will be emphasized, to prepare you for the practical servicing instructions that come later.

Station Frequencies. The diagram in Fig. 2 shows all the frequencies that you will deal with in servicing television and radio receivers.

The regular a-m radio broadcast band extends from 540,000 to 1,600,000 cycles. To change these values to kilocycles, drop the last three zeros. The broadcast band thus is 540 to 1,600 kc. This is the range over which ordinary home radio receivers can be tuned.

Tuning-dial scales of a-m broadcast receivers are seldom marked with the full frequency value in kc, because of lack of room for the zeros on the dial. Instead, one or even two zeros are dropped from each value.

Above the broadcast-band frequencies are the ranges covered by two-band, three-band, and all-wave receivers. Here you can pick up international broadcast stations, police calls, code stations sending messages in dot-and-dash form, and amateur radio conversations. This is commonly known as the *short-wave region*. It extends from 1,600 to 54,000 kc, or up to 54 mc, which is the first television station frequency.

Television station frequencies are in three bands. The first, called the *low vhf band*, or simply the low band, contains television channels 2, 3, 4, 5, and 6. The second, called the *high vhf band*, or high band, contains channels 7, 8, 9, 10, 11, 12, and 13. The third is called the *uhf band*, and contains 70 more television channels. The abbreviation vhf means very-high frequency; uhf means ultrahigh frequency.

Originally, 13 television channels were assigned in the vhf band, but the

How Television and Radio Sets Work

first one has since been dropped without renumbering the other 12 channels. The vhf television channel numbers and their frequency ranges are as follows:

	Channel No.	Frequency, mc
Low VHF Television Band	2	54–60
	3	60–66
	4	66–72
	5	76–82
	6	82–88
F-M Broadcasting		88–108
High VHF Television Band	7	174–180
	8	180–186
	9	186–192
	10	192–198
	11	198–204
	12	204–210
	13	210–216
UHF Television	14–83	470–890

Between the f-m band and the high band for television are communication station frequencies of no interest to servicemen. Similar communication stations operate between the high vhf band and the uhf television band. Above the uhf television band are more communication stations, radar stations, and other special services that have no importance to home receiver servicing.

Frequency values alone are not enough to identify stations, as two or more stations may be assigned to the same frequency. Call letters are therefore assigned to each station for identification.

A-M Broadcasting System. The major units of an a-m broadcasting system are shown in Fig. 3. When the announcer speaks before the microphone in the studio, he produces *sound waves* that act on the microphone.

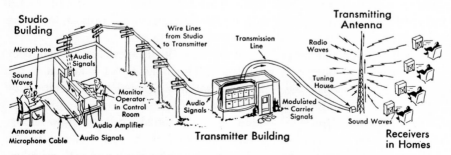

Fig. 3. Signal paths from microphone to receivers in a typical a-m broadcasting system

Microphone. A thin metal sheet or diaphragm in the microphone moves back and forth when a sound wave hits it. Attached to this moving diaphragm is an electrical device that changes the movements into a corresponding electrical signal called the *audio signal*.

The device inside the microphone may be a piece of crystal, a coil moving between the ends of a permanent magnet, or simply a corrugated metal ribbon moving between the poles of a permanent magnet. All types of microphones produce the same result—they change the sound waves into audio signals.

Microphone Cable. The audio signal produced by a microphone is an alternating electric voltage. To transfer the voltage from one place to another, wires are needed.

The wire that comes out from a microphone is called the *microphone cable*. This cable provides a path for transferring the audio signal to the audio amplifier that is the next part of the a-m broadcasting system shown in Fig. 3.

Audio Amplifier. The audio signals coming from a microphone need to be strengthened many thousands of times. This electrical strengthening process is called *amplification*. It is done with radio tubes and radio parts in an audio amplifier located in the control room.

A meter on the amplifier panel shows the strength of the audio signals. There is also a volume-control knob for changing the strength of the signals. The man at the control desk must keep the signal strength nearly constant, even though the singer or speaker in the studio moves around a lot.

From the control room, the strengthened audio signal travels over telephone lines for miles to the transmitter building. This transmitter is usually located outside the city.

A-M Transmitter. As soon as the audio signal gets inside the transmitter building, it is boosted in strength some more. The audio amplifier that does this is much like the one in the control room, only a lot more powerful.

The amplified audio signal is combined with the special radio-frequency signal needed to carry it through space. This radio signal is known as the *carrier signal*, and has the frequency that was assigned to that station by the Federal Communications Commission.

Why a Carrier Signal Is Needed. An audio signal, no matter how strong, would not travel more than a few hundred feet through space if fed to an antenna. This is why the audio signal must be combined with a higher-frequency signal that is capable of traveling great distances through space.

The carrier signal is generated in a vibrating quartz crystal, and is amplified by radio tubes to build up its strength or power. Next, both audio signal and carrier signal are fed into the same tube and combined. The result is known as the *modulated carrier signal*.

In most stations the modulated carrier signal is fed directly to the antenna tower. This signal is still an electric current, so it travels through wires to the transmitting antenna. These wires can be one inside the other to give what is known as a coaxial cable. Occasionally the two wires are side by side, forming a transmission line.

A-M Antenna Towers. Many different kinds of antenna towers are used by a-m broadcast stations. Some are tall and some are short. Some are thick and some are thin. Some stand up by themselves, while others require guy wires to hold them erect.

The height of the tower for an a-m station is carefully tailored to match the frequency assigned to the station. The higher the assigned frequency, the shorter is the tower needed. You will find that this simple rule holds true for everything in radio and television. Remember—the higher the frequency, the shorter is the antenna needed. Both f-m and television stations use much higher frequencies than a-m stations, so their transmitting and receiving antennas are much shorter than for a-m stations.

A transmitting antenna tower changes the modulated carrier-signal current into *radio waves* that travel away from the tower in all directions. These radio waves travel with the speed of lightning but are invisible. What they actually are does not matter now, because even scientists are not too sure about it today.

A-M Receiving Antenna. The radio waves that carry an a-m radio program away from the transmitting antenna are changed back to sound waves by a receiver. The first part of a receiver is the antenna on which the radio waves act.

Many different kinds of receiving antennas are found in homes. In modern radios, ferrite-rod antennas and loop antennas mounted at the back of the cabinet are the two commonest types. No matter what the receiving antenna looks like, however, it has just one job. It changes the radio wave back into an electric current. This carrier-signal current is exactly like that which flowed up the transmitting antenna tower, except for being much weaker.

A-M Superheterodyne Receiver. All modern radio receivers are superheterodyne receivers, usually called *superhets*. These sets have four main sections between the receiving antenna and the loudspeaker, as shown in Fig. 4—the converter, i-f amplifier, detector, and audio amplifier.

The converter in a superhet changes the carrier frequencies of all incoming signals to a fixed lower carrier frequency known as the intermediate frequency, or i-f. In modern home and portable radio sets the fixed i-f value is usually 455 kc. In auto radios the i-f value may be either 262.5 kc or 455 kc.

Most of the amplification in superhets is done in i-f amplifiers. The i-f carrier signal still contains the original audio signal. The detector separates this audio signal from the now-unwanted i-f carrier.

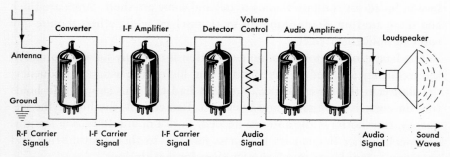

Fig. 4. Circuit arrangement in a typical a-m superheterodyne radio receiver. A transistor may be used in place of a tube in each stage

Do not be surprised if you encounter a radio that has no i-f amplifier tube. In some radios that are designed only for local reception, the converter provides all the amplification that is needed. Here there is just an i-f (intermediate-frequency) transformer between the converter and the detector.

In the audio amplifier are tubes that boost the strength of the audio signal enough to operate the loudspeaker. The first tube in this section is called the first a-f tube. The last tube is called the a-f output tube or the power tube. In some large radios there are two output tubes and a phase inverter tube, to provide greater output power.

The loudspeaker, more commonly called a speaker, has a large paper cone or diaphragm that moves back and forth to produce sound waves when a strong audio signal is fed into it. In this way a speaker gives back the original sounds produced in the studio.

Transistor radios work the same way, using transistors and crystal diodes in place of tubes.

How Television and Radio Sets Work   51

How a Superhet Converter Works. The converter section of a superhet receiver contains two separate stages, the *r-f oscillator* and the *mixer–first detector*. These two stages are usually in a single tube, and have tuning circuits that are controlled by a single tuning knob on the front panel of the receiver.

Whenever you tune a broadcast-band superhet to a new station, you are tuning the r-f oscillator so it generates a frequency 455 kc higher than that station's frequency. The mixer–first detector combines these two signal frequencies to produce a new carrier signal whose frequency is exactly the desired 455-kc i-f value.

The mixer stage in a superhet receiver is also called the first detector because it has the job of detecting or separating out the i-f signal. The detector stage that does the real job of separating the audio signal from the rest in a superhet is therefore usually called the second detector. You will be using all these terms a lot in your work, so it is a good idea to get familiar with them now.

Tuning Action. A receiving antenna intercepts radio waves from many different stations and changes them all into carrier-signal currents, each having a different carrier frequency. Tuning action is needed to separate the one desired station frequency from all the others.

When a coil and capacitor are connected together, the combination works best at one particular frequency. Changing the electrical value of either part changes the frequency at which they work best. This is tuning action, found in every television and radio receiver. Usually the capacitor value is changed by means of a variable capacitor like that shown in Fig. 5.

When the plates of a variable capacitor are fully closed or meshed, the combination tunes to the low-frequency end of the band. When the plates are fully unmeshed, the combination tunes to the high-frequency end of the band. This is illustrated for the a-m broadcast band in Fig. 5.

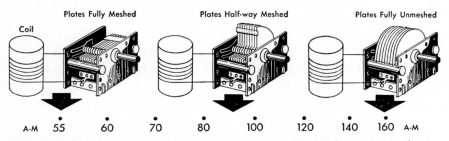

Fig. 5. Positions of tuning-capacitor rotor plates for three different frequencies in the a-m broadcast band. Remember that fully meshed plates give the lowest frequency

Modern receivers use two or more variable capacitor sections mounted on the same shaft. Each section has its own coil. With a two-section or two-gang tuning capacitor, one section tunes the mixer circuit to the desired station frequency. The other section then tunes the oscillator to a frequency 455 kc higher.

With a three-gang tuning capacitor, the extra section tunes an r-f amplifier stage located between the antenna and the mixer. This one-tube stage has two jobs. First, it amplifies a weak desired signal before this signal gets changed to the i-f value. Second, the r-f stage provides additional tuning action (improved selectivity) to keep undesired signals out of the receiver circuits.

Getting Power for Receivers. When radio tubes or transistors are used to boost the strength of a signal, power is added to the signal. This power has to come from somewhere—either from power lines or from batteries. Television receivers, radio receivers, and audio systems that use tubes get their power from the power lines that serve homes, while those using transistors generally get their power from batteries.

In the early days of radio, batteries provided the power required to operate the tubes. When engineers learned how to make radio sets operate directly from power lines, batteries went out of the picture for a while. Today batteries are back again in portable receivers, in auto radios, in tiny portable radios, and in radios for homes that do not yet have electricity.

Transistor Radios. The most popular portable radios of today are transistor radios that operate from easily replaceable flashlight cells or equally small special batteries. These tiny sets almost always use printed circuits. The transistors are soldered directly into the printed wiring since they rarely give trouble.

Transistors operate from very low voltages, so most transistor radios have power-supply voltages somewhere between 1½ and 9 volts. Battery replacement is the chief problem here. The transistor radio circuits themselves are generally very much the same as in ordinary superheterodyne radio receivers.

Clock Radios. Many radios today have built-in electric clocks. The combination is logical because these clocks usually have a built-in time switch that can be set to turn the radio on or off at preset times. This ensures that desired programs will not be missed. The time-switch arrangement can also be used in place of an alarm clock, so a person is awakened by music in the morning rather than by the harsh clatter of a gong.

Even battery-operated transistor portable radios are now made with time clocks. Here the clock is driven by a tiny d-c motor that operates from a separate flashlight cell, so the clock runs even when the radio is turned off. Common troubles in these radio time clocks are covered in the chapter dealing with controls and switches.

Auto Radios. The radio receivers used in automobiles are generally high-quality a-m superheterodyne receivers having a number of special features. First of all, they operate from the storage battery in the automobile, which provides either 6 or 12 volts d-c depending on the car.

In older auto radios, the storage-battery voltage is stepped up by a power supply that includes a vibrator and transformer, to provide the higher d-c voltages required by tubes. In newer auto radios, power transistors are used in the output stage that drives the speaker. These transistors, along with special low-voltage tubes in other stages, operate directly from the 12-volt storage battery that is used today in practically all cars.

Auto radios are generally much more sensitive than table-model radios, to permit good reception when driving in the country far beyond the local range of broadcast stations.

The antenna for an auto radio must be located outside the steel body of the car, because steel serves as a shield that blocks radio waves. Practically all autos today use a whip antenna mounted on a fender or on the roof. Some cars have two such antennas, but this is usually just a design feature. One of them may be a dummy, not even connected.

A few auto radios are complete transistor radios. These are sometimes constructed so there is one removable portable transistor radio containing its own dry cells and its own midget speaker. When this transistor radio section is pushed into its recess in the dash of the car, it automatically makes connections with the auto radio antenna and with a power-transistor output stage that is required to drive the larger speakers mounted in the car. This output stage obtains its power from the 12-volt automobile storage battery, since it would quickly run down ordinary dry cells.

Auto-radio Repair Problems. From a servicing standpoint, auto radios differ very little from other radios. The whip antenna mounted outside the car gives no particular troubles that cannot readily be fixed. The power supply is different, but its troubles are easily located and fixed. The chief problem with auto radios is getting them out of the car so you can work on them. This usually means lying on the floor or seat with your head under the dash, as in Fig. 6.

Sometimes you are pleasantly surprised to find that removing one nut on the engine side of the fire wall allows the auto radio to drop out. At the other extreme, however, you will occasionally have to remove the entire glove compartment or the heater in order to get out the set.

You will soon learn to recognize the models of cars on which you will have to add a few dollars to your repair estimate to cover such factors as extra time, adhesive bandages, liniment for a stiff neck, and damages to

Fig. 6. Working on an auto radio. A flashlight helps to locate mounting bolts under the dash. (Sylvania photo)

work clothes. These hard-to-believe factors are almost certain to become a part of your overhead expense when you work upside down with your head under the dash, trying to loosen hard-to-reach bolts deep in a rat's nest of hoses, wires, control cables, and sharp metal edges. This is why some servicemen prefer to leave auto-radio work for the specialists who have the required garage facilities, bench storage batteries for testing, and all the special wrenches that speed up the work.

Fortunately for servicemen, the auto-radio picture is not quite so gloomy as it sounds. In three out of four sets, the trouble is simply a bad tube or a bad vibrator. The vibrator is replaced just as easily as a tube because it plugs into its own socket. Thoughtful manufacturers now mount the tubes upside down, and make the bottom of the auto-radio housing removable. By taking off this cover, you can remove and replace tubes and the vibrator quite easily with the aid of a flashlight, without taking the set out of the car.

F-M Broadcasting System. Just before World War II an entirely new system of radio broadcasting, called *frequency modulation* or simply f-m, was introduced. This system uses transmitters of its own, operating on an entirely different principle from a-m transmitters, and requires special receivers. Older sets intended for reception of standard a-m broadcasts cannot pick up f-m programs. Many radio receivers have an extra band for f-m so that they can pick up either type of program.

Frequency-modulation broadcast stations operate in the frequency band between 88 and 108 mc. This is very much higher in frequency than ordinary all-wave a-m receivers tune to, so special antennas are needed for f-m. The signals of an f-m station cannot ordinarily be heard farther than about 100 miles from the station.

Within the service radius of an f-m station, the signal obtained is almost entirely free from static. The f-m signal is usually also better in fidelity (naturalness) than that of an a-m station. An f-m station can legally broadcast a much wider range of audio frequencies than a-m broadcast stations are allowed to do. This means that f-m stations can broadcast such high-frequency sounds as the tinkle of bells, the squeak of a door, and the highest notes of string musical instruments, such as the violin. Of course, f-m receivers must be better technically to take advantage of this higher fidelity. Poorly designed or poorly adjusted f-m sets can sound horrible.

F-M Transmitter. Up to the point where the audio signal arrives at the transmitter building, f-m and a-m systems differ only in quality of equipment. Higher-quality amplifiers and microphones are used in the f-m studio to provide the higher-fidelity program. Now let us see how f-m really differs from a-m.

In an a-m transmitter, the process of modulation (mixing the carrier signal with the audio signal) is done in such a way that the *strength* of the carrier signal is varied by the audio signal. In f-m, however, it is the *frequency* of the carrier signal that is varied or modulated by the audio signal. This is where the name frequency modulation comes from.

F-M Transmitting Antennas. The range of an f-m station is limited essentially to the line-of-sight distance from the transmitting antenna to the horizon. The higher a tower is, the farther you can see from it. The higher an f-m transmitting antenna is, the farther the f-m radio wave can travel in a straight line to points on the earth. For this reason, the antenna of an f-m transmitter is located at the highest conveniently available point in a community. Sometimes the antenna is on top of a mountain or hill. Sometimes it is on top of a tall steel tower that is also used by an a-m transmitter.

Line of sight means that, if you can see the f-m transmitting antenna with your eyes or even with a telescope on a clear day when standing alongside your f-m receiving antenna, you can receive the program with an f-m receiver. The higher the transmitting antenna, the greater is the distance from which it can be seen.

There is some bending of f-m radio waves along the curved surface of the earth, so good reception can often be obtained beyond line of sight.

The higher and better the f-m receiving antenna and the more sensitive the f-m receiver, the better will reception be. Similarly, the higher the power of the f-m transmitter and the higher its transmitting antenna, the greater will be the distance at which its signals can be regularly picked up.

Many f-m sets will work with built-in antennas if located within about 10 miles of stations. Some f-m sets have a simple connection to the power line for signal pickup. Other sets have built-in loops containing only one or two turns of wire, or small dipole antennas attached to the inside walls of the cabinet. Best f-m reception is obtained with a television-type antenna located in the attic or on the roof of the house, however.

F-M Receiver. Except for just one stage, an f-m receiver is about the same as a superheterodyne a-m receiver. This different stage is the discriminator, which replaces the second detector as shown in Fig. 7. The r-f oscillator signal still combines with the incoming signal in the mixer to produce an i-f signal, just as in an a-m superhet. The i-f value will usually be 10.7 mc, which is much higher than the standard 455-kc value for a-m radios.

The job of the discriminator is to remove the audio signal from an i-f carrier signal that is swinging back and forth in frequency at an audio rate. The discriminator delivers the desired audio signal to the audio amplifier, just as does the second detector of an a-m receiver. From there on, f-m and a-m receivers are identical in operation.

Some f-m receivers use a ratio detector in place of the discriminator. The circuits are quite similar, and servicing problems are essentially the same for both.

An f-m superheterodyne receiver tunes over a band ranging from 88 to 108 mc, and f-m tuning dials are generally marked directly in these megacycle values.

Small variable capacitors are used for tuning in most f-m sets. Other f-m sets employ unusual tuning arrangements involving sliding bars, moving plungers, or adjustable coils. You will find, however, that all are quite easy to fix once you get acquainted with them.

There will usually be more i-f amplifier tubes in f-m receivers than in a-m receivers. Sometimes the last i-f amplifier tube is called a *limiter*. This tube passes only signals that are *below* a certain strength. It is adjusted to pass the desired constant-amplitude f-m signal. It then blocks or limits any undesired noise peaks that are stronger than the signal. This limiting action

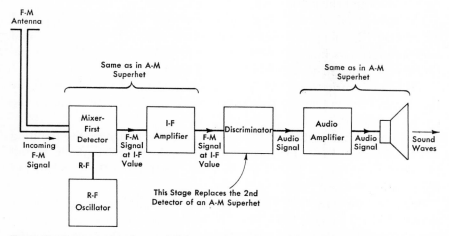

Fig. 7. Circuit arrangement in a typical f-m superheterodyne receiver. The mixer–first detector and the r-f oscillator together serve as the converter. In a-m sets a single tube provides all converter functions, but at the higher frequencies of f-m sets a separate r-f oscillator tube is generally used

is what removes static crashes from f-m programs. An f-m receiver gives enjoyable listening even during thunderstorms, whereas a-m radios are so noisy in storms that they must usually be turned off. Sets that have no limiter use a ratio detector in place of the discriminator.

Television Broadcasting System. Television is a combination of the two broadcasting systems you have just covered, along with some new techniques. Television uses f-m to bring in the sound portion of a program and uses a-m for the picture portion.

Sound System for Television. A complete television broadcasting system is really two systems in one. The sound portion is already familiar to you, so a brief review will take care of it for now.

In the television studio, the microphone on the boom over the performers picks up the sounds and converts them to audio signals. These signals travel through the microphone cable to the control room where they are amplified. The audio signals then go over wires to the transmitter. Here the audio signals are fed to the f-m transmitter that broadcasts the sound portion of the program. This transmitter operates near the high-frequency end of the 6-mc-wide channel assigned to that television station.

The picture transmitter operates near the low-frequency end of the channel. Thus, for a station on channel 2 (54–60 mc), the sound carrier is on 59.75 mc and the picture carrier is on 55.25 mc.

The sound and picture carriers in a television channel are always 4.5 mc apart. By keeping all channels alike in this way, it is possible to build television receivers in which both sound and picture of any television station can be tuned in accurately with one knob.

The sound carrier for a television station is changed to radio waves by the transmitting antenna. These radio waves travel through space to the receiving antenna.

The f-m receiver portion of the television receiver changes the resulting f-m signal current back to the sounds picked up by the microphone. Sound for television programs is thus handled very much as in the standard f-m broadcasting system.

Television Camera. The picture portion of a television program starts at the television camera. Here, lenses focus the desired scene onto a coated plate inside a large camera tube. This tube is known as the iconoscope, the orthicon, or the image orthicon, depending on which particular type is used.

The camera tube changes the picture at the studio into a corresponding picture signal. The job of the television camera thus corresponds to that of the microphone in a sound system.

Many elaborate electronic circuits are needed along with the camera tube. Each tiny speck of the projected picture image inside the camera tube must be changed into an appropriately weak or strong electrical signal. This must be done in the correct sequence so that every part of the picture is transmitted. This orderly sequence is also necessary so that the picture tube in a television receiver can reverse the process and paint the picture back on the viewing screen electrically.

Here are a few technical figures. The picture is transmitted as 525 separate horizontal lines. The process is repeated 30 times a second, which means that 30 complete snapshots of the scene are transmitted per second. The human eye sees these 30 glimpses of the scene as a continuously moving picture, because the eye is able to retain the impression of each glimpse for a fraction of a second. In medical language, this is called persistence of vision.

There is actually only one tiny dot of brightness on a television receiver screen at any given moment. This moving dot is varying continually in brightness. It moves along the lines of the screen so fast, however, that our eyes see only the complete picture.

From the television camera, the picture signal goes to the master control desk in an adjoining room. Here operators watch viewing tubes and switch from one camera to another as called for by the program director.

The picture signal must be accompanied by various synchronizing signals that will keep receivers in step with the transmitter. These synchronizing signals are broadcast along with the picture signal. They are sent out during the intervals when the screen is momentarily dark after the end of each line and after the end of each complete picture.

From the control room, the picture signal goes through special electrical cable to the picture transmitter in another room nearby. Here the picture signal is amplified and then combined with its own carrier signal.

After more amplification, the resulting modulated carrier signal travels over another cable to the transmitting antenna high above the transmitter building. This antenna serves for both the sound and picture transmitters.

Television Receiving Antennas. The same receiving antenna serves for both the sound and picture portions of a television program. In locations close to television stations a built-in loop antenna in the set, a small V-shaped rod antenna on top of the set, or a simple two-rod antenna in the attic can be used. Usually a more complicated directional antenna arrangement above the roof is needed for good television reception, however. A later chapter in this book tells how to install and aim television antennas.

The reliable maximum reception range of television signals is limited to about 150 miles because of the high frequencies involved, and is normally about 50 miles. This is much the same as for f-m. In hilly country, the range is less if hills block the straight line-of-sight path between transmitting and receiving antennas. Highly directional antennas, aimed accurately at the transmitter, are needed to pick up a good television signal at distances over 50 miles.

Television Receiver. The picture carrier signal from the television antenna is boosted in strength and tuned in much as for an ordinary a-m superheterodyne receiver. Because of the higher frequencies involved, the circuits and parts look different, but they work in the same way.

The second detector in the picture portion of a television receiver is often called the video detector, because it separates the picture or video signal from the picture carrier. This video signal is boosted in strength by the video amplifier tubes and fed to the television picture tube, as indicated in Fig. 8. There it causes the spot on the screen to vary in brightness from instant to instant. This spot is produced by an electron beam that is produced by an electron gun in the narrow neck of the picture tube.

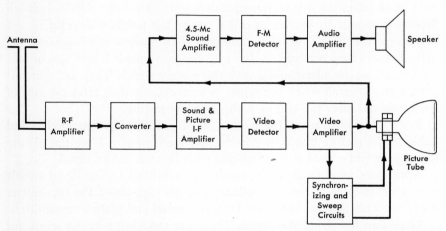

Fig. 8. Circuit arrangement used in modern television receivers. This is known as the intercarrier system. Sound and picture signals take the same path up to the picture tube. The r-f amplifier stage builds up the strength of incoming signals before feeding them to the converter

The synchronizing signals that are transmitted to keep receivers in step come out from the video amplifier. These sync signals are used to drive horizontal and vertical sweep circuits.

The horizontal sweep circuit feeds a coil that makes the spot on the screen move back and forth over each of the 525 horizontal lines into which the picture is divided.

The vertical sweep circuit feeds a coil that makes the spot move down to the next line when it reaches the end of one line. The vertical sweep also makes the spot move up to the top of the screen for a fresh start each time it reaches the bottom of the picture. Both the horizontal and vertical deflection coils are mounted around the neck of the picture tube.

Intercarrier Sound System. In the arrangement of Fig. 8, the television sound carrier signal travels with the picture carrier signal through the r-f amplifier, converter, i-f amplifier, video detector, and video amplifier. In passing through the video detector, the sound and picture i-f carriers mix with each other to produce a new sound i-f carrier signal of 4.5 mc. The picture signal goes to the picture tube, and the sound i-f carrier signal goes through its own 4.5-mc i-f amplifier, f-m detector, audio amplifier, and loudspeaker just as in an f-m receiver. This is known as the intercarrier television receiver system.

Separate Sound Channel. Older television receivers use a different arrangement for the sound channel, as shown in Fig. 9. At the output of the

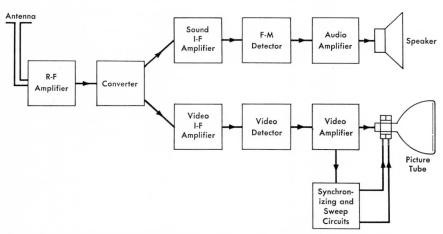

Fig. 9. Circuit arrangement used in older television receivers, in which sound and picture signals take separate paths after the converter

converter tube the picture and sound i-f carrier signals are separated. The sound i-f carrier signal then goes through its own i-f amplifier, its own f-m detector, an audio amplifier, and a loudspeaker just as in an f-m receiver. Many sets using this separate sound-channel arrangement are still in use, because it was in all sets made before 1948. After 1952 all sets used the intercarrier arrangement. In sets made between 1948 and 1952 you can expect to find either arrangement.

Remote Controls for Television Sets. Many television receivers can be bought with remote controls. These permit changing channels, changing volume, and turning the set on or off from anywhere in the room. The various systems may use ultrasonic waves, radio waves, a light beam, a cable, an air hose, or power-line wires to link the remote-control unit with the television set.

One of the most fascinating remote-control systems is the one that uses ultrasonic sounds, having frequencies above the human hearing range. The

62 Television and Radio Repairing

control unit is small enough to fit in the palm of the hand. It may contain only a number of short aluminum rods, each producing a different ultrasonic frequency when tapped. Pressing a pushbutton for a particular control function causes a hammer in the control to tap the corresponding ultrasonic gong. The resulting inaudible sound is picked up by a microphone concealed in the cabinet of the receiver. After amplification and rectification, the signal operates an electromagnet or motor that advances a station selector or performs some other desired control function.

Manufacturers' service manuals are generally required for adjusting and repairing remote-control mechanisms, because the procedures are different for almost every set. With these instructions, the repair jobs are no more difficult than those on other home-entertainment equipment.

Television Troubleshooting. Television receivers usually have somewhere between 15 and 20 tubes. Despite all these tubes, television-receiver troubles are very much the same as those of f-m and a-m receivers. Actually, it is often easier to find the defective part in a television receiver because the picture itself gives clues to the trouble.

UHF Television Reception. Most television receivers can be obtained with or without a separate uhf tuner for channels 14 to 83. This tuner is not needed in localities that are served only by vhf stations, and is therefore left off to keep the cost of the set down.

In television receivers having turret-type tuners, uhf tuning strips can be installed in the tuner to give any desired combination of uhf and vhf channels. A special uhf antenna is usually required for good uhf reception.

Color Television Reception. The screen of a black-and-white television picture tube gives off white light when hit by the electron beam inside the tube. This white color is really a mixture of many different colors. Thus, when red, green, and blue primary colors are mixed in the proper proportions, the human eye blends them all together and sees white.

Color television receivers take advantage of this basic fact by having groups of tiny red, blue, and green phosphor dots on the screen of the color picture tube. Three electron beams, produced by three separate electron guns in the neck of the picture tube, are focused close together on the screen of the tube. Each beam hits only its own assigned color of dot when the trio of beams is swept back and forth across the screen by the single deflection yoke.

For color television, the television transmitter broadcasts a special signal that tells how much of each primary color is required at each point being scanned. The color television receiver has special circuits that extract this color information and use it to change the strengths of the red, green, and blue electron beams from instant to instant.

When a color television receiver is used to receive black-and-white programs, the receiver circuits automatically set themselves to feed to the three guns the correct signal proportions for white light. The strengths of all three beams are then varied simultaneously in accordance with the brightness of the scene being scanned by the camera just as beam strength is varied in a black-and-white receiver to paint the picture electronically on the screen in shades of gray.

Color Television Servicing. Except for the color picture tube and a few special parts, you will see familiar parts when you take the back cover off a color television set. The back-of-set adjustments are more critical, however, and the set is more complex than a black-and-white set. Furthermore, the color picture tube voltage is about 25,000 volts, as compared with around 15,000 volts for black-and-white sets. This means that color television sets are best left alone until you become more familiar with the special adjustment procedures, high-voltage precautions, and special servicing techniques required on individual makes of sets. Do not try to do everything at once. Start with the easiest jobs, on which you can do good work at a good profit, before complicating your life by tackling the tough ones.

Getting Acquainted with Consoles. A receiver that rests directly on the floor rather than on a table is known as a console. The simplest consoles have a television or radio chassis exactly like that used in table models, but generally have a larger loudspeaker that is attached to the cabinet rather than to the chassis. Larger consoles may also have a control unit for remote tuning, along with a radio receiver, record changer, and perhaps also a magnetic-tape recorder.

High-fidelity Home Audio Systems. When the audio amplifier of a radio or television set is built as a separate unit, using better parts and providing higher audio output power, it becomes the main unit of a high-fidelity home audio system. This amplifier can be used with higher-quality loudspeakers mounted in special cabinets, to give reproduction approaching that heard in the concert hall. The input to the amplifier may be an automatic record changer, a high-quality single-play turntable for long-playing $33\frac{1}{3}$-rpm records, a magnetic-tape recorder, an a-m radio tuner, an f-m radio tuner, or a microphone.

Between the input signal source and the main amplifier is usually a preamplifier having jacks for the various inputs, separate bass and treble tone controls, and a volume control. There may also be a switch for selecting the desired input and providing the correct extra amplification for that input. More elaborate preamplifiers have still more controls for playing the different makes and types of phonograph records with maximum fidelity, for suppressing needle-scratch noise on worn records, and for activating other special control circuits.

Since home audio systems use the same parts as home radio-phono combinations, most of the instructions in this book for testing and replacing parts apply to these systems also. It should be pointed out, however, that owners of expensive home audio systems are usually fussy about the quality of the sound reproduction. Until you have acquired more experience with the special circuits used in these systems, it may be better to call in an audio expert when checking and replacement of tubes does not clear up the trouble.

Stereo Home Audio Systems. When you place two ordinary home audio systems in the same room and feed them with a stereo recording or broadcast, you have a stereo home audio system. Except for the input, servicing problems are the same as for a single audio channel.

Stereo Records. Stereo $33\frac{1}{3}$-rpm records are the commonest stereo sound sources today. The grooves in these records are at an angle of exactly 45 degrees to the vertical. One side of the groove contains the sounds picked up by one microphone during recording. The other side of the groove contains sounds picked up by another microphone that is spaced several feet away. The special stereo pickup used for playing these records has a single needle that drives two elements, each delivering the signal from one side of the groove. A stereo pickup can thus feed two separate amplifiers simultaneously.

Stereo Tapes. Stereo magnetic tape is another signal source for stereo home audio systems. Two tracks are recorded side by side on the tape. Sound signals recorded in these tracks actuate two magnetic playback heads mounted side by side on the stereo tape machine, each feeding one stereo amplifier input.

Stereo Loudspeaker System. The two loudspeakers or two sets of loudspeakers for a stereo system are spaced several feet apart, approximating the spacing of the microphones at the recording studio. Listeners seated in front of these loudspeakers hear the various instruments of an orchestra from various directions, just as does a listener in front of the orchestra itself.

When playing a stereo demonstration record of a pingpong game, the sound of the ball hitting the bat comes first from one speaker and then the other. With a railroad sound-effects record, the train is first heard whistling far off to the left. The engine roar begins faintly as the train approaches, builds up to a peak as the locomotive passes the microphones, then gradually fades away to the right. In the background is heard the steady clickety-clack of car wheels as they pass the rail joint in front of the microphones.

Recognizing Radio Parts. Your first repair jobs will involve locating and replacing bad parts. The sooner you learn to recognize these parts by appearance, the faster you will be able to locate the trouble. For this reason, the following paragraphs show you what each type of part looks like, tell how to distinguish one part from another, and give a general idea of what each part does in a set.

Getting Acquainted with Tubes. You will work with tubes far more than with any other part, because a tube is the cause of the trouble about three out of four times when an average receiver goes bad. This is why tubes are made with plug-in terminals for easy removal and replacement.

One important job of a tube is boosting the strength of a weak signal. Some tubes amplify r-f signals, while others are better for amplifying a-f signals.

Tubes are designed for specific amplifying jobs in specific circuits. This is why there are so many different tube types. You will find that each type is identified by a combination of numbers and letters stamped on the glass or metal wall of the tube.

Other jobs for tubes include rectification, frequency conversion, and detection. Cathode-ray tuning-indicator tubes show when a receiver is correctly tuned. Larger cathode-ray tubes serve as the picture tubes in television receivers. The fact that tubes have so many different jobs in receivers makes it easier to locate the bad tube in many sets just by noting the nature of the trouble.

Tubes require heat. This is obtained by sending current through a filament wire to make it red-hot, just as in an electric lamp. The heated wire may give off electrons directly, as in a tube for battery-operated portable radios. More often, the filament provides heat for a chemical coating on the surrounding cathode electrode, to provide much greater emission of electrons. This means that a properly operating tube must feel warm or hot when touched.

Getting Acquainted with Transistors and Crystal Diodes. Study carefully the typical transistor and crystal-diode shapes shown in Fig. 10. Both

66 Television and Radio Repairing

parts will usually be soldered directly to circuit terminals, since they seldom go bad. You need to be able to recognize these parts at a glance right from the start, because they are easily damaged by carelessly applied soldering heat. You will learn later how transistors and crystal diodes can be unsoldered and soldered safely.

Transistors do the same jobs as amplifier and oscillator tubes in receiver circuits, but have many advantages over tubes. Thus, transistors last longer than tubes, require no filament or heater current, and operate on very low

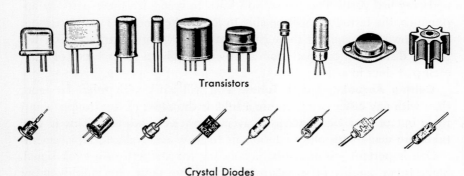

Fig. 10. Examples of transistors and crystal diodes used in radio and television receivers

voltages. A few flashlight cells in series provide all the voltage and power needed by all the transistors in the average portable receiver.

Crystal diodes do the same jobs as vacuum-tube diodes, but are smaller, cheaper, and require no filament or heater current. Crystal diodes are most often used as second detectors in superheterodyne television and radio receivers.

Getting Acquainted with Resistors. The three basic types of resistors used in home-entertainment equipment are shown in Fig. 11. These resistors serve to limit and control the flow of electric current in circuits. The higher the resistance value of a resistor, the more opposition it presents to the flow of current. You will learn more about resistance in the chapter dealing with electricity and magnetism.

Carbon Resistors. Some carbon resistors are molded from a mixture of carbon and clay, then baked until hard. Others consist of a carbon coating on a glass rod, covered with a protective plastic insulating coating. When a carbon resistor burns out, it turns black and usually gives off a pungent odor of overheated plastic.

Most carbon resistors have tinned copper wires coming out from their ends. These wires are called axial leads. Only in older sets will you occasionally find carbon resistors with radial leads, coming out at right angles to the ends.

Carbon resistors usually have three or more bands or dots of color. These colors give the value of the resistor according to the EIA color code, created many years ago by the Electronic Industries Association. In the chapter on resistors you will learn how to read these color-coded values at a glance.

Wirewound Resistors. A length of high-resistance wire wound on a ceramic or other insulating form serves as a wirewound resistor. The turns of

Fig. 11. Examples of the three commonest types of resistors. Left—carbon resistors; center—wirewound resistor; right—potentiometer

wire are spaced just far enough apart so they cannot touch each other. The ends of the wire are fastened to terminal lugs or to tinned copper leads. The resistance value may be printed directly on the resistor or given by the color code.

Wirewound resistors are used to provide extremely low resistance values, to give more precise values than can be obtained with carbon resistors, and to handle higher power values. They are more expensive than carbon resistors, and are therefore used less often.

Rheostats and Potentiometers. In some circuits it is necessary to change the amount of current flow from time to time. Here a different type of resistor is used, in which a contact arm is moved on the resistance element by rotating a knob. A volume-control circuit in a radio receiver is a familiar example. This type of resistor is called a rheostat if it has two terminals, one for the movable contact arm and the other for one end of the resistance element. Practically all controls have three terminals, however—one for each end and one for the movable arm; these are called potentiometers.

When designed to be adjusted from the front of the set, a potentiometer has a long shaft that projects through the cabinet of the set, with a knob on its end. When designed for only occasional adjustment, it will usually be at the rear of the set or behind a panel at the front, with a short metal shaft that is knurled or has a screwdriver slot for adjustment purposes. One entire chapter in this book deals with the testing, repair, and replacement of controls like these.

In television sets, potentiometers and rheostats are widely used in circuits for adjusting picture height, width, centering, linearity, and synchronism.

You will often encounter dual potentiometers, mounted one behind the other. Here the shaft for the rear potentiometer goes through the hollow shaft of the front potentiometer. Each shaft has its own knob, one behind the other. The controls can thus be adjusted independently, yet they take no more room than would a single control on the front panel of the set.

Defective potentiometers and rheostats are easy to locate, because their misbehavior when adjusted pinpoints their guilt. You simply order an exact-duplicate replacement from your jobber and install it in place of the bad unit. The entire repair job can be completed in well under half an hour, once the new part is on hand.

Getting Acquainted with Switches. A switch is nothing more than a device for making or breaking a connection between two points. The construction of the switch is generally determined by the amount of current that passes between its contacts when they are closed, by the operating method desired, and by the number of circuits that are to be opened or closed at the same time.

Switches can be operated by turning a knob, flipping a lever, pushing a button, or performing some other type of movement. One of the commonest types of switches is mounted at the rear of a volume control, to turn on a set automatically when the volume control is turned up. Here the set goes off when the volume is turned down.

A television tuner generally contains a switch that is controlled by the vhf station selector knob. This is called a selector switch, because it switches the correct combination of parts into the tuner circuit for reception of the vhf channel selected by the user. Switches serve as controls, so you will find them covered along with controls in a later chapter.

Getting Acquainted with Capacitors. The commonest types of capacitors that you will work with are shown in Fig. 12. Each serves to block the flow of direct current while controlling the flow of alternating current.

Paper capacitors are the commonest. They will usually have a waxed tubular paper housing or a colorful molded plastic housing. The electrical value will be printed directly on the housing or indicated by means of a

How Television and Radio Sets Work 69

color code. Axial leads serve as terminals. If you cut open an old paper capacitor, you will find in it long strips of metal foil separated by insulating paper. The commonest trouble is a short, wherein the insulation goes bad at one point and opposite layers of foil touch each other. The capacitor must then be replaced.

Ceramic capacitors look somewhat like buttons, but have two parallel leads coming out from the edge. They consist simply of thin metal layers separated by an insulating ceramic. Values are usually printed directly on the capacitor.

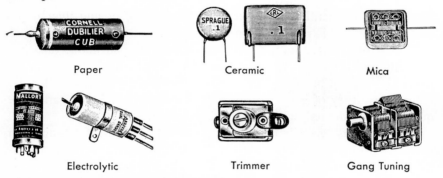

Fig. 12. Examples of fixed and variable capacitors

Ceramic capacitors are used chiefly in signal circuits, where they provide low capacitance values for tuning a circuit or for keeping r-f signals in their proper paths. Ceramic capacitors seldom go bad, so there is no need to carry a stock of them in your toolbox. In the chapter on capacitors, you will find that you can get along with just a few standard sizes on your workbench. Others can be ordered as needed.

Mica capacitors do the same jobs as ceramic capacitors, and are equally good. You will find them chiefly in older sets, because recently perfected ceramic capacitors are much cheaper in a given size. Mica capacitors consist of alternating layers of sheet mica and metal foil, encased in molded plastic. They have two leads, each connecting to alternate layers of the foil.

Electrolytic capacitors often resemble large paper capacitors, and have axial leads for soldering to circuit terminals under the chassis. Larger electrolytic capacitors are usually in tubular metal cans mounted on top of the chassis. These have terminals or insulated leads that project through a hole in the chassis, for making connections to receiver circuits. When two

70 *Television and Radio Repairing*

or more electrolytic capacitors are in the same housing, there will be three or more leads or terminals.

Electrolytic capacitors have polarity markings. This means that one terminal of a section will be marked minus and the other plus. A new capacitor must always be installed with the same polarity as the old one.

Electrolytic capacitors are used chiefly in power supplies, to provide a filtering action that removes hum voltages from signal circuits. To do this, they have very large capacitance values, obtained by using chemicals to form a thin film of insulating gas on the surface of the metal foil inside. Electrolytics frequently go bad because of natural aging, and thereby provide many profitable repair jobs for servicemen.

Getting Acquainted with Tuning Capacitors. As you know, capacitors act with coils to provide tuning action. With fixed coil and capacitor values, only one station would be received. With a variable capacitor, the natural frequency of a circuit can be changed to match the carrier frequency of the desired radio or television station.

Since receivers generally have two or more tuned circuits for tuning purposes, gang tuning capacitors are widely used. Here two or more sections are mounted on the same shaft for rotation by a single knob.

Many receiver circuits require only occasional adjustments. Here trimmer capacitors are used, having slotted or nut-shaped adjusting screws designed for use with special insulated-shaft screwdrivers or tools. These trimmers are adjusted correctly at the factory and rarely need attention thereafter. Since they can be readjusted properly only with the aid of signal generators and other test instruments, be careful not to change their settings when working on a set.

Getting Acquainted with Resistor-Capacitor Units. Combinations of two or more resistors and capacitors wired together as a subassembly as in

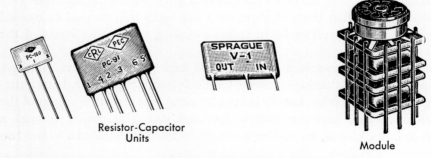

Fig. 13. Examples of resistor-capacitor units. Modules are found only in a few sets

Fig. 13 are widely used in modern television and radio sets. These may have separate ordinary resistors and capacitors assembled on a small printed-circuit board, or may have resistor elements printed directly on a ceramic plate, with capacitor elements soldered to the plate.

Another type of combination unit, called a module, contains all the printed resistors and capacitors that make up a single stage. Small coils and a tube socket are also included in some modules. The parts are printed on individual plastic or ceramic slabs, with the tube socket on the top slab. Connections between the slabs are made with stiff wires that are dip-soldered into position at the factory to serve also as the frame for the module. Some of these wires are longer and serve as terminals that can be inserted in holes in printed-circuit boards. An example is shown in Fig. 13.

Getting Acquainted with Coils. A coil is simply a number of turns of insulated copper wire, used to pass direct current while offering a desired amount of opposition to the flow of alternating current. The smallest coils are no larger than carbon resistors, and are used extensively in television sets to keep signals in desired paths.

Small coils are widely used in television tuners to provide tuning action. Here the turns of wire support themselves in air, with nothing inside the coil. For the higher vhf channels, such as channel 13, the tuning coil may have only a fraction of one turn, made by punching an arc out of a sheet of metal.

When coils have a large number of turns of fine wire, they are usually wound on a cardboard or plastic insulating form. Coils do not have to be circular; those used in the deflection yoke that surrounds the neck of a television picture tube are pressed into an odd shape so they interlock with each other and fit around the neck of the tube.

The electrical performance of a coil is increased by using a metal core inside the coil. At radio frequencies a powdered-metal core is used, as in loopstick antennas. For lower frequencies, as in audio chokes and filter chokes, the core consists of sheet-metal laminations bolted tightly together.

Getting Acquainted with Transformers. When two or more coils are wound close together on the same insulating form or iron core, the result is known as a transformer. This part is used to transfer alternating current from one circuit to another through space by means of a magnetic field. A transformer thus passes alternating current without transferring direct current.

In radio and television sets, an r-f transformer will be found between the antenna and the converter stage, where it handles the carrier frequencies picked up by the antenna. An i-f transformer will be found between the converter and the second detector, where it handles the lower intermediate frequency to which the signal has been changed by the converter. An output transformer is located between the output stage and the loudspeaker, and serves to transfer audio signals. The power transformer, found only in a-c sets, is located in the power supply where it serves to transform the a-c power-line voltage to the higher and lower values required by the set.

You can recognize an i-f transformer by the fact that it is in a round or square aluminum shield can mounted on top of the chassis, with one or more holes on top to provide access to its adjusting screws. The terminals will project under the chassis through a hole.

An r-f transformer does not usually have a shield, and may be either above or under the chassis. Both r-f and i-f transformers have very fine wire that is easily broken if bumped with a sharp tool.

An output transformer is most often mounted directly on the loudspeaker. Occasionally you may find it on top of or underneath the chassis, particularly in transistor portables. The iron core is a clue to its identity.

A power transformer has more leads or terminals than other transformers, and is also larger and heavier. For this reason, it will always be found mounted on top of the chassis. A power transformer will usually have a heavy shaped-metal shield as its housing, to prevent its magnetic field from acting on nearby parts.

Although coils and transformers seldom go bad, they can be damaged by the failure of other parts or even by a stroke of lightning that hits somewhere near the set. For this reason, one entire chapter in this book is devoted to these parts.

Getting Acquainted with Metallic Rectifiers. In place of a rectifier tube, more and more receivers are using a metallic rectifier in the power supply. This may be either a selenium, silicon, or germanium rectifier. Metallic rectifiers are usually soldered directly to receiver circuits, because they last much longer than ordinary rectifier tubes.

Getting Acquainted with Pilot Lamps. You will find two distinct types of pilot lamps in sets. One has a screw base and the other a bayonet base. Both require sockets. The same type of lamp should always be used for a replacement. When the identifying number cannot be read, use the color of the internal glass bead, the size and shape of bulb, and the size and type of base as your clues for ordering the right replacement.

Getting Acquainted with Fuses. Television and radio fuses usually consist of glass tubing through which the fuse wire can be seen. Metal end caps serve as terminals, just as in automobile fuses. These fuses fit into spring-clip holders so they can easily be replaced.

You will also find fuses that have wire leads and are soldered directly into receiver circuits. There is a good reason for this, beyond that of lower cost; a fuse usually blows only when some other part in the set goes bad. If the fuse is too easy to replace, some customers may be tempted to do it themselves and damage the set still more. Fuse troubles are covered in detail in the power-supply chapter, since most fuses are located in power circuits.

Which Parts Go Bad Most Often? After tubes, paper and electrolytic capacitors go bad most often. Next come volume controls, fixed resistors, and switches.

The remaining parts, such as metallic rectifiers, power transformers, r-f and i-f transformers, coils, loudspeakers, ceramic capacitors, wirewound resistors, tuning capacitors, transistors, crystal diodes, and fuses, need replacement in less than 1 out of every 100 jobs.

Pilot lamps are ignored in this failure-expectancy picture. They may burn out as often as tubes but are easy to spot, easy to replace, and do not usually affect receiver performance.

4

Making a Service Call

Nature of a Service Call. When a television set or console radio goes bad, the average customer will pick up the telephone and call a serviceman. A few will do this also for table-model sets, but most of these small sets will be brought into your shop. In both cases, however, your goals will be the same—to make a good impression on the customer, confirm the customer's complaint by trying out the set, check it over carefully for obvious defects, test the tubes, and then remove the chassis from the cabinet for troubleshooting if these preliminary checks fail to locate the trouble.

This chapter takes you through this entire sequence, from your first contact with the customer right up to the time you start troubleshooting underneath the removed chassis. You learn what to ask and say when a phone call comes in for service, what to take with you on a call, how to look your best, how to act to make a good impression, and how to carry on conversation with a customer while you work.

You learn to look and listen first, so you will not overlook a simple and obvious defect that does not even require removing the chassis. You learn how to try out a television or radio set you have never seen before, without letting the customer know that the knobs are all strange to you. You learn what troubles to look for in each type of set before removing the chassis.

Finally, you get practical step-by-step instructions for removing the chassis from even the most complex television-radio-phono console. The entire chapter is thus in exactly the same order as the events of a typical service call, right up to the final paragraph on collecting the money. So—let us assume now that you hear the telephone ringing in your shop.

Handling Telephone Calls. Always answer your telephone with a cheery hello, as if greeting the prospective customer personally with a smile on

Making a Service Call 75

your face. If doing spare-time work from your home, you will not know whether the call is a personal or business one until the other person asks, "Is this the man who fixes radios?" Your answer would then be, "Yes, this is John Jones. Can you tell me a little about your set?" This, of course, is what the customer wants to do most of all.

Listen patiently and carefully even though the detailed description is technically worthless. Just remember that from this phone conversation

Fig. 1. Answer your phone with a cheery greeting before it rings the second time, and you will have a good start for another profitable service job. (Sylvania photo)

you must obtain the customer's name, address, a definite day-and-hour appointment for looking over the set, and if possible the make and model number of the set. This last information is desirable because you can then take along spare tubes of the types used in that set, by looking up the circuit in your service manuals. Keep a pad or order book alongside your telephone for this purpose, as in Fig. 1.

Equipment for Home Calls. Your toolbox, tube caddy, tube tester, and multimeter are all that you ordinarily need on home calls. Your own car will be entirely all right for business use at the start, but eventually you will want a business car or truck of some kind.

Take only your toolbox and multimeter out of the car on your first trip. Set them down before ringing the doorbell. When someone opens the door, take off your hat and identify yourself by giving your name and the purpose of your call. Keep the hat off from then on. You can easily carry

76 Television and Radio Repairing

it in the same hand as your toolbox when you go into the house. Of course, you are not smoking when you go in.

The next question is: What sort of an impression do you make as the customer opens the door in answer to your ring? Your appearance, your professional attitude, and your technical ability all determine whether people will call you regularly or just use you as a last resort when their set goes bad.

How to Make a Good Impression. What you do in the customer's home has a great deal to do with the impression you make, as Fig. 2 clearly

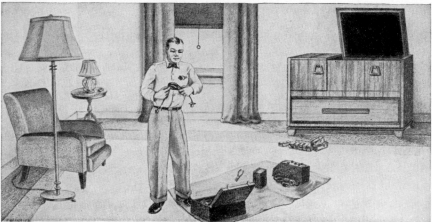

Fig. 2. Top—How to make a bad impression in a customer's home while working on high-fidelity home audio set. Bottom—How to make a good impression. Note use of cloth under toolbox to protect rug while working on home audio system. (Philco illustrations)

shows. Your customer regards his home as his most valuable possession. Any damage that you do through carelessness—any disrespect you show toward his home and his furnishings—will not be overlooked or forgotten.

Here is one way to make a good impression immediately when you enter a home. Spread out a large, clean piece of cloth on the rug or floor alongside the console before you even touch it. Place your toolbox on this cloth, never on the furniture. The cloth will prevent the toolbox and tools from soiling the rug or scratching the floor.

Whenever it is necessary to move furniture in order to get at the television or radio set, be careful not to scratch either the furniture or the floor. If any floor lamps are near the set, ask the customer's permission to move them away temporarily, since they are so easily knocked over.

Conversation. In answering questions, never give a lecture on technical subjects just to show how much you know. The customer always appreciates a simple and definite answer. Let him ask more questions if he wishes. Answer each one briefly as best you can. Remember that you are going to have to charge somebody for the time you spend in idle conversation with a customer if you are going to earn a good living from servicing. On the other hand, no customer is going to pay willingly for talking time. Therefore, you yourself have to keep the lost time at a reasonable minimum.

In conversation with a customer, avoid mentioning politics, religion, or obvious *personal* matters. Many servicemen have got into trouble and lost customers by commenting on these three dangerous topics. Let the customer give his own opinions if he wishes, but stay neutral yourself always.

Look and Listen First. Records show that in five to seven calls out of ten there is no need to remove the chassis of a television or radio set from its cabinet to find the trouble. Bad tubes are the commonest trouble. Other troubles that can be fixed without removing the chassis are:

1. Improper adjustment of the operating controls for the receiver.

2. Power-supply trouble, such as a defect in the line cord, faulty house wiring, blown fuses in the house or in the set, dead batteries and loose battery connections in portable radios, and a bad vibrator in an auto radio. All these are covered in the chapter dealing with power supplies.

3. Trouble in the antenna system that picks up signals and brings them to the set.

4. Obvious above-chassis trouble, such as tubes loose or out of their sockets, loose or missing tube shields, broken wires, or toys jammed in the set by children.

5. Man-made or natural interference that is noted for the first time and blamed on the set by the customer.

All these troubles can be recognized and eliminated or explained without removing the chassis. Many can be checked while you are trying out the set to confirm the customer's description of the complaint.

Trying Out a Radio Set. Whenever possible, *let the customer turn on the set and demonstrate it for you.* Watch closely and note what each knob is for, so you can operate the set with confidence yourself even if you have never seen that particular model before. Only a few sets have lettering on the cabinet or knobs to identify each knob.

No serviceman is able to identify all the control knobs at a glance on a strange set, since controls are arranged in so many different ways. Despite this, customers often think their set is the only one in the world and feel insulted if you ask which knob turns on the set. Watch carefully while the customer tries the set. Ask him to try certain things, such as the tone control and band-changing switch, if you cannot tell which they are.

Sometimes you will have to find out for yourself what each knob does. On table-model sets there will usually be only two knobs. One will be for tuning, and can be identified by a glance at the tuning dial while turning it. The other will be a combination on-off switch and volume control. Turning this knob clockwise until a click is heard turns on the set. Further clockwise turning then increases the volume or loudness of the program. On some sets, the volume-control knob must be pushed in or pulled out to turn on the set. This allows the customer to leave the volume control at the desired setting all the time.

Radio sets having an f-m band or a short-wave band in addition to the broadcast band will also have a band-changing switch or knob. Check performance on each tuning band. If the set works on one band, you know immediately that most of the set is O.K. and the trouble is in the parts for the other band.

Many radios have a tone-control knob. This can be in any position when trying out the set.

Larger table-model radios and most consoles will also have a radio-phono switch. A receiver that is set to the phono position will sound just as dead as if there were trouble, unless it is connected to a phonograph and a record is playing.

When trying out a radio set, your ears will tell you whether performance is satisfactory. Local stations should come in loud and clear. Larger sets should bring in a few distant stations as well. On local stations there should be no hissing or static noise.

Hum heard along with the program should not be objectionably loud. On most sets some hum is normally noticeable during moments of silence in a program if your ear is close to the loudspeaker.

If no stations can be heard, turn the volume control up and down a few times. If the hum increases when the volume is turned up, you will know that the audio amplifier and loudspeaker are O.K. You then look for the trouble between the antenna and the volume control, in the r-f and i-f stages. If the hum is quite loud and not affected by the volume control, look for trouble between the volume control and the loudspeakers in the audio amplifier and the power-supply filters.

There should be no raspiness or severe distortion. The amount of distortion will depend on the size, quality, and cost of the set, so listen to new sets whenever possible and note how they sound. You will quickly learn what to expect in tone quality from various sets.

Trying Out a Television Set. With television sets it is particularly desirable to let the customer show you how the set behaves or misbehaves. By watching, you can identify all the controls and at the same time see if the customer is tuning in the set properly.

The controls found most often on the front, side, or top of a modern television receiver are the on-off switch, channel-selector switch, fine-tuning control, contrast control, brightness control, and volume control, as shown in Fig. 3. These will have large knobs, generally arranged in pairs one behind the other on concentric shafts. The on-off switch is usually combined with the volume control on one knob just as in radio sets, and may often be the push-pull type. There may even be a separate knob in the center of the volume control that is pushed once to turn on the set, and pushed again to turn the set off.

Other controls occasionally found with large knobs on the front panel will have such names as horizontal hold control, vertical hold control, focus control, and tone control. When not visible on the front panel, some or all of these additional controls have small knobs, and may be concealed behind a hinged or sliding portion of the front panel or may be at the rear of the chassis along with other semiadjustable controls. When a receiver is working properly, only the large-knob front-panel controls are needed for tuning in and adjusting picture and sound.

Many different special controls will be found at the rear of a television set on top of the chassis, behind a small hinged door, or on the front or side of the cabinet. Sometimes these have small knobs or have knurled shafts that can be turned with the fingers like a knob. More often the ends of the control shafts are slotted so they can be turned with a screwdriver.

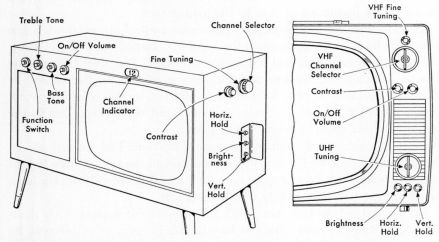

Fig. 3. Locations and functions of controls on typical console and portable television receivers. Function switch on console has four positions; the TV position gives television reception, while the PHONO, TAPE, and STEREO positions permit plugging home audio equipment into jacks at the rear for use with the audio amplifier and speaker of the set

Before adjusting a screwdriver control, mark or measure its original position as shown in Fig. 4, so you can return exactly to the original setting if changes have no effect on the trouble.

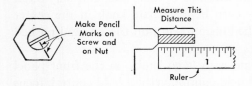

Fig. 4. Methods of marking positions of screwdriver controls so you can go back to their original settings if necessary

Circuit Breakers and Fuses. Some television sets have a circuit breaker to provide protection against current overload. There will usually be a manual reset button at the back of the chassis. If the set is completely dead, try pushing in this reset button. Sometimes a circuit breaker will open of its own accord, particularly when the a-c line voltage in the home is higher than normal.

If the set works when the button is pushed, wait at least 10 minutes to see if the circuit breaker opens again. If it does open, the set needs trouble-

Making a Service Call   81

shooting and should be taken to your shop. Tubes seldom cause a circuit breaker to open, so you will have to look for a shorted capacitor or some other defective part.

In place of a circuit breaker, some television sets will have replaceable fuses. If you see one that is blown and have the correct replacement, try inserting it. If the new fuse blows, troubleshooting is required just as for a circuit breaker that opens again.

What Television Controls Do. Picture trouble in a television receiver can often be cleared up by adjusting one of the small-knob controls at the

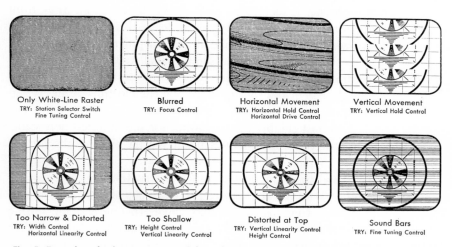

Fig. 5. Examples of television picture defects that can be due to improperly adjusted controls. For an entirely dark picture, try the brightness and contrast controls. For an off-center or tilted picture, adjust the deflection yoke.

front or rear of the set. Once you learn what each of these controls does to the picture, you will be able to select and adjust the correct control for the particular trouble. This knowledge of control action is important, because careless tuning of some of the controls can throw the set way out of adjustment.

Effects of some of the controls on the picture are shown in Fig. 5. For each of these picture troubles, the controls that may need adjusting are listed. In each case, the first control listed is the one that ordinarily fixes the trouble, but the others listed may need slight adjustments because they interact with the first one.

The following paragraphs give in alphabetical order the common names of the various controls found at both the front and rear of television sets,

82   Television and Radio Repairing

tell what each control does, and give adjusting instructions where needed. Study this list carefully now, and use it for reference whenever in doubt about some control.

Television Controls

agc control   Back-of-set control that varies the level at which the automatic-gain-control (agc) circuit works. Used only on a few sets. Adjust for clearest picture when tuned to the strongest local television station. Picture may vanish when this control is way out of adjustment.

agc switch   Permits disconnecting the automatic-gain-control circuit when maximum sensitivity is required, as for receiver locations far away from television transmitters. For most locations, leave this agc switch in the ON position, where it serves to make all station programs come in with approximately the same picture and sound signal strength.

brightness control   Varies the picture-tube operating voltage that determines the brightness of the spot produced by the electron beam. When the brightness control is adjusted, the contrast control will usually need adjusting also. Adjust the brightness control for the dimmest acceptable picture, as excessively bright pictures mean shorter picture-tube life.

brilliance control   Same as brightness control.

channel-selector switch   Same as station-selector switch.

contrast control   Varies the strength of the video signal, just as a volume control varies the strength of an audio signal. In one extreme setting the contrast control makes the picture all harsh blacks and bright whites like pen-and-ink drawings. In the other extreme position it makes the picture a dim, muddy gray all over. The correct setting gives blacks, whites, and many tones of gray. When the contrast-control setting is changed, it is usually necessary to readjust the brightness control also. In fringe areas where signals are very weak, pictures can often be improved by cutting back the contrast and brightness controls. This gives a dimmer but clearer picture.

fine-tuning control   Often provided on front panel of sets having a station-selector switch, to permit correcting for tube and circuit variations when the coil and capacitor values placed in the circuit by the switch are not exactly right for a particular station. The fine-tuning control generally adjusts a small variable capacitor, and sometimes needs readjusting when the station-selector switch is set to a different channel. Adjust the fine-tuning control for maximum sound, then turn away from this point enough to eliminate the sound bar pattern and ragged appearance of the picture on the screen. Some receivers with switch-type tuners have 12

Television Controls (Continued)

fine-tuning adjustments, one for each vhf station. Once each is set for clear picture and sound, they do not ordinarily need to be changed for months. The controls may become accessible for adjustment with a screwdriver when the tuning dial is removed, or the set may have an extra knob that is pushed in and turned at each station-selector position to adjust the fine-tuning control. You will learn more about preset fine-tuning controls in the chapter on tuning devices.

focus control   Changes area of spot made by electron beam on screen. May be either a chassis control, a mechanical means for moving a ring-type permanent magnet located on the neck of the picture tube, a movable connecting link on the base pins of the picture tube, or a plug that can be inserted in one of three jacks on the chassis. Adjust the focus control for sharpest possible horizontal line structure.

height control   Determines the vertical distance occupied by the picture on the screen. Adjust so the height of the picture is just a little greater than the height of the mask opening on the screen, tune to all channels in turn, and readjust if top or bottom edges of picture show on any channel. The vertical centering control may need readjustment after the height control is changed.

horizontal afc control   Generally serves to assist the horizontal hold control by adjusting an automatic-frequency-control circuit that acts on the horizontal sync system. Adjust only when you cannot stop tearing or zigzagging of the picture with the horizontal hold control.

horizontal centering control   Moves entire picture sidewise on screen. If it is a chassis control, it applies centering voltage to the horizontal deflection coils. More often, horizontal centering is obtained by rotating magnetic centering rings on the neck of the picture tube, or by adjusting the position of the focus coil. If the first adjustment made moves the picture in the wrong direction, adjust in the opposite direction.

horizontal drive control   Affects strength of output voltage of horizontal sync system. Generally needs adjustment only after a new tube is put in set. When out of adjustment, diagonal zigzag line pattern cannot be eliminated by any horizontal hold control adjustment. If set has no horizontal linearity control, the horizontal drive control may be used for the same purpose of improving the shape of a circular test pattern. If the set has a linearity control, try it first before disturbing the drive control. If the horizontal drive control makes the picture brighter, do not turn it any farther than absolutely necessary.

Television Controls (*Continued*)

horizontal frequency control A secondary control that assists the horizontal hold control, usually by adjusting an automatic-frequency-control circuit acting on the horizontal sync system. Adjust the frequency control only when you cannot stop tearing or zigzagging of the picture with the horizontal hold control.

horizontal hold control Makes the horizontal sync system operate at a frequency close to that of the horizontal sync pulses that are put in the television signal by all station transmitters, so that these pulses can more easily control the receiver. If incorrectly set, weird zigzag diagonal moving patterns appear on the screen. Adjust the control so the complete picture is restored, then check all channels. Adjust further if the zigzag patterns reappear on any channel. Best position for this control is usually about midway between the two points at which the picture locks in and becomes steady, when turning toward the lock-in region of the control first from one direction and then from the other. If the final setting leaves the picture off center, make the necessary correction with the horizontal centering control.

horizontal linearity control Makes spot move across screen at uniform speed when properly adjusted. Rarely needs adjustment. Adjust so vertical lines in picture or test pattern are exactly straight.

horizontal sync control Same as horizontal hold control.

local-fringe switch Permits choice of best possible picture quality for reception of nearby stations, or greater sensitivity but slightly poorer picture quality for locations over about 25 miles away from television stations.

on-off switch Turns television set on and off. Generally combined with the volume control, just as in radio sets. On many new sets it is operated by pulling out or pushing in the volume control knob, without changing the volume setting.

picture control Same as contrast control.

station-selector switch Serves to switch into the receiver circuits the pretuned coil and capacitor combination required for reception of the particular channel desired.

sync control Same as horizontal hold control.

tone control Changes tone of sound portion of television program, just as in radio sets.

tv-phono switch Serves to remove power from all but the audio amplifier tubes when using a phonograph with the television set.

Making a Service Call **85**

Television Controls (*Continued*)

vertical centering control Moves entire picture up or down on screen. If it is a chassis control, it applies centering voltage to the vertical deflection coils. More often, vertical centering is obtained by rotating magnetic centering rings on the neck of the picture tube, or by adjusting the position of the focus coil.

vertical hold control Adjusts frequency of vertical sync system. When this control is far out of adjustment, several pictures are seen overlapping each other vertically and moving up or down on the screen. When the control is slightly off in one direction, the picture will roll upward. When the control is slightly off in the other direction, the picture will roll downward. The rolling movement may be faster in one direction than in the other. Best adjustment is obtained by approaching the correct setting from the direction in which the picture movement is slowest. Adjust until the picture is stationary. After adjusting, reduce the contrast control setting until the picture is just barely visible; the picture should be stationary if the vertical hold control is correctly set.

vertical linearity control Affects vertical spacing between horizontal lines of raster. Adjusts so circles are true circles, with no flattened or peaked portions at top or bottom. Easiest to adjust when station is broadcasting test pattern. It is usually done just before first scheduled program each day. Improper adjustment makes people's faces or legs look either too long or too short. The height control may require readjustment.

vertical size control Same as height control.

volume control Adjusts loudness of sound portion of television program.

width control Determines how far electron beam moves horizontally from side to side on screen. Adjust so width of picture is just a little more than width of mask opening on screen. Check all channels, and readjust if sides of picture show on any channel. The horizontal centering control and the horizontal linearity control may need readjustment also when the width control is changed.

Next Step on Service Call. So far in your service call, you have found out exactly how the radio or television set is misbehaving. You have noted whether the trouble occurs on some or all channels or band-switch settings. You have become familiar with the receiver controls and have found that adjustment of receiver controls does not clear up the trouble at hand. You have also made sure the trouble is actually in the set and not due to transmitter trouble or interference. The next step, then, is to remove the back cover of the set so you can see the top of the chassis clearly.

Removing Back Cover of Radio Set. The cardboard or wood back cover of a table-model radio set will usually be fastened with either screws or snap fasteners. The screws are easily removed with an appropriate size of screwdriver if they have slotted heads, or with a socket wrench if they have hexagonal heads. The snap fasteners are pried out with a small screwdriver.

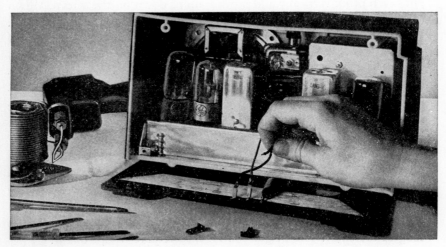

Fig. 6. Leads going to back-cover loop antennas of small radios break easily, as shown here, so handle loops carefully. Resolder the joints when leads are frayed or broken

Sometimes the back cover simply snaps into position inside the plastic cabinet. Spread apart the cabinet slightly with your hands to remove such covers. The cover may also serve to hold the chassis in the cabinet, so the chassis can be slid out as soon as the cover is removed.

Modern radios use a ferrite-rod antenna that is mounted above the tubes on the chassis. There are no antenna terminals or connections on the back covers of these sets.

Sometimes the power-line cord runs into a simple socket fastened to the back cover. This socket mates with a corresponding plug on the chassis only when the back cover is in its correct position. Removal of the back cover disconnects power automatically from the chassis, as a safety precaution.

In many older radio sets, a loop antenna is attached to the back cover on the inside, with leads going from the loop to the chassis. Be careful not to break these connecting leads. Leave the loop connected to the chassis as shown in Fig. 6, because the loop is needed for testing the set while working on it. Loop antennas are an actual part of the radio circuit, so a broken wire here will cause trouble.

If the loop antenna is connected by means of plugs and jacks, make a diagram showing where each antenna lead goes, before removing any of them. Accidental reversal of loop-antenna leads is a trouble that is very difficult to locate. Play safe and make a connection diagram first. This precaution also applies when the loop-antenna leads go to terminal screws, unless both leads and terminals are color-coded.

Back covers with loops are sometimes fastened to the chassis as well as to the cabinet. Just remove the cabinet fastenings, leaving the loop attached to the chassis.

Removing Back Cover of Television Set. If adjustments of the television-receiver controls do not clear up the trouble, return all controls to their original settings, pull the line cord out of the wall outlet, and remove the back cover of the set. A screwdriver will usually serve for loosening the screws in the cover. On a few sets, a spin-type socket wrench of the correct size may be needed. The cover will stay in position after the cover screws are removed, because of the line-cord socket that is attached to the cover.

On some sets the interlock socket has a bracket that is fastened to the cabinet with screws. These screws will have to be removed also, before the back cover can be taken off.

To remove the back cover, pull it back firmly by gripping near where the line cord goes through the cover. This disconnects power from the set.

The back cover can usually be removed without disconnecting the antenna twin-line. If the twin-line goes through a hole in the cover, just slide the cover down the line so it is out of the way. You can disconnect the twin-line later if necessary.

Do not start testing tubes and do not remove the chassis from the cabinet of a radio or television set until you have carefully looked over the top of the chassis for obvious defects such as loose parts, tubes out of sockets, or broken wires.

Hot-chassis Sets. Whenever you see a line-cord socket on the back cover of a television or radio set, you can be pretty sure the chassis of the set is hot. This means the chassis is connected to one side of the power line, so touching it with power on is just like touching one wire of the power line. If some part of your body is grounded when doing this, you can get a dangerous shock. Even standing on a damp concrete basement floor provides such a ground connection through the leather soles of your shoes. Radiators, water pipes, plumbing fixtures, gas pipes, BX wiring, metal switch or outlet plates, and metal boxes of home wiring are other examples of grounds found in homes.

How to Stay Alive. When you make operating tests on hot-chassis radio and television sets at your workbench, you will always plug the set into your isolation transformer. This allows you to handle the exposed chassis without getting a dangerous shock.

An isolation transformer is too heavy to carry on service calls, yet there will be times when you need to work on a hot radio or television chassis in a home. Here you first move the set toward the center of the room where it is well out of arm or leg reach of metal wall outlet plates, radiators, and other grounded objects. Next, put one hand in your pocket. Now you can touch the chassis while changing tubes or doing other work, without providing a dangerous path for electricity through your body to ground. The situation is the same as that of a bird sitting on a bare high-voltage power line; the bird is safe because no part of his body is touching a grounded object.

Of course, you can get a shock even with one hand in your pocket, if you put a finger on a grounded terminal in the chassis while your hand is touching the chassis. The current then passes through only your hand, and will be unpleasant but not necessarily dangerous unless you have a weak heart. Keep your fingers where they belong when working on a hot chassis without an isolation transformer.

Always pull the line-cord plug out of the wall outlet when doing work that does not require power in the set. Turning off the set is not enough, because in many hot-chassis sets the chassis is connected to the dangerous hot side of the a-c line even when the set is turned off.

If the customer is watching while you work on a hot-chassis set, be sure to point out the danger. If you do not, the customer may try to do the same thing himself later without realizing that you were following important safety rules.

The types of trouble that can be fixed in television and radio sets without removing the chassis will now be taken up one by one. Examples of common causes for each trouble will be given, along with suggestions for making the necessary repairs. This practical information applies to all types of sets, including home audio as well as radio and television, from the tiniest table models to huge consoles.

Obvious Above-chassis Troubles. When you look over the top of the chassis, make sure all tubes are firmly pushed down in their sockets. Make sure all tube shields are in position and pushed down. If any of the tubes have top caps, make sure the connecting leads are not touching anything else. Examine the loudspeaker for damage to the cone. Be sure the loudspeaker connecting plug and all other plugs are firmly pushed in. Look carefully for broken wires.

Look also for things that should not be in the chassis, such as children's toys poked through holes in the back cover. Be on the lookout for damage by insects and mice. These pests are attracted by the warmth of a set.

One serviceman found 17 pieces of metal play money in a television set. They had been pushed through a slot in the back cover by a child. Several coins fell between exposed terminals and made the set go bad.

Finding Bad Tubes. About 75 per cent of all failures in radio, television, and home audio sets are due to bad tubes. In most sets you can reach all the tubes from the rear of the cabinet, without removing the chassis. Instructions for locating and replacing bad tubes are given in following chapters.

With radios particularly, it is a good idea to tap each tube a few times with a finger or pencil. A loud crashing sound or intermittent reception of a station generally means that the tube being tapped is shorted and needs replacement.

Tube checking is one of the most important things you do in the presence of the customer. When the trouble is only a bad tube, you can complete the repair job quickly by putting in the required new tube. After you have proved that all tubes are good and the trouble is elsewhere, you are in a much better position to get paid for a thorough troubleshooting and servicing job. The customer feels that the trouble must be serious and hard to find, and becomes resigned to the larger bill that troubleshooting jobs deserve.

Identifying Power-supply Trouble. Look for trouble in the power-supply circuit, the line cord, and wall outlet when the set is stone-cold-dead, with no hum, no pilot lamps lit, no tubes warm or glowing, and nothing at all on the screen if a television set. If the tube filaments are in series, as in transformerless a-c radio and television sets, one burned-out tube filament can make the entire set dead, so check also for bad tubes. An entire later chapter covers these power-supply troubles.

Identifying Antenna-system Trouble. With all types of sets, form the habit of checking antenna connections to the set while waiting for tubes to warm up after you have turned the set on. Look for loose terminal screws. Look for bare loop antenna leads touching the chassis. Look for a lead-in wire that has slipped off the antenna terminal of the set. Look for a break in one of the tiny wires coming from a ferrite-rod antenna, as well as for other obvious defects in the antenna system.

If there are two lead-in wires as on f-m, television, and short-wave sets, be sure the bare ends of the wires are not touching each other at the set. Follow the lead-in wire around the room as far as you can conveniently do at the time, to see if heavy furniture has damaged the insulation in such a way that the wires touch each other or touch a grounded object.

In a television set, if the sound is weak and noisy and the picture is gray, with a lot of white spots called snow on all stations or on all but the strongest, a possible cause is antenna trouble. Of course, ask the customer first if all the stations came in clearly before. Some sets are so poorly made or so carelessly installed that they bring in only one or two of the strongest television stations properly.

In an auto radio, few stations and a lot of noise may mean that the shield of the lead-in wire is poorly grounded to the frame of the car. It can also mean that the antenna itself has become grounded to the frame of the car. Examine the antenna mounting carefully. Look also for breaks in the lead-in wire that runs from the antenna to the radio set.

In battery-powered portable radios, ferrite-rod, loop, or shoulder-strap antennas are used. They have the same types of trouble as home radio antennas.

More information on troubleshooting and repair of antennas is given later in this book.

Identifying Television Interference. Television-receiver interference troubles are easy to recognize but hard to cure. Each has its own characteristic pattern on the screen, as shown in Fig. 7. With experience you will know which troubles must be endured.

A common type of r-f interference encountered in cities is the diagonal-line pattern produced by the local oscillator of an old television receiver that has a 25.75-mc i-f value. Table 1 shows that this interference

Table 1. Local-oscillator Interference

Channel to which old set is tuned	2	3	7	8	9
Channel on which interference is seen	5	6	11	12	13

can be seen only on a television set that is tuned to channel 5, 6, 11, 12, or 13, and only when the old set is tuned to a particular channel.

Making a Service Call **91**

The oscillator of an old set is always 25.75 mc higher in frequency than the channel to which the set is tuned. The oscillator acts as a miniature transmitter at this higher frequency, and thus radiates a signal on a higher channel. The interference will be seen on all nearby sets, both old and new.

This table can help to locate the offending receiver. For example, if the characteristic diagonal-bar interference is seen on channel 11, make tactful inquiries to find out who has an old set that is tuned to channel 7.

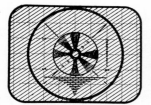

R-F Interference
Commonest Type of Interference. Usually Produced by Radiation from Oscillator of Nearby Television Receiver. Can Also Be Produced by Police, Amateur, or Other Radio Transmitter

Auto Ignition
Seen as Flashing White Horizontal Dashes Only When Auto Without Ignition Noise Suppressors is Passing Nearby. Dashes are Shown in Black Here for Clarity

Ghost
Two or More Patterns are Seen, the Fainter One Being Due to a Signal Arriving Over a Longer Path After Reflection from a Hill, Building, Other Large Object. Turning Antenna May Help

Fig. 7. Examples of interference troubles most often seen on television pictures

Ask the owner of the set to change channels for a moment. If the interference stops, you have located the guilty set.

Nothing can be done to prevent this interference from entering a receiver, because the interference is on the same channel as the desired station. Your customer will have to endure the interference, tune to another channel, or induce the offending neighbor to change channels or get a new set.

Signals from police radio stations, amateur radio stations, and other types of radio service can produce r-f interference patterns because of inferior circuit design in the customer's own television set. Here you can try connecting an interference filter to the antenna terminals of the television set. These filters are sold by jobbers, and serve to block signals of all stations operating on frequencies below that of television channel 2.

You will learn later how to adjust the direction and position of a television antenna to reduce or eliminate the effect of a particularly troublesome ghost or other interfering signal.

Removing the Chassis. When all tubes check O.K. and no troubles can be found above the chassis, the next step is getting that chassis out of

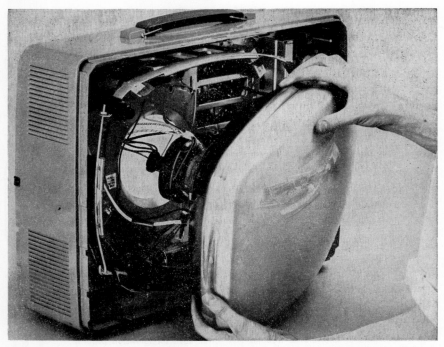

Fig. 8. The picture tube in this portable television receiver can be loosened and pulled forward far enough to permit work on the printed-circuit boards without removing the chassis. (RCA photo)

its cabinet as quickly as possible, because time means money to you. Equally important is getting the chassis back again correctly after the repairs are made.

In the customer's home you must handle a chassis with confidence and assurance, even though you have never seen that particular set before. You have to make a good impression so the customer will allow you to take the chassis to your shop for repairs if this becomes necessary.

This chapter covers the commonest mounting methods and chassis-removal procedures. You will learn what to look for and what to do on any set. You will acquire the knowledge and confidence to tackle big consoles. On these, it often takes longer to get out the chassis than to make the actual repair.

Some sets are designed so you can work on both sides of the chassis without removing the chassis. The top or sides of the cabinet may be removable for this purpose. In one portable television set, the picture tube can easily be loosened and moved forward as in Fig. 8 without disconnecting wires, for work on the front of the printed-circuit boards used in the set.

Unplug Line Cord First. Always pull the receiver line-cord plug out of the power outlet before removing or putting back a chassis. This eliminates the possibility of getting a bad electrical shock. The ends of the line cord under the chassis are soldered to exposed terminals, and your fingers can easily slip under the chassis and touch one of these.

More important still, the chassis is connected directly to one side of the a-c power line in many universal a-c/d-c television and radio sets. In one position of the line-cord plug in the wall outlet, the entire chassis may be electrically energized or hot, even when the set itself is turned off. In the other position of the plug, the chassis may be electrically hot when the set is on.

How to Remove Control Knobs. Having unplugged the line cord, the next step in getting a chassis out of its cabinet is removing the knobs that fit over the ends of the control shafts. These are usually held on their shafts by friction, and are simply pulled off. Sometimes these knobs are jammed on and require quite a bit of force. Only rarely will you encounter older sets in which knobs were fastened on shafts with setscrews.

In some television sets the channel-selector knob has a spring lock. This must be released to remove the channel-selector and fine-tuning knobs. The procedure for one set is shown in Fig. 9. It involves turning the channel knob to channel 7, turning the fine-tuning knob so its slot faces

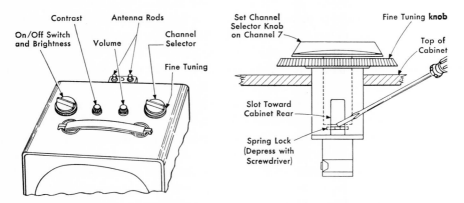

Fig. 9. Example of portable television set having spring lock on channel-selector knob, and method of inserting screwdriver from back of set to release lock. Other knobs pull off conventionally

the rear of the set, then reaching in from the rear of the set with a small-shank screwdriver to release the lock while pulling up on the channel-selector knob.

Some older radios have mechanical pushbutton tuning systems in which the tuning knob is fastened by a screw going through the end of the knob. This screw must be removed to get the knob off so the chassis can be slid out. Removing this screw releases the locking mechanism for the pushbuttons, making it necessary to reset the pushbuttons after replacing the chassis in its cabinet again. This is a fairly easy job, however. Instructions for setting up pushbuttons are given in the chapter covering tuning devices.

Removing Friction-grip Knobs. The friction for holding a knob on its shaft is obtained either with flat springs or with serrations or grooves inside the knob and on the shaft. All friction-grip knobs are removed by a firm, steady pull. Be careful not to lose any of the knob springs.

Double controls on television sets have two knobs. The smaller knob is on the inner shaft. The larger knob is next to the front panel and is on the outer metal sleeve. These knobs are removed one at a time, the smaller knob first, by pulling straight off.

Sometimes small knobs are difficult to grip tightly enough to pull off. For these, a knob puller is worth many times the few dimes it costs. The prongs of the puller fit between the knob and the cabinet, giving a firm grip on the knob without risk of damaging the brittle plastic knob or the equally fragile cabinet.

Prying off a knob with a screwdriver is risky business because of possible damage to an expensive and oftentimes irreplaceable cabinet. If a knob will not come off, it is far better to crush it with pliers and then put on a whole new sets of knobs. The new knobs need not be exactly the same shape as the old ones, as long as they look good on the set and are made to fit the type of shaft at hand.

Pushbuttons for automatic tuning usually slip through the holes in the cabinet and come out with the chassis. Only in a few older sets is it necessary to remove the pushbutton caps before sliding out the chassis. Pushbutton caps are generally held on by friction, and come off with a firm pull.

What to Do with Knobs. To avoid losing knobs, always put them in an empty compartment of your toolbox or some other equally safe place as soon as you take them off. If at your bench, use jars, trays, cans, boxes, or dishes for the same purpose. If the chassis will be out of the cabinet more than a few minutes or is scheduled for a trip to your shop, put the knobs back on their shafts after removing the chassis.

You will realize the importance of using containers the first time you have to hunt on the floor for a knob that rolled off the bench. Neatness both at your bench and in customers' homes can really save you time and money. A cloth mailing bag with tie strings is ideal for holding knobs and hardware. Tie the bag to the chassis after all loose parts have been put in.

Removing Dial Pointer and Scale. In a few table-model radio sets the dial pointer is outside the cabinet and must be removed before the chassis can be slid out of the cabinet. The pointer is held in place on its shaft by friction. It can usually be pulled off easily by grasping the center cap with your fingers. Sometimes gentle prying with a small screwdriver will help get the pointer loose. Be careful not to scratch the dial face.

Some sets have a slide-rule dial with a pointer that comes through a slot in the cabinet. This pointer rides on the dial cord, and must be disconnected from the cord before the chassis can be removed. Examine the pointer to see how it comes off. Usually the pointer will have spring clips that grip the dial cord. This type of pointer can be pulled off the cord with your fingers.

If the pointer is fastened to the cord of an old set with cement, apply speaker cement solvent or acetone and wait a few minutes for the cement to soften before pulling the pointer off the cord. Cement solvent may dissolve nylon, so do not use it on modern nylon dial cords.

Disconnecting the Speaker. In some sets the speaker is attached to the chassis, while in others it is fastened to the cabinet, with wires running to the chassis. Usually these wires are long enough so you can get the chassis out of the cabinet and turn it upside down for troubleshooting without removing the speaker.

On some sets you will find a connecting plug in the speaker cable, either at the speaker or on the chassis. This plug is removed by a firm, steady pull. It disconnects the speaker for convenience in working on the chassis.

If the speaker wires are not long enough to permit removal of the chassis and there is no disconnect plug, remove the nuts around the back of the speaker rim with a socket wrench, leaving one of the top nuts for the last. Hold the speaker in place with your hand while spinning off this last nut with your fingers, then grasp the speaker magnet frame and pull straight back to remove the loudspeaker. Place the speaker face down on a clean piece of paper on your workbench Be careful not to puncture the fragile paper cone with any sharp object. Be sure there are no iron filings on your bench near the speaker.

Do not turn on a set when its loudspeaker is disconnected, as this may burn out a part. The speaker is the electrical load for the audio output stage, and holds voltages down to their normal values. When the load is removed, the voltages shoot up and may cause breakdown of a capacitor or some other part.

Taking Speakers to Your Shop. It is seldom necessary to bring separately mounted speakers into your shop along with a radio or television chassis. Just keep a spare speaker on your bench for test purposes. It is a good idea to check the speakers for opens or rubbing voice coils first, however, as explained later in the chapter covering speakers.

If you do not have a suitable spare speaker in your shop or if there is a possibility that the speaker may be defective, take the speaker to your shop along with the chassis. If there are two or more speakers, take all of them.

Be careful not to damage the speaker cone when handling it. Keep the speaker *face down* in a clean location to minimize chances of damage. Prop up the speaker on one side about half an inch if testing out the set, to permit air movement. A speaker that is down flat on a bench may sound distorted.

Handling the Picture Tube. The commonest method of mounting the picture tube of a television receiver is by means of brackets and strips attached to the chassis. The chassis and picture tube then come out together.

In some sets, the picture tube is held by brackets attached to the cabinet. Before removing the chassis here, carefully pull the socket off from the base of the picture tube. If the socket is tight, pry it off gently by inserting a thin screwdriver between the socket and the base of the picture tube. Disconnect the plugs in the wires going to the deflection yoke on the neck of the picture tube, and pull the high-voltage lead out from its button-type terminal on the funnel of the picture tube.

Disconnecting Record-changer Leads. Always try to avoid cutting a wire or cable that passes through a hole in the cabinet framework. Look carefully for a disconnecting plug or terminal screw somewhere along the length of that wire.

The power leads to the phonograph motor are soldered to terminals inside the chassis in a few receivers, instead of being plugged into a socket. If there is no plug for these leads on the chassis or at the motor, you will have to unsolder or cut the leads before you can remove the chassis from the cabinet. Stagger the cuts, so the joints cannot touch each other even if your insulating tape falls off.

In some sets the phonograph motor plugs into an outlet on the chassis.

This outlet is wired so power to the phonograph motor is cut off when the set is turned off.

In the commonest type of cable used to connect the phono pickup of a record changer to a radio receiver, the insulated wire is connected to a metal prong. The braided metal shield around the wire is connected to a metal cup that surrounds the prong. This plugs into a single-hole jack that is generally located at the back of the chassis. If there is only one such jack, you can safely pull out the lead without any worry about getting it back right. If there are two or more identical jacks, make a sketch first before disconnecting the record changer. Extra jacks are sometimes provided for plugging in a microphone, recording equipment, a television tuner, or an f-m tuner.

Pilot-lamp Connections. Consoles often have one or more pilot lamps in the record-changer compartment. In sets having solid doors that close to hide all controls, there is usually a pilot lamp behind a red glass button or jewel above or below the doors. This glows to indicate that the set is on.

Sometimes the pilot lamps that illuminate elaborate tuning dials are mounted on brackets attached to the cabinet rather than to the chassis. Pilot-lamp sockets in consoles are usually screwed or bolted to the cabinet and have soldered wire leads running to the chassis. Leave the wires connected and remove the sockets when preparing to remove the chassis.

Some pilot-lamp sockets are held in place by spring clips that fit over brackets attached to the cabinet. To remove a spring-type pilot-lamp socket from its bracket, grasp the socket firmly with your fingers and pull.

Use your flashlight for examining pilot-lamp socket-mounting arrangements deep inside the cabinet. One good look is usually enough to tell you which way a socket comes off.

Rubber bands are handy for fastening pilot-lamp wires and other cables to some part of the chassis, to prevent the wires and sockets from flopping around while carrying the chassis to your shop.

Removing Chassis Bolts. The chassis is usually bolted to the cabinet in some way. If the set has a hot chassis, all chassis bolts will be inside the cabinet. In transformer-type sets the bolts will usually go through the bottom of the cabinet since there is no shock hazard.

There are dozens of different ways of mounting a chassis in the cabinet of a television set. Fortunately, the mounting bolts can usually be located at a glance. A flashlight will help to locate those in dark corners. Remove the bolts or screws with the appropriate screwdriver or socket wrench, and store them in a bag or in an empty compartment of your toolbox.

When a television set has a separate control panel, the leads running to it are usually long enough to permit working on the chassis without removing the panel. If the leads are too short for this, remove the knobs first, then remove the screws that fasten the control panel to the cabinet.

If a table-model set has to go to your shop, you will take the cabinet along anyway, so do not remove the control-panel assembly in the home unless absolutely necessary. With consoles where you leave the cabinet in the home, you will of course have to take the control panel to your shop along with the chassis.

Sliding Chassis out of Cabinet. With the cabinet right side up and its back toward you, hold the cabinet in position with one hand, and pull the chassis slowly out of the cabinet by grasping the power transformer or some other rigidly mounted part. Be careful not to damage the loop antenna of a radio set or the picture tube of a television set as you take out the chassis.

Rubber vibration-absorbing washers will usually be found under the chassis on big sets. Lift the chassis to get clear of these, then slide it out through the back. Get a firm grip on the chassis when doing this, and pause for inspection whenever something sticks. Be prepared to tilt the chassis to get it past the back framework on some sets.

The chassis is sometimes mounted on a square of ½-inch plywood. This baseboard in turn is screwed to the cabinet. Here you remove the baseboard mounting screws first, lift out the chassis and baseboard as a unit, then lay the chassis on its side or back so you can get out the bolts that fasten the board to the bottom of the chassis.

Tuning dials of consoles are oftentimes mounted on the chassis and also screwed to the front of the cabinet for greater rigidity. If the chassis does not slide out easily, look for screws at the back of the dial. On some sets the chassis must be tilted to free the dial from the cabinet.

Do Not Use Force. A chassis should slide out easily. If it does not, stop pulling and look carefully to see what is still holding it. The chassis may have feet that rest in holes or depressions in the bottom of the cabinet. There may be rubber shock-absorbing washers under the chassis that have become stuck to the cabinet. Gently lift or pry the chassis upward and then slide it out.

Sometimes the tuning dial may be fastened to the back of the cabinet. You may even find a loudspeaker that is fastened to both the chassis and the cabinet.

In the years to come, receiver designers will undoubtedly figure out still different ways to arrange and mount parts, so do not worry about details of these things now. By using your eyes as you gently pull on the chassis, you will always be able to figure out where it is still holding and what to do about it. Remember—never use force.

Console-cabinet Problems. A console cabinet is ordinarily too large to be carried conveniently by one man. The handling involved in hauling a console to your shop involves risk of scratching the cabinet finish, so it is

Fig. 10. Only the chassis and loudspeaker of an old console radio were brought to this shop for repairs. The chassis jack, available from jobbers for around $5, holds a chassis at any convenient angle. Note use of heavy-duty electric soldering iron instead of soldering gun. A few servicemen prefer to use irons. (Centralab photo)

safer to leave the cabinet in the customer's home. This is especially true for television sets. If troubleshooting and major repairs seem necessary, take to your shop only the chassis and other parts needed to operate the set on your bench. For radio sets the chassis and loudspeaker are all you need, as shown in Fig. 10.

Chassis-removing Precautions for Consoles. A console is usually an expensive set, so do your work even more carefully than on table-model sets. Be thinking always about getting each part back again correctly. It may be days or even weeks before you will be putting the chassis back. You will find from sad experience, just as every other serviceman has, that your memory cannot be trusted that long when working on a lot of different sets.

Make careful notes of the positions of knobs, pointers, screws, wires, and everything else you loosen or remove, whenever there is the slightest possibility of getting things back wrong. Thus, if some knobs are different from others yet all shafts are alike, make a sketch showing where each knob goes.

If there is any chance of getting leads mixed up, as there usually is, make a sketch before disconnecting a single lead. Scotch tape and chinamarking crayon are good to carry in your toolbox for labeling leads. Keep a supply of 3- by 5-inch index cards and a pencil in your pocket for making diagrams and jotting down information. Put the customer's name or the make and model number of the set on each card at the same time, because loose cards easily get mixed in a shop.

Cabinet Scratches May Cost You Money. Never lay your tools or even knobs and screws on top of the cabinet. You run a serious risk of scratching the cabinet. There is also the risk that screws and knobs will roll off the cabinet and get lost under furniture without your realizing it. Remember that you are responsible for any damage that you do to the cabinet, as well as to any other furniture in the customer's home. Even a small scratch that you are unable to fix satisfactorily may become the customer's excuse, and rightfully, for not paying your bill.

Handling Built-in Antennas. A loop antenna is an actual part of the radio circuit. It should therefore be taken to your shop along with the chassis and loudspeaker.

When a television or radio antenna in actual use is stapled to the inside of the cabinet or to its back edges, make a sketch of the stapled antenna and put estimated dimensions on the sketch. You can then roughly duplicate the antenna on your bench with ordinary hookup wire or twin-line.

Removing Record Changers. If poor radio or television reception in a phono console is the complaint of your customer, there is no reason to touch the record changer once you have disconnected its leads. Only where the customer mentions record-changer trouble would you take this to the shop.

Removal of the record changer usually involves lifting it directly out of the cabinet, because the changer floats on springs. Look first for shipping bolts inside the springs, however. Usually these bolts will be loosened rather than completely removed. Shipping bolts should never be tightened except for shipping purposes. Fasten the pickup arm to its support with Scotch tape before lifting out a record changer.

Making a Service Call 101

Inspecting the Chassis. After removing any chassis in a customer's home, place it on a cloth or newspapers that you have spread out to protect the rug and floor. Now turn the chassis upside down or on a side so that you can inspect the parts underneath. Look for obvious defects, such as burned-out resistors, paper capacitors from which all the wax has run out, bad wiring, and bare leads or terminals touching each other.

If you want to go a bit farther and can operate the set outside of its

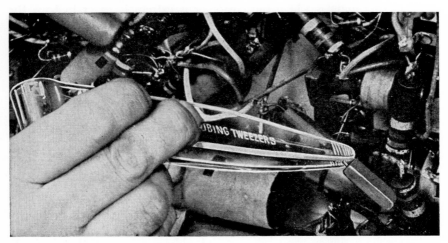

Fig. 11. Tweezers like these, made of insulating plastic, are handy for wiggling suspected parts and wires while a set is on

cabinet, plug in the set and turn it on. Poke or wiggle each under-chassis part in turn with a stick of wood, with the handle of a small screwdriver, or with plastic tongs made for this purpose, as shown in Fig. 11. Do this also for tube-socket terminals and other soldered joints. If proper operation is restored or noise is produced when a particular part is tapped, that part is probably intermittently defective and needs replacement. Many simple defects can be located in this way.

Tap adjacent parts before removing the suspected part, however. Sometimes a heavy tap almost anywhere on the chassis will give the same effect. Tap lightly to narrow down the trouble to the part or joint that is most sensitive to tapping.

Troubleshooting. It is usually impractical to operate the chassis of a large console on the floor, because antenna and speaker leads are not long enough. Also, the home is by no means an ideal place for troubleshooting. Once you have gone this far, your best bet is to take the chassis and all other needed parts to your shop. Customers will invariably permit this when you explain that you will call them for approval of your estimate of the charges before you go ahead with the actual repair work.

As you acquire experience, you will occasionally do some chassis troubleshooting in a customer's home. Usually there will be special circumstances, such as a rural location 25 miles or more from your shop where the mileage cost of an extra trip would be pretty high. Another situation justifying work in the home is where the customer is extremely eager to have the set fixed immediately for use at a party or other event.

Do not tackle a repair job in a customer's home unless you are sure you can finish it, however. Your workbench is the only place where you can really do justice to the customer's receiver under normal circumstances.

Taking the Chassis to Your Shop. When leaving the customer's home with a chassis, make several trips from the house to the car, to avoid the risk of dropping something. Put the chassis on the floor of the car rather than on the seat. Put the loudspeaker face down on the floor, on newspapers.

The chassis is always safest on the floor of the car. On spring seats or on a cushion there is more danger that it will flip over when hitting a bump. Rotate tuning capacitor plates to the fully meshed position, so they cannot readily be bent accidentally.

Always have the customer's name and address on a card that is firmly wedged between parts on a set or is tied to the set. This is particularly important when you pick up several sets on the same trip. Otherwise, sets can easily get mixed up.

Be just as careful with each chassis in your shop. Shelves keep sets out of danger. Keep your bench clear except for the set on which you are working.

Cleaning the Chassis. Always expect to find a lot of dust and dirt on the chassis and inside the cabinet of a receiver. A housewife is not supposed to clean the inside because of the possibility of damage and shock. Therefore, never comment on the amount of dust that you find inside a set. Consider a thorough cleaning as a regular part of every job that comes on your service bench. Do not do this cleaning in the customer's living room, however. Leave the dust on or take the chassis outside for cleaning.

Clean the chassis before you start troubleshooting. If you keep your eyes open while doing this, you may even come across the trouble, or at least find things that need fixing to prevent future troubles.

A cloth is used for removing dust. A few drops of liquid furniture polish on the cloth will make it hold dust better. A small, clean paint brush with soft bristles is handy for getting between crowded parts on top of the chassis. Later you may wish to get a small vacuum cleaner or use an attachment with your regular home vacuum cleaner.

Use a system when wiping the chassis of an old radio. Remove a tube and wipe it, wipe the chassis area and socket made accessible by removal of that tube, then replace the tube. Repeat for each other tube in turn. This is better than removing all the tubes at once, unless you are going to do this anyway for tube-testing purposes. The important thing is to avoid getting the tubes mixed up and replaced incorrectly.

After wiping off the tubes and chassis, take a good deep breath and blow out the dust from between the gang tuning capacitor plates. A feather or pipe cleaner can be used between individual pairs of plates if the dust sticks stubbornly.

Any grease or dirt on top of the chassis that cannot be removed by simple wiping should be rubbed with a cloth dipped in cleaning fluid such as carbon tetrachloride.

Do not use cleaning fluid on any plastic material such as cabinets, dial windows, or plastic safety windows of television sets, as cleaning fluid will soften and ruin some plastics. Wipe the tuning dial and its transparent window, if present, with a soft dry cloth. There is nothing underneath the chassis that needs to be cleaned.

Caution. In looking over a set and cleaning it, do not touch the adjusting screws for coils and capacitors even if their bolts or nuts appear to be loose. These adjustments are often critical, and special test equipment is needed for realigning the receiver if the adjustments are changed.

Cleaning Television Glass Surfaces. With television sets, the face of the picture tube and the plate-glass or plastic safety window in front of it should be cleaned carefully without scratching the highly polished surfaces. First dust each surface lightly with a soft dry cloth to get off as much dust as possible without grinding it in. Next, soak a chamois or soft cloth in warm water and wring nearly dry, for use in cleaning the glass or plastic surfaces thoroughly.

Putting the Chassis Back. Replacing a chassis is the reversal of the steps followed in removing the chassis. Replacement is even easier than removal if you have been observant and have made careful notes.

When replacing a speaker on stud bolts, place the speaker on one stud first, then rotate it on this bolt slowly until the other three speaker holes line up with the remaining studs.

If any screw holes in a wood or plastic cabinet are stripped or too large, insert a small unmelted piece of rosin-core solder in the hole before putting in the screw.

Final Check of Performance. When the entire job is finished, plug in the set and check its performance yourself carefully. Tune slowly over the entire range of each band. If something is wrong, check the installation again step by step to see what you have overlooked. Remember that band-changing and radio-phono switches must be set for radio reception. Set the tone control for maximum bass, or ask the customer to set it at his favorite position. When you are satisfied with the performance of the set, ask the customer to check it also because this is good business.

Final Safety Check. Before leaving a set on which you have worked, make certain that it is completely safe to operate. This means there must be no danger of electric shock when any exposed metal part on the set is touched. Do not do this by hanging onto a cold-water pipe with one hand while running your other hand over the set, because this may mean business for the undertaker. Instead, use the a-c voltmeter section of your multimeter. Complete instructions for making this safety check are given in the chapter on multimeters. If the check indicates that there is a dangerous situation, look up the chapter on power-supply troubles.

A safety check is particularly important for sets having metal cabinets, because it is so easy for a screw or piece of solder to drop unnoticed to a position where it connects the metal cabinet to the hot chassis. Omission of insulating strips or use of damaged strips when replacing the chassis can be just as dangerous. Insulation is important for safety, so replace every piece exactly where it was before.

Do Not Leave Tools in Customer's Home. Before you leave the customer's home after repairing a set or returning a bench-repaired set, make one quick search to see if you have left any tools. Next, clean up all dirt and scraps of insulation that are left on the rug or floor. Put back in position any lamps or chairs that you moved. Do everything else needed to make the room as neat and clean as when you came into the house.

Collecting the Money. To complete the service call after returning and installing the repaired chassis, you would next hand the customer a bill that was carefully prepared beforehand in your shop. To make sure the customer has the money ready for you, in check or cash, call the customer beforehand and ask for an appointment to install the chassis at a time when it would be convenient to pay the bill.

On home calls where the repair can be made without taking the set to your shop, make out your bill after completing the job, as in Fig. 12. Itemize your bill and list the new tubes and parts separately, just as is done

Fig. 12. Making out a bill after completing a television home service call, while the customer reads a copy of the guarantee covering the repair work. (Electrical Merchandising Week photo)

by garages. It is good business psychology to let your customer read a copy of your guarantee while you are making out the bill.

Television and radio repairs must be on a cash-and-carry basis if you are to stay in business for long, because the time spent on extra calls to collect money is an expensive loss to you.

When you put your first chassis back in a console in a customer's home, when you hear him say it looks and sounds good, when you get paid—then you will have completed your first real service job. It is one of which you can well be proud—and the forerunner of many more because you will be on the right track to success in servicing.

5

Getting Acquainted with Electricity and Magnetism

A Foundation for Your Training. All television and radio circuits depend on simple basic things like voltage, current, resistance, power, and magnetism for correct operation. More important still are the volt, ampere, ohm, and watt units for measuring these things in circuits. You will use one or more of these units practically every time you look for bad parts or order new parts. Thus even the foundation training in this chapter is highly practical.

Introducing the Electron. When you poke a finger into an electric lamp socket, that unpleasant throbbing feeling is due to electrons flowing through the fingertip. What are these electrons? Where do they come from? Where do they go? Why do they go? And most important of all, what can they do? The simple answers to these questions will be for you the practical story of electricity at work in television and radio.

Electrons are everywhere, in everything. The electron is one of the smallest particles of matter, so small that it has never been seen by human eyes. Despite this smallness, an electron's behavior is well known. Think of the electron as a tiny ball, smaller than the smallest ball bearing you ever saw. Imagine it to be so small that it can travel through the invisibly small pores of metal objects, just as easily as water flows through the pores and holes in a sponge.

Electrons are everywhere because they are in each one of the atoms that make up everything in the world. An atom is bigger than an electron but still is too small a particle to be seen even in the most powerful new microscope.

An atom has a central core around which are whirling one or more electrons. Over 100 different kinds of atoms are known today, each having a different number and arrangement of electrons.

Some atoms are able to hang onto their own electrons pretty well. Other atoms have trouble keeping their full quota of electrons. An electron that gets away and wanders from atom to atom is called a *free electron*.

An atom is normal when it has its full quota of electrons—no more and no less—whirling around it. When an electron gets away, the atom has a way of attracting another electron to complete its quota again.

Negative Charges. An electron is technically known as the *unit negative charge of electricity*. The minus sign of arithmetic is used to represent the

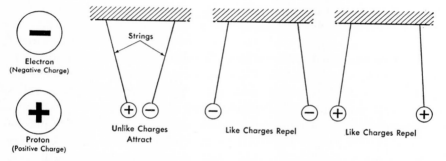

Fig. 1. Kernels of puffed cereal hung from silk threads will act like this if charged as indicated, demonstrating the law of electric charges on which the action of every tube depends

negative charge on the electron, as shown in Fig. 1. The same minus sign is also used to indicate a negative charge produced at some point in a circuit by a group of electrons. An example of such a point is the negative terminal of a battery.

Nature seeks to maintain a balance at all times. Therefore, for every unit negative charge there must be an equal and opposite positive charge. This positive charge is represented by the plus sign of arithmetic. The unit positive charge is called a *proton*.

A normal or neutral atom has the same number of electrons as it has protons. If an electron wanders away from a neutral atom for any reason, there will be one positive charge in the atom that is no longer canceled by a negative charge. The atom will then have a positive charge. This is all it needs to attract the next stray electron that comes along.

It is a basic law of nature that *unlike charges attract each other*. The diagrams in Fig. 1 will help you remember this. When objects with unlike

108   Television and Radio Repairing

charges, one positive and the other negative, hang side by side on strings they are attracted to each other.

When both objects have positive charges, they repel each other. Likewise, when both objects have negative charges, they repel each other. These repelling actions illustrate the other part of this basic law of electric charges, that *like charges repel each other*.

The operation of every tube and transistor depends on this law of electric charges, so memorize these six words: *Unlike charges attract; like charges repel*. An easy way to remember is to think about the same natural law in the moonlight, where girls attract boys.

What Is Electricity? Whenever electrons travel toward a positive charge, they form a current of electricity. With the lamp socket as an example again, the action of the power company's generators is such as to produce at one instant a strong positive charge at one terminal in the socket. At the same instant, there is an oversupply of electrons (negative charges) at the other terminal. The electrons cannot travel through air to get to the positively charged terminal, but they can and do travel through the path provided by your finger inserted in the socket. What you feel is electrons in motion—a current of electricity.

Voltage Sources. Anything having the ability to produce a surplus of electrons at a terminal is a source of electricity, usually called a *voltage source*. Such a source will always have two terminals. There will be a surplus of electrons at the negative terminal and a corresponding shortage

Fig. 2. When the switch is closed, the electron flow around this simple circuit makes the bell ring. The outside terminal of the standard No. 6 dry cell is negative, so electrons flow out from it through the wire as indicated by the arrows. (General Electric photo)

Electricity and Magnetism **109**

of electrons to provide a positive charge at the positive terminal. The greater the surplus of electrons, the higher is the voltage between the terminals.

Whenever an electrical path is completed between the terminals of a voltage source, the surplus electrons at the negative terminal will travel over this path to the positive terminal, as indicated by the arrows in Fig. 2. This travel of electrons along a path is called *electric current*. The current will flow just as long as the voltage source can continue replenishing the supply of electrons at the negative terminal.

Batteries are voltage sources that use up their supply of electrons and hence run down. Power-line voltage sources will not run down because the huge generators at the power station can keep up the supply of electrons.

Voltage Is Electrical Pressure. The higher the voltage, the more electrical pressure there is at the negative terminal of a source. This electrical pressure urges the electrons to take some path—any path—from the negative terminal to the positive terminal. If the voltage gets high enough it can even make the electrons jump through air. This can be seen as a spark, as an electric arc, or as lightning in nature.

Volts and Millivolts. The answer to any question about how much voltage exists between two terminals is expressed in *volts*. Voltages of some of the devices you will encounter in your work are indicated in Fig. 3.

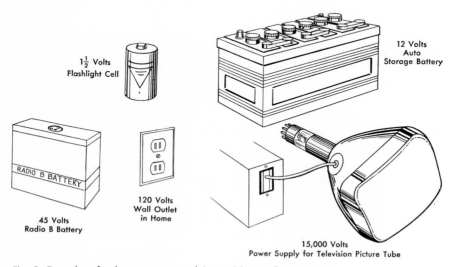

Fig. 3. Examples of voltages encountered in servicing work

The shock that you feel becomes more violent as the voltage goes up.

Voltages below about 50 volts cannot usually be felt, but all voltages encountered in servicing can be measured with meters called voltmeters. In servicing work, you will be measuring voltages frequently because they are an important clue to causes of trouble. The abbreviation used for volts is *v*.

The smallest battery used in portable radios is rated at about 1½ volts. Much smaller voltages are quite common in circuits, however. To eliminate the decimals required to express small voltages in volts, the *millivolt* unit is used. One millivolt is equal to a thousandth of a volt; instead of writing 0.001 volt, you write 1 millivolt. The abbreviation for millivolts is *mv*.

Voltages of signals in receiver circuits are often even lower than 1 millivolt. For such small voltages a still smaller unit called the *microvolt* is used. One microvolt is equal to a millionth of a volt; this means that 1 microvolt is the same as 0.000001 volt.

In television receivers, voltages of thousands of volts are needed for the picture tubes. To eliminate writing zeros for these high voltage values, a larger unit called the *kilovolt* is often used. One kilovolt is equal to 1,000 volts. The abbreviation for kilovolts is *kv*.

What Is Current? Voltage is a measure of the strength of electricity. But what can that strength do? Just one thing: It can push a certain number of electrons over a given path during each second of time after the path is provided. This flow of electrons is called *current*.

The more electrons flowing per second, the higher is the current. More electrons can do more work, so higher currents mean greater results. Current, electric current, electron flow, and current flow all mean the same thing—all are the result of a voltage that is making electrons move through a wire or other path.

The strength of a current could be expressed in terms of the actual number of electrons flowing per second. This is such a tremendously large number, however, that several more convenient units of current flow are used in television and radio servicing.

Amperes and Milliamperes. The basic unit for measuring current flow is the *ampere*. One ampere of current is flowing when 6.28 million million million electrons are flowing past a given point per second. This figure is given merely as a matter of interest, because you will never bother to count electrons. You will always deal with amperes and related smaller units for current flow. Ampere is abbreviated as *amp* or *a*.

In television and radio work, the unit of current that you will use even more than the ampere is the *milliampere,* abbreviated *ma*. One milliampere is equal to one-thousandth of an ampere. A still smaller unit is the *microampere*, equal to a millionth of an ampere.

The meters used for measuring current will always indicate which unit is intended. A meter that reads in amperes is called an *ammeter*. One that reads in milliamperes is called a *milliammeter*. One that reads in microamperes is called a *microammeter*.

It is unnecessary to have a separate meter for each current range. Instead, you will use a single instrument called a *multimeter*. This has one meter and a switching arrangement that permits changing the meter circuit so it can be used to measure several different current ranges.

What Is a Conductor? Electricity can flow freely only through materials known as *conductors*. In a good conductor, like copper, silver, or aluminum, there are many free electrons wandering aimlessly from atom to atom at all times. When a voltage source is connected to a conductor, these free electrons are attracted to the positive terminal of the source. This movement of electrons is the current that flows whenever a conductor is used as a path for electricity.

The more free electrons there are in a material, the greater is the current that flows when a voltage is applied. Most metals are good conductors, though some are better than others. The human body is a conductor, especially when the skin is wet.

What Is an Insulator? A material that has only a few free electrons is called an *insulator*. Examples of good insulators are glass, plastics, mica, dry air, and a vacuum.

There is no definite dividing line between conductors and insulators. Everything will conduct electricity to a certain extent, particularly if the voltage is made sufficiently high. Thus, with voltages of about 15,000 volts, electricity will flow through air a short distance between the points of spark plugs in automobiles. At the lower voltages encountered in radio sets, however, air is a good insulator.

Insulating materials are used between the terminals of all television and radio parts, to prevent electrons from taking short cuts. Sheets of brown plastic are perhaps the commonest insulating material. This plastic insulation has a peculiar characteristic odor when cold. When seared by a hot soldering iron, its pungent smell is unforgettable.

What Is Resistance? You have learned that voltage is the force which makes electrons move. Current is the movement of electrons resulting from this force. You also know that materials called insulators offer almost complete opposition to the movements of electrons. This opposition to the flow of electrons is called *resistance*.

The resistance of an electron path through a good insulating material is extremely high. The resistance of a path through a copper wire or other good conductor is very low.

In television and radio circuits, in-between resistances are frequently needed for certain electron paths. Special parts called resistors are inserted in the electron path to provide a desired amount of resistance to current flow.

Ohms and Megohms. The unit used to specify the amount of resistance offered by a particular path for electron flow is called the *ohm*. Just to give an idea of its meaning, a 1-foot-long piece of copper wire about as fine as a human hair has a resistance of about one ohm. This is the kind of wire used in some types of television coils.

Thick copper wire, such as that used in wiring houses, has almost zero ohms of resistance per foot. The No. 18 copper wire that is so widely used for making connections in television and radio sets has a resistance of 0.006 ohm per foot. This resistance is so small that for connecting purposes we think of it as zero ohms.

Wire made from a metal alloy called Nichrome has a much higher resistance than copper wire. For this reason, Nichrome wire is often used to make resistors.

By winding a long length of fine Nichrome wire on an insulating form, resistance values as high as 10,000 ohms and even more are readily obtained. By using less wire or a thicker wire, any desired value of resistance between zero ohms and 10,000 ohms can be obtained.

High values of resistance, as high as 10,000,000 ohms, are widely used in television and radio sets. It would take too much Nichrome wire to get such high resistance values. This is why carbon is used in resistors instead of wire when higher values are needed.

It takes a lot of zeros to specify resistance value above 100,000 ohms. Therefore, a larger unit of resistance called a *megohm* is also used. One megohm is equal to 1,000,000 ohms. You will often find megohm abbreviated as meg or as M. Another common abbreviation seen on diagrams is K, which means thousands of ohms; thus, 47K is 47,000 ohms.

What Is an Electric Circuit? Whenever a complete electron path contains the three things shown in Fig. 4A, it is a useful electric circuit.

First, a circuit must have a voltage source. This source can be a battery, a power line, a transformer, or a generator.

Second, a circuit must have a load through which electrons can flow to do useful work. This load can be a loudspeaker in which electrons can produce sound, a picture tube in which electrons trace out a picture, or simply a resistor that passes on a signal to the next tube.

Finally, a circuit must have connecting wires. These provide a complete path through which electrons can travel from the negative terminal of the voltage source through the load and back to the positive terminal. The voltage source can be thought of as a pump that keeps the electrons circulating around this path.

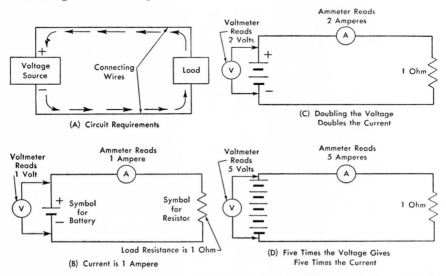

Fig. 4. These diagrams show that when you increase the voltage in a circuit, the current increases correspondingly. Study each note on each diagram

Predicting How Much Current Flows. You already know that a high resistance cuts down the amount of current flowing in a circuit. Zero resistance allows the largest possible current to flow. Also, the higher the voltage, the more force there is to push electrons around a circuit, and hence the larger is the current.

There is actually a simple relationship between voltage, current, and resistance. This is illustrated in Fig. 4B. Here, a voltage of 1 volt produces 1 ampere of current in a circuit containing 1 ohm of resistance.

When the voltage is increased to 2 volts in Fig. 4C, the ammeter shows 2 amperes. Similarly, 5 volts gives five amperes, as in Fig. 4D, and 35 volts gives 35 amperes. The important thing to remember is that, when you double the voltage in a given circuit, you double the current.

Now suppose the voltage is left at 1 volt and the resistance of the load is doubled, as in Fig. 5. Again an ammeter will show how much current is flowing. Note that doubling the resistance cuts the current in half, to 0.5 ampere. In like manner, a resistance of 10 ohms passes only one-tenth of the current that 1 ohm passes.

To find what other values of voltage and other values of resistance would do to the current in this simple circuit, just divide the voltage value *in volts* by the resistance value *in ohms*. This gives the current value *in amperes*. Engineers call this rule Ohm's law, in honor of the man who first figured it out. You do not need to remember the law, however, as such figuring is not necessary for the practical servicing techniques covered in this book.

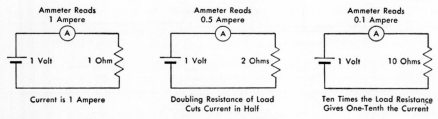

Fig. 5. Increasing the resistance in a circuit makes the current go down

Heat and Power. Whenever electrons flow through something, it gets hot. For example, an electric lamp gets hot because electrons are flowing through its filament. An electric stove gets hot because electrons are flowing through the Nichrome wire in its heating element. Similarly, many resistors in a television set get hot or at least slightly warm when electrons flow through them. In each case, the heat means that electric power is being used.

There is a limit to how much electric power a device can take. Too much power means too much heat, and too much heat can make any device go bad. For this reason, resistors and many other parts in a television or radio set have a power rating.

The larger the physical dimensions of a resistor or other part, the higher will be its power rating because it has more surface area for getting rid of heat. The units used for expressing a power rating are already familiar to you if you have electricity in your home.

Watts and Kilowatts. The unit of electrical power is the *watt*. Electric lamps have ratings ranging from 15 to 300 watts in the commonest sizes. The resistors used in television and radio sets have much lower ratings, ranging from 0.1 to 3 watts for carbon resistors and on up to about 20 watts for wirewound resistors.

In servicing you will not need to use the larger and more familiar unit of power called the *kilowatt*, which is equal to 1,000 watts. It is good to know this relationship between watts and kilowatts, however. Your customers pay their electric bills on the basis of how many kilowatts of power they use each hour of the month.

Power is charged for on the basis of *kilowatthours*, abbreviated kwh. A special meter called a watthour meter is needed to measure this. Every home with electricity has such a meter. The meter is read by a power-company employee once a month to determine how much power was used. A 1,000-watt electric heater that is on for 1 hour draws exactly 1 kilowatt-hour of power. For 2 hours it would be 2 kwh; for 5 hours it would be 5 kwh. A 100-watt heater could be run for 10 hours to use up 1 kwh, however, because you multiply power in kilowatts by hours of use to get kwh.

You may sometimes wish to explain that television and radio sets cost very little to operate. For example, suppose a television set draws 200 watts, as indicated in the service manual for the set. In an hour, then, the power used is 200 watthours, or $\frac{1}{5}$ kilowatthour. If the customer pays 5 cents per kilowatthour for electricity, the operating cost per hour is $\frac{1}{5}$ of 5 cents or 1 cent per hour.

In servicing, you will rarely if ever have to figure out power values. You simply order the same size of resistor that was in before.

The audio power output of a radio set or home audio system is also measured in watts. This represents the a-f power that is fed to the loudspeaker, and will always be much less than the input power drawn from the a-c power line. A small radio that draws 30 watts from the a-c line may have an output power of only 3 watts. A 20-watt audio amplifier delivers 20 watts of a-f power, and may draw 100 watts from the a-c line.

Voltage Drops. Whenever current flows through a resistor, it produces a voltage drop across the resistor. The resistor value in ohms multiplied by the current value in amperes gives you the amount of this voltage drop, as illustrated in Fig. 6A.

As you get farther along in your study of circuits, you will find that resistors are often inserted intentionally in circuits to produce voltage drops. These voltage drops are often used to act on following circuits. The thing to remember now is that, whenever current flows through a resistor, there is a voltage drop across that resistor.

When there is only one resistor in a circuit, as in Fig. 6B, the voltage drop across the resistor is equal to the voltage of the source or power supply. When there are two or more resistors in the circuit, the voltage drops across the individual resistors always add up to the voltage of the source, as in Fig. 6C. This is useful to know when troubleshooting in circuits.

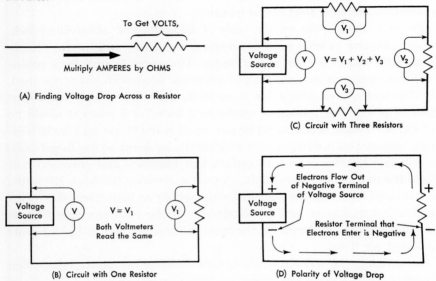

Fig. 6. Voltage drops across resistors in a circuit are useful troubleshooting clues. Study each diagram in turn carefully as you read the associated text

Polarity of Voltages. You already know that a voltage source has polarity, wherein one terminal is negative with respect to the other source terminal. Electrons, which are negative charges, flow out of the negative terminal when the circuit is completed. Once you locate the marked negative terminal of any voltage source, you know that the other terminal is positive.

The voltage *drop* across a resistor likewise has polarity. The direction of electron flow determines what this polarity is. The resistor terminal that electrons enter is *negative* with respect to the other resistor terminal, as indicated in Fig. 6D.

Polarity must always be expressed with respect to something. Thus, the polarity of one resistor terminal is expressed with respect to the other resistor terminal. The polarity of one battery terminal is expressed with respect to the other battery terminal. The polarity of one point in a circuit is usually meant to be with respect to ground, unless otherwise specified.

You do not even need to figure out the direction of electron flow through the resistor if you can trace the circuit from the resistor to the voltage source. The resistor terminal that connects to the negative terminal of the voltage source is always negative. Study Fig. 6D and you will see that this is quite logical.

Direct Current. Voltage sources that have a definite polarity at all times are known as direct-current sources. Batteries are an example. Direct-current sources make electrons flow in one certain direction through a circuit. This direction of electron flow, you will remember, is away from the negative terminal of the voltage source.

The term *direct-current source* is often abbreviated as d-c source, or simply as d-c. The voltage of a d-c source is similarly abbreviated as d-c voltage, or simply d-c. Thus, d-c can stand for direct current, for the voltage that sends direct current through the circuit, or for the source that has the voltage that sends the current through the circuit.

This sounds complicated, but do not worry about it; from the way the abbreviation d-c is used, you will always know what is intended. Most generally, you will find that d-c is used to indicate that electrons are flowing in the same direction at all times.

There are still quite a few thousand homes in this country, chiefly in the older sections of large cities, that have d-c power lines. The number is gradually dwindling, however.

Alternating Current. In most homes today, the current flowing through the power lines is *alternating current*, abbreviated a-c. This alternating current is produced by an alternating voltage, abbreviated a-c voltage or simply a-c.

An alternating voltage is produced by a generator that reverses its polarity many times a second. Each time the polarity reverses, the electrons flowing in all connected circuits must change their direction, too. The current is changing or alternating in direction many times a second, and is thus logically known as an alternating current.

In the United States, practically all a-c power stations generate power at a frequency of 60 cycles per second. During each cycle a given terminal will change from negative to positive in polarity and go back to negative again.

Most of the tubes and transistors in television and radio receivers require several d-c voltages. For this reason, each a-c set will have a power-supply circuit that changes the a-c voltage to a d-c voltage by a process known as rectification. As a result, you will be working mostly with d-c voltages. You will not have to worry a bit about the rapidly changing polarities of the a-c power-line voltage.

Alternating Voltage. At first thought it might seem impossible to specify the value of a voltage that is continually increasing or decreasing and even reversing in polarity regularly. Actually, this is no problem at all. A d-c voltage of 1 volt will produce a certain amount of heat when applied to a given resistor. The a-c voltage which produces *the same amount of heat* in that resistor is also 1 volt.

The a-c voltage value that produces this equivalent heating effect is called the *effective value*. Unless otherwise specified, a-c voltage values are assumed to be effective values. The meters you will use in troubleshooting read these effective values.

Controlling Alternating Current. You will recall that direct current is controlled by two factors—voltage and resistance. For alternating current the corresponding factors are voltage and impedance. Think of impedance as the opposition that a coil or capacitor offers to the flow of alternating current.

Coils have an impedance that increases with frequency. At zero frequency, corresponding to direct current, the impedance of a coil is the same as the resistance of the copper wire in its winding. An iron core increases the impedance of a coil at a given frequency. At radio frequencies a coil can completely block the flow of signal currents.

Capacitors are the opposite of coils, because the impedance of a capacitor decreases with frequency. At zero frequency, corresponding to direct current, the impedance of a capacitor is very high and essentially no current flows through it. At radio frequencies the impedance of a capacitor is very low, so it provides a good path for signal currents while blocking direct currents.

A coil thus passes direct current while blocking signal currents. A capacitor passes signal currents while blocking direct current. This is why coils and capacitors are used together so often to keep signal currents in desired paths, while providing for the flow of direct current from the power supply to the tubes or transistors of a television or radio set.

Magnets and Magnetism. Anything that attracts iron or steel is called a *magnet,* and this attracting ability is called magnetism. The operation of every transformer, practically all television picture tubes, every standard loudspeaker, many types of phonograph pickups, and all types of electric phonograph motors depends on magnetism.

In service work you will be dealing with two types of magnets. A *per-*

manent magnet holds its own magnetism and cannot be turned off. An *electromagnet* has magnetism only when current is flowing through its coil. An electromagnet can therefore be turned off by stopping the current flow. A few practical facts about these two types of magnets will definitely prove useful.

Poles of Magnets. The attracting power of any magnet is generally greatest at its two ends. These ends are called the poles of the magnet. Iron filings sprinkled over a permanent magnet will cling to these two poles.

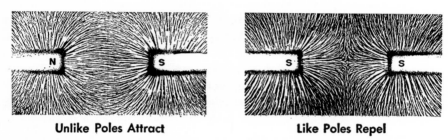

Unlike Poles Attract **Like Poles Repel**

Fig. 7. This is what happens when iron filings are sprinkled on cardboard placed over permanent magnets. The filings arrange themselves in lines corresponding to magnetic lines of force. Here the magnets are arranged to demonstrate the rules of magnetism

Both poles of a magnet have equal ability to attract iron filings and hence have equal magnetism, but actually the two poles are different. This difference can be demonstrated with two straight bar magnets by covering them with cardboard and sprinkling iron filings on the cardboard. If identical poles are brought together, the magnets will repel each other, and the pattern of iron filings will show a repelling action. On the other hand, if unlike poles are brought together, they will attract each other, and the iron filings will form lines going straight across between the two poles. A simple basic rule expresses this: *Unlike poles attract, and like poles repel.* This rule, illustrated in Fig. 7, is exactly the same as for electric charges.

If a straight permanent magnet is hung or pivoted at its center so that it can rotate readily, one pole will always point in the direction of north. This is because the earth itself is a huge magnet. For convenience in identifying the poles of a magnet, the pole that swings to the north is called the *north pole*. The other magnet pole is called the *south pole*, because it is then pointing in a southerly direction. Every magnetic compass is based on this simple principle.

The letter N is often used in place of the word north to identify the north pole of a magnet. The letter S is similarly used for the south pole.

Electromagnets. Any ordinary piece of iron or steel can be made into a magnet by winding insulated copper wire around it and sending direct current through the wire. When current flows, the ends of the iron or steel rod will be poles and will attract iron filings. As soon as the current is interrupted, the rod will lose its magnetism, and the filings will drop off. Any magnet that depends on electricity in this way is called an electromagnet.

The more current that flows through the coil of an electromagnet, the stronger will be the magnetism and the more iron will be attracted to the poles. Electromagnets used in some large loudspeakers can easily hold a pair of pliers or a screwdriver. These powerful magnets therefore attract iron filings like honey attracts bees.

Even a single iron filing in a speaker can ruin the moving coil and cause distortion. It is extremely important to keep iron filings away from speakers. Good modern speakers are carefully sealed, and the magnets are covered with dust caps to keep out filings, but the magnets of many older speakers and cheaper modern units are exposed. You will learn more about what iron filings do to speakers later. You will also learn how to get out the filings.

Permanent Magnets. Certain types of steel can hold magnetism permanently. These pieces of steel continue being magnets after the magnetizing force of the current-carrying coil is removed. Such magnets are called permanent magnets and are now used extensively in speakers.

Modern permanent magnets are made from alloys that hold much more magnetism than steel alone. The most widely used of these alloys for speakers is Alnico V, which contains ALuminum, NIckel, and CObalt.

Magnetic Fields. Magnetism is the same, no matter whether produced by permanent magnets or electromagnets. Magnetism is considered to be a magnetic *field* made up of magnetic *lines of force*. All magnets are surrounded by magnetic lines of force, arranged much like the lines formed by the iron filings in Fig. 7.

Magnetic lines of force occur for any coil that is carrying current. Even a wire has magnetic lines of force around it when carrying current. The more complete loops or turns there are in a coil, the more magnetic lines of force there are, as shown in Fig. 8. This means that more turns of wire give more magnetic effect.

The actions of all transformers depend on the magnetic field around a current-carrying coil. If another coil encloses part or all of the magnetic

Electricity and Magnetism

field, the two coils can act on each other even if not touching. Any change in current through the first coil can make a current flow through the second coil when the two coils are connected into a circuit.

Since alternating current is continually changing, it can easily be transferred from one coil to another by a magnetic field, as in Fig. 8. You thus already know the operating principle of a transformer, even before studying this important part.

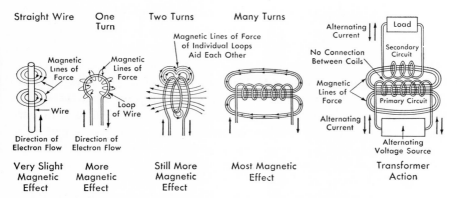

Fig. 8. Coil and transformer actions. The more turns of wire there are in a coil, the greater is the magnetic effect

Interference from Magnetic Fields. The power transformer in an a-c receiver and the motor of a record changer both have coils that carry fairly large alternating currents. These coils therefore produce strong and continually changing magnetic fields that can affect other nearby parts. For this reason, parts whose performance might be affected by these magnetic fields must be kept well away from the troublemaking coils or shielded from them by iron or steel housings or partitions.

Even a single wire can pick up hum from the magnetic field of a motor or transformer, if this wire is in a part of the receiver circuit that is followed by a number of amplifier tubes. Here the wire itself must be enclosed in a shield. Insulated wires inside a braided wire shield will often be found in receivers.

Iron or steel is required to shield parts from the magnetic field of coils that carry 60-cycle alternating current. Aluminum or copper shields are used for coils and transformers carrying signal currents at higher frequencies. Even tubes are sometimes shielded, to prevent stray magnetic fields from affecting them.

6

How to Use a Multimeter

Getting Acquainted with a Multimeter. In servicing, you need to measure d-c volts, a-c volts, ohms, and occasionally amperes when testing to locate the defective part in a receiver. The instrument used for this purpose is usually called a *multimeter*. With it, practically every part in a television or radio receiver can be checked for internal faults, such as breaks in wires, changes in resistance, short-circuits between wires, and grounds.

A multimeter is sometimes called a volt-ohm-milliammeter, a VOM, a multitester, a set tester, or a multirange tester. Multimeters having only volt and ohm ranges are also called volt-ohmmeters.

A multimeter has a number of different ranges for each type of measurement, to cover all testing and troubleshooting requirements in servicing. These voltage, resistance, and current ranges are obtained by combining special switching circuits and plug-in terminal jacks with a single meter.

To use a multimeter, set the range switch of the multimeter to the kind of reading you want, insert the test-lead plugs into the correct pair of jacks, clip or hold the test prods on the terminals of the part being checked as in Fig. 1, and read the meter.

In troubleshooting, you measure voltages while the set is on or measure resistances while the set is off. By comparing your multimeter readings with the normal values given in service manuals, you get clues to the location of the trouble.

Consider the multimeter as a tool for indicating what your eyes, ears, and nose cannot do. For example, if you suspect the defective part to be a resistor yet the resistor is not charred or burned in any way, use the ohmmeter range of the multimeter to test the resistor.

How to Use a Multimeter 123

As another example, suppose all the tubes in a receiver test good and all the parts look good, yet the set is dead. Here you would use the d-c voltmeter ranges of the multimeter to measure voltages in the set. You then compare these values with those given in the service manual for the set and look for trouble in the section where measured and printed voltages do not agree.

In this chapter you will learn how to use a multimeter correctly for practical servicing tests. You will find that reading the meter is just as

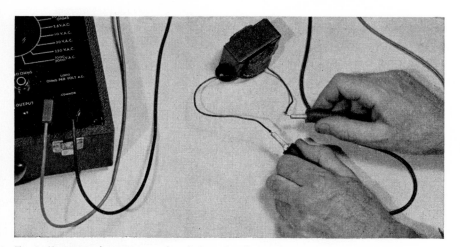

Fig. 1. Measuring the resistance of a choke coil with a multimeter

easy as reading length on a ruler. You will get simple instructions for choosing the correct multimeter range for any test. Most important of all, you will get advice on how to choose a good multimeter for your servicing work.

Choosing a Multimeter. Many different sizes and styles of multimeters are available at an equally great variety of prices. Choosing the best one for your particular requirements can be a real problem if you have only catalogs to go by.

Examples of good multimeters for servicing use in both the shop and home are shown in Fig. 2. If you choose one of these, there is no need to worry about what to look for in a multimeter. Both are designed and constructed properly for practical television and radio servicing, and will get you started right. Each has a convenient handle for carrying into homes and at the same time meets all requirements for bench use in your shop.

Each has the required high sensitivity for television testing. Prices of both are about the same, in the range of $40 to $50.

Other multimeters that deserve consideration in this price range include the Hickok model 457, RCA model WV-38A, and Weston model 980. Look also at the Hickok model 455A, which costs around $70 but has a special fuse system that protects the meter and parts against overload and burnout.

Fig. 2. Two excellent high-sensitivity multimeters recommended for anyone getting started in servicing work. Left—Triplett model 630-PL. Right—Simpson model 260. Each comes with a pair of test leads

The construction and design information in the following sections will help you to get acquainted with the important parts and features of your multimeter. This information will also serve as a buying guide if you want to consider other makes and models.

Meter Sensitivity. Multimeters are rated according to their sensitivity in ohms per volt when used as voltmeters. The higher the sensitivity, the more accurate is the multimeter.

Multimeters with a sensitivity of 20,000 ohms per volt are widely used by experienced servicemen. The recommended multimeters in Fig. 2 both have this sensitivity. Pocket-size multimeters with sensitivities of 1,000 or 2,000 ohms per volt may be cheaper but are no good for you, because they cannot measure the higher resistance values encountered in modern receivers.

Meter Scales. A neat and simple arrangement of scales on the meter of a multimeter helps prevent mistakes in reading. The three most important scales are for ohms and for a-c and d-c values.

The ohms scale is usually at the top on a multimeter, because its divisions are unequally spaced, and require more identifying numerals for easy reading. The d-c scale comes next, and the a-c scale is under it.

The a-c scale is often printed in red to distinguish it from the d-c scale. Extra sets of numerals are usually printed below the a-c and d-c scales, to make it easier to read other ranges of values on these scales. Sometimes the same numbers or even the same scales serve for both the a-c and d-c scales. In a good multimeter the scales are carefully planned for easiest possible reading on all ranges.

One requirement for easy reading of resistance values is the use of *multiplying factors* for the higher ohms ranges. With this system, the ohms scale is read directly when the switch on the instrument is set to $R \times 1$. The designation $\times 1$ means multiply by one. When the switch is set to $R \times 10$, the ohms scale readings are multiplied by ten. Similarly, for settings like $R \times 100$, $R \times 1,000$, $R \times 10,000$, and $R \times 100,000$ the scale values are multiplied by the designated number.

Why Extra Ranges Are Needed. Several ranges are needed on a multimeter for each type of measurement so the pointer can be read more accurately. The effect of using just one high-range scale would be like using a 300-pound bathroom scale to weigh a letter. You cannot read small fractions of a pound on a scale made for hundreds of pounds. Likewise, you cannot read a 1½-volt battery voltage on a 5,000-volt scale designed for television power-supply measurements. For such low measurements a 0-to-3-volt range, 0-to-5-volt range, or 0-to-15-volt range is needed to give the accuracy required. The same holds true for resistance and current measurements.

Controls. Simplicity of operation is important in a multimeter. It helps to speed up testing and reduces chances of damage by using the wrong range.

Most multimeters require only one operation to change a range for most of the settings. The extra jacks for high voltage and for high current (amperes) are not serious objections, as these ranges are not used very often. These extra jacks are required because the range-selector switch is not ordinarily made to handle such high voltages and currents.

Test Leads. Two leads, one red and the other black, are generally furnished with a multimeter. Each lead has a short plug at one end for plugging into a jack on the multimeter panel. The long test prod at the other end of the lead is held in the hand when measuring. In addition, two removable alligator clips are often furnished with the test leads. These clips can be pushed onto the metal tips of the prods when semipermanent connections are desired. Some prods have permanently attached clips.

Carrying Case. Although you can get a leather or plastic carrying case for most multimeters, the extra cost is rarely justified. Multimeters are made well enough to withstand ordinary handling on your bench, in your car, and in homes. The case can actually be a nuisance, since it takes time to remove the multimeter from the case and put it back.

Electronic Multimeters. Another type of multimeter used by servicemen is the electronic multimeter, also called a vacuum-tube voltmeter, VTVM, and vacuum-tube volt-ohmmeter. These instruments have one or more vacuum tubes, and therefore operate from the a-c line or from higher-voltage batteries than are used in ordinary multimeters.

The advantage of the electronic instrument is that it has practically no effect on the circuit to which it is connected. This feature is desirable when hunting for trouble in certain critical circuits, but is not needed for the simple repair jobs covered in this book. There is no need to buy an electronic multimeter until you have mastered this entire book and are ready for more advanced troubleshooting techniques.

Multimeter Kits. Although it is possible to save a few dollars by buying a kit of multimeter parts and assembling the instrument yourself, this is definitely not recommended for a beginning serviceman. A multimeter is your most important instrument, and you want to be able to rely on the accuracy of its readings. If you put it together yourself and have no way to calibrate the readings on each scale and range, you may always be questioning its accuracy and reliability. Get a factory-calibrated instrument, to get started right in professional servicing.

Instruction Booklet. With each multimeter comes the manufacturer's booklet of operating instructions. The general instructions in this chapter cover all common precautions against damage, but multimeters differ greatly in design. Do not take chances in operating your multimeter—*read its instruction book first*, before trying to use the meter. Remember that you can burn up a ten-dollar bill (the usual minimum price of a multimeter-repair job) faster with improper use of a multimeter than with a match!

If a registration card is included with your new multimeter, fill it in and send it back to the manufacturer as soon as possible. This will give you full benefit of the manufacturer's guarantee on your instrument.

Measuring D-C Voltages. Measurements of d-c voltages in receivers provide extremely valuable information when hunting for the trouble-making section. Learn the correct procedure now, to avoid possible damage to your new multimeter.

1. Observe safety rules. If working on a hot-chassis set in a customer's home, move the set away from any grounds. If at your bench, plug the set into your isolation transformer.

2. If the multimeter has an a-c/d-c switch, turn it to the d-c position. If no such switch is used, proceed directly to the next step.

3. If the multimeter has a polarity switch, set it to the correct position for ordinary d-c voltage measurements, as specified in the instruction manual for the instrument.

4. Set the range switch to a d-c voltage range, usually marked D-C V. If you know about how many volts to expect at the place you intend to check, set the range switch to the range next higher in voltage. For example, if you expect to measure a voltage around 180 volts, then set the range switch to 250 or 300. If you have no idea how many volts to expect, set the range switch to 1,000 or 1,200.

5. Put the short black test-lead plug in the common black jack, generally marked COM, —, or COMMON.

6. Put the short red plug in the red jack identified by V-O-MA or a similar voltage designation.

7. Hold the long-handled black prod on the negative voltage terminal of the set. This is not always the chassis, so it is a good idea to check the service manual first. If you are making several voltage tests, push an alligator clip over the end of the prod and clip it to the chassis.

8. Turn on the set. Touch the tip of the long red prod to the terminal you want to check. Make this connection lightly, so you can remove the prod quickly if the pointer goes off scale or reads backward.

9. If the meter reads backward, your assumption as to polarity was wrong. Reverse the positions of the test leads either at the multimeter or at the set. If your multimeter has a polarity-reversing switch, just flip this switch to make the meter read up-scale. Be sure to set the switch back to its normal position before making the next reading.

10. Read the meter on the correct scale for the range being used.

11. If the reading is way down near zero for the range you start with, lift the red prod off the terminal and switch to a lower range, then repeat the measurement. Always disconnect the multimeter in this way before changing ranges, to prevent arcing at the range-switch contacts inside the multimeter.

Reading D-C Scales. Reading the d-c scale of a meter is very much like reading a 1-foot ruler. The zero end of the meter scale is at the left,

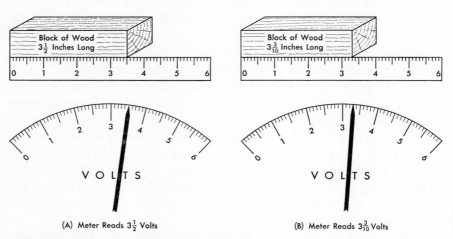

(A) Meter Reads $3\frac{1}{2}$ Volts   (B) Meter Reads $3\frac{3}{10}$ Volts

Fig. 3. Reading a meter is exactly like reading a ruler

just as on the ruler. The full-scale end of the meter scale corresponds to the other end of the ruler.

The place at which the meter pointer comes to rest corresponds to the point at which you read a ruler when measuring something. For example, in Fig. 3A the block measures $3\frac{1}{2}$ inches long, and the meter pointer indicates $3\frac{1}{2}$ volts. If the block of wood were $3\frac{3}{10}$ inches long and the voltage were changed to $3\frac{3}{10}$ volts, the two indications would again match as shown in Fig. 3B.

The ruler used in the illustrations is one commonly used by machinists. It differs from a conventional ruler in that it is divided into tenths of an inch instead of eighths or sixteenths.

When reading a meter, look straight down on the pointer. If your head is off to one side, you may see an entirely different division line under the pointer and thus get a wrong reading. This is called a parallax error.

A horizontal position, flat on the bench, is best for a multimeter. A

How to Use a Multimeter 129

meter resting on end or propped at an angle will give essentially the same accuracy, but is more easily knocked over or damaged.

Meter Scale Divisions. The space between two adjacent lines on a ruler or meter is called a division. Several different lengths and thicknesses are used for division lines to make them easier to read.

If the pointer is exactly halfway between division lines, as in Fig. 4A,

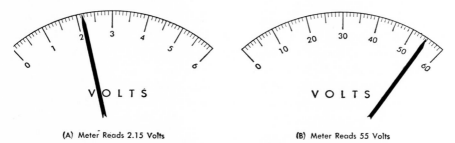

(A) Meter Reads 2.15 Volts   (B) Meter Reads 55 Volts

Fig. 4. Examples of how meters are read when the pointer is between division lines

you would read 2 plus 1½ tenths. This is expressed as 2.15. Actually, a reading of 2.1 is close enough for troubleshooting.

In many tests the pointer does not fall right on a division line nor does it fall exactly halfway between two lines. When this happens, read the value of the nearest division line, as shown in Fig. 4B.

Reading Higher D-C Ranges. A meter in a multimeter will usually have two or more rows of numbers for the d-c scale. Thus, in the example of Fig. 5A there are numbers for a 300-volt range as well as for a 60-volt range. Read only the row of numbers for the range being used, just as if the meter had only that one scale.

As an example, the pointer indication is read as 165 volts when using the 300-volt scale in Fig. 5A. This reading is figured out by noting first

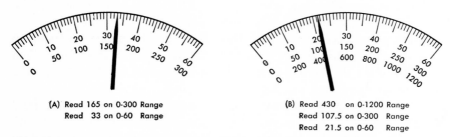

(A) Read 165 on 0-300 Range
Read 33 on 0-60 Range

(B) Read 430 on 0-1200 Range
Read 107.5 on 0-300 Range
Read 21.5 on 0-60 Range

Fig. 5. Multimeter scales are easy to read if you look only at the row of numbers for the particular multimeter range being used for troubleshooting or testing

that the pointer is three divisions past 150. Since there are ten divisions between 150 and 200, each division is 5 volts. Three divisions past 150 is therefore 15 past 150, or 165.

If you were using the 60-volt scale for the pointer setting of Fig. 5A, you would ignore the 0–300 row of numbers and simply read 33 volts directly.

An example of reading a three-range scale is given in Fig. 5B. Here again, you read only the numerals for the range being used. Note that for the 60-volt range the smallest scale division represents 1 volt. For the 300-volt range this smallest division represents 5 volts. For the 1,200-volt range the smallest division represents 20 volts.

In most servicing measurements it is not necessary to read more than three digits. A reading of 107.5 volts would therefore be called 107 volts or 108 volts. Accurate readings are not necessary, because in troubleshooting you are looking for a very large change from the value given in the service manual.

The number opposite each switch position on the multimeter is the right-hand or full-scale value for voltage and current ranges.

Reading Extra D-C Ranges. Multimeters often have voltage ranges for which there are no rows of scale numbers. However, there will always be a row of numbers that can be multiplied mentally by 10 or occasionally by 100 to get the range of values required. For example, when a 6,000-volt range setting is used, you can read the 60-volt scale and multiply the reading by 100. A reading of 33 volts on a 60-volt meter scale would then indicate a voltage of 33 multiplied by 100, which is 3,300 volts. Here, 100 is the multiplying factor.

Off-scale D-C Voltage Readings. If the pointer goes off scale to the left, the polarity of the d-c voltage is opposite to what you thought it would be. Disconnect the test leads quickly and reverse them at the receiver. The pointer cannot go far to the left of zero because of stops inside the meter, so ordinarily no damage is done.

If the pointer goes off scale to the right, the voltage you are trying to measure is more than the full-scale value of the range you are using. When this occurs, lift off the test prod, turn the multimeter range switch to the next higher range, and repeat the test.

If the pointer moves violently across the scale, lift off the test prod quickly and check your multimeter settings. You may have the multimeter in a milliampere or ohms position or in way too low a d-c voltage range.

If the meter pointer jumps around intermittently, you probably have a defective part in the circuit being measured.

If the pointer vibrates on a d-c voltage measurement, there probably is a-c present along with the d-c voltage being measured. For some circuits, this may be normal, but in most cases it is a clue to trouble.

Accuracy of D-C Voltage Readings. When a multimeter is used to measure d-c voltages in a receiver, some current is drawn from the receiver circuit by the meter. In getting to the meter, this current flows through

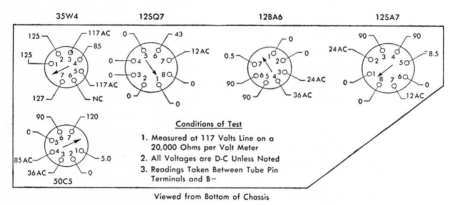

Fig. 6. Example of voltage data given in a service manual for a table-model clock radio

resistors in the receiver and produces voltage drops across them that rob from the voltage being measured. The voltage indicated by the meter will therefore be slightly lower than the actual voltage.

With a 20,000-ohm-per-volt multimeter, the meter current is so small that its effect cannot be noticed for most measurements. Only with lower-sensitivity 1,000-ohm-per-volt multimeters does the error in voltage readings become appreciable.

When voltage values are given in service manuals, the ohms-per-volt sensitivity of the meter used is generally specified. If using the same meter sensitivity, you can forget about errors due to meter current when measuring voltages. Most service manuals today give the voltages obtained with a 20,000-ohm-per-volt multimeter sensitivity. An example of one way used in service manuals to specify correct voltages is shown in Fig. 6.

Precautions When Measuring High Voltages. Observe the following precautions when measuring the high voltages of television receivers and even the plate circuit voltages of radio sets:

132 Television and Radio Repairing

1. Use extreme care in working with all voltages over 100 volts! Do not let the first few low-voltage shocks fool you into thinking you can take it! You may get by with just a sharp, tingling sensation when you lightly touch a 250-volt terminal on a dry day. However, if your hands are wet or you happen to grab these same wires with more firmness on a damp day, *you can get killed!*

2. Do not work on high-voltage circuits in a place where the floor is wet, such as a basement workshop, unless you are wearing a good pair of rubbers or rubber-soled shoes. A wooden platform spaced off the floor helps in wet or damp locations.

3. Do not work on a set if you are overly tired. You can easily ruin yourself or your test equipment or both! No job is important enough to take chances. Get some rest and come back to it in the morning.

4. Keep one hand in your pocket when working on dangerous high-voltage circuits. This leaves only five fingers, instead of ten, that can carelessly or accidentally slip onto a high-voltage terminal. Having one hand in your pocket also eliminates the dangerous hand-to-hand path through your heart for electricity.

5. Keep your fingers well away from the metal tip of the test prod.

Measuring A-C Voltages. To make a-c voltage measurements, follow the same procedure as for d-c voltages, except switch to the a-c volt positions on the range switch and read on the a-c meter scale (generally printed in red).

Some multimeters have separate rows of numbers for the a-c scale, as shown in Fig. 7A. Others use the d-c range numbers, as in Fig. 7B, for the 300-volt a-c range. Examples of readings are given for each type.

The a-c scale divisions are spaced closer together at the left end of the

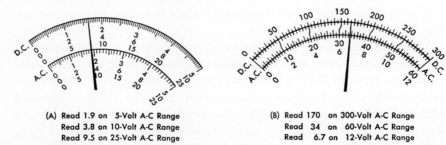

(A) Read 1.9 on 5-Volt A-C Range
Read 3.8 on 10-Volt A-C Range
Read 9.5 on 25-Volt A-C Range

(B) Read 170 on 300-Volt A-C Range
Read 34 on 60-Volt A-C Range
Read 6.7 on 12-Volt A-C Range

Fig. 7. Examples of a-c and d-c scales of multimeters, with the pointer reading for each a-c range. If you can read these scales, any other meter will be easy

scale. This is due to the rectifying characteristics of the metallic rectifier used in the multimeter to change the a-c to d-c so it can be measured by the meter.

To read the value of the a-c voltage at a particular pointer indication, follow the same procedure as for reading a d-c scale, except use the a-c scale.

Safety Check. Before returning a repaired receiver to its owner, be certain that it is completely safe to operate, without danger of electric shock. You can do this with the a-c voltmeter ranges of your multimeter.

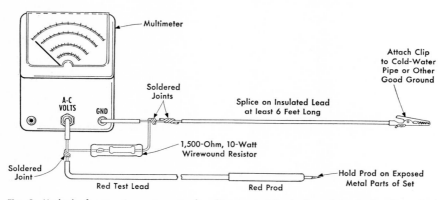

Fig. 8. Method of preparing extra set of multimeter test leads for safety checks of repaired receivers. Wrap No. 33 Scotch electrical tape around resistor body, resistor leads, and soldered joints

Set the multimeter to 150 volts a-c or the nearest other range that will measure a-c line voltages. Get an extra set of multimeter leads from your jobber and fix them up as shown in Fig. 8. This involves lengthening the black ground lead and connecting a 1,500-ohm, 10-watt wirewound resistor between the two leads close to the multimeter.

The safety test is made by clipping the ground lead to a cold-water pipe or other good ground, then holding the other multimeter lead in turn on each exposed metal part of the receiver after the chassis is installed in its cabinet.

The a-c voltage indicated by the meter should not be higher than 8 volts. A reading between 8 volts and full line voltage indicates that a potentially dangerous shock hazard exists. Switch to a lower a-c voltage range if the meter reading is near zero, to get a more accurate measurement.

If the receiver passes the first check, switch back to the 150-volt a-c range, reverse the position of the line-cord plug in the wall outlet, and repeat the test. Here also the voltage should be below 8 volts.

Make this safety test on all receivers, even if they have a power transformer. It is particularly important for sets having metal cabinets. If the receiver does not pass the test, look for a defective capacitor, poor insulation, or a wrong connection to the metal cabinet.

Measuring Audio Output Voltages. An important use for a multimeter is measuring the strength of the audio signal that a receiver feeds to its loudspeaker. This is done when aligning the receiver; hence you will not need to know the procedure for a while.

Output measurements are made with a-c voltage ranges by moving the red test lead to the OUTPUT jack on the multimeter panel. Some multimeters have a switch setting for OUTPUT which make it unnecessary to move the test lead. Switch the output ranges the same as a-c volt ranges, using the highest range first.

The a-c audio voltage measured across a loudspeaker changes continually, in step with the volume and frequency of the sound coming from the loudspeaker. The louder the sound, the higher the voltage indicated by the pointer. For music, the pointer moves up and down in time with the music.

Measuring Resistance. The ohmmeter ranges of a multimeter are used to test parts that may have become defective because of overload, natural aging, moisture, or mechanical failure of the mounting. Practically every television and radio receiver part can be tested quickly and easily with an ohmmeter. Detailed instructions for making these ohmmeter tests are given in later chapters of this book; only general instructions for measuring resistance will be given here.

The ohmmeter section of a multimeter uses different combinations of resistors and a battery with the meter to obtain the different ranges of resistance. The battery, located inside the multimeter, makes the meter pointer swing over to zero on the ohms scale when the test leads are touched together.

A rheostat in the ohmmeter circuit provides an adjustment to compensate for the drop in battery voltage as the battery is used. A well-designed ohmmeter requires very little if any readjustment of this control in going from one range to the next. With poorly designed ohmmeters, the ohms-adjust control must be adjusted every time you change to another ohms range.

How to Use a Multimeter 135

Procedure for Measuring Resistance. Several precautions must be observed when using the ohmmeter, to prevent damage to the multimeter or excessive drain on the ohms range battery.

1. Be sure the receiver is turned off whenever measuring resistance. As a further precaution, unplug the line cord. In transistor radios and other battery portables, disconnect or remove all batteries before making resistor checks. The on-off switch for battery sets generally controls only the filament circuit, leaving the plate voltage still connected.

2. Discharge any large or high-voltage capacitors before testing them for shorts, because good capacitors can store energy. An easy way to do this is to grasp a screwdriver by its insulated plastic handle and hold the metal blade against the capacitor terminals. This usually produces a loud but harmless spark as the capacitor discharges. It needs to be done chiefly to the high-voltage capacitors in television power supplies, for otherwise these capacitors may discharge through the meter and damage the meter coil.

3. If you know the approximate value of the resistor to be checked, set the range switch to an ohms range which has that value somewhere near the center of the scale. If you do not know the approximate resistance, set the range switch to the highest ohms range.

4. Plug the test leads into the ohmmeter jacks of the multimeter. These are generally the same jacks used for d-c voltage measurements. One is the COMMON or − jack, and the other is marked +, V-O-MA, or V-Ω. The symbol Ω here means ohms.

5. Hold the tips of the long prods together and check the zero-ohms adjustment. If the lowest ohms range is used, it is a good idea to clip the leads together to ensure a good low-resistance contact. If the meter pointer is not at zero on the ohms scale, adjust the Ω ADJUST or ZERO OHMS knob to bring the pointer to zero.

6. Hold or clip the long test prods on the leads of the part under test, and read the ohms scale. If the pointer indicates too close to the highest-resistance mark to give a good reading, switch to the next lower range. For resistance measurements it is not necessary to disconnect one test prod when changing ranges. Switch to a still-lower range if necessary, until the reading is near the center of the scale.

If you cannot get a reading (if the pointer does not move noticeably from the highest-resistance or infinite end of the scale), either the part you are testing is open or its resistance is higher than the full-scale ohms range of your tester.

136 Television and Radio Repairing

Reading an Ohmmeter Scale. The zero is at the right end of an ohms scale on most multimeters, so that values increase to the left.

The divisions of an ohms scale are definitely not of equal value across the scale. You must therefore figure out the value of the scale divisions each time the pointer rests in between numbered lines on the scale. For example, in Fig. 9A there are five divisions between 0 and 1. The value of each of these divisions is therefore 1 divided by 5, which is 0.2 ohm. Between 5 and 10 on the scale there are also five divisions, but here they cover a scale distance of 5 ohms, so each of these divisions is 1 ohm.

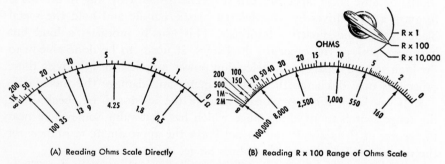

(A) Reading Ohms Scale Directly   (B) Reading R x 100 Range of Ohms Scale

Fig. 9. Examples of ohms-scale readings on two different multimeters. Check each reading

The same procedure is followed for all other portions of the ohms scale, as indicated by the reading for each pointer position in Fig 9A. This procedure applies for the three divisions between 50 and 200, even though the divisions are very unequal at this crowded end of the scale. The scale distance between 50 and 200 is 150, which means that each of the three divisions is 50. The first division line to the left of 50 is therefore 100.

Some ohms scales, chiefly on electronic volt-ohmmeters, have the zero at the left. This ohms scale is read the same as ordinary ohms scales, remembering that numbers increase to the right rather than to the left.

Reading Ohmmeter-scale Abbreviations. The ohms scale of a multimeter often has special markings and abbreviations not found on other scales. A few of these are listed below for your reference.

Ω is the symbol for ohms. It is the Greek capital letter omega. The lower-case letter omega, written as ω, is also used, mostly in wiring diagrams, as an abbreviation for the word ohms.

∞ is the symbol for infinity. It is usually used at the high-resistance end of the ohms scale. It means an infinite amount of ohms or an open circuit. The meter pointer indicates at this mark when the ohmmeter leads are not connected to anything.

K is the symbol for 1,000. It is often used after a number to indicate that the number has three more zeros after it. For example, 500K means 500,000 ohms.

M is often used in place of meg to designate that the number is in megohms. Add six zeros to change it to ohms.

M is sometimes used instead of K for 1,000. Its position with respect to other numbers on the ohms scale or the range switch will usually tell you at a glance what M stands for, so do not let this confusion of abbreviations worry you.

Reading Higher Ohms Ranges. Most multimeters have two or more ohms ranges. The commonest and most convenient method of reading higher ohms ranges is with multiplying factors that only involve adding zeros to the scale value, as in Fig. 9B. Here there is only one easy-to-read set of numbers on the ohms scale, and the range switch setting gives the multiplying factor. Just multiply the scale reading by the switch setting. Thus, a scale reading of 15 is 15 ohms for $R \times 1$, 1,500 ohms for $R \times 100$, and 150,000 ohms for $R \times 10,000$.

Getting Accurate Resistance Readings. Several cautions must be observed in using the ohms ranges to get the most accurate reading and to protect your multimeter.

1. Disconnect one lead of the resistor, capacitor, or coil being checked, whenever other parts are also connected between the two measuring points. Any shunt (parallel) path around the part being checked could cause a lower resistance reading than normal or even hide a defect. A shunt path may also indicate a false leakage resistance, a short in an actually good capacitor, or some other nonexistent defect.

2. Keep your fingers off the bare test-prod tips, and keep your fingers off the terminals of the part being tested, because your own body provides a shunt resistance path. This body resistance can cause a considerably lower resistance reading than normal, especially if your hands are moist and the resistance being measured is around a megohm or higher. You can easily measure your body resistance by holding one test-prod tip in each hand. Body resistance can be as low as 10,000 ohms if your fingers are moist, though it is usually around 100,000 ohms.

3. Turn the multimeter range switch to an off position or a high-voltage range position when you are through using the ohms ranges. This prevents

damaging the ohmmeter if the test prods accidentally touch voltage-supply terminals. In addition, this precaution prevents the ohmmeter battery from being run down if the test prods touch each other.

4. Remove worn-out batteries from the multimeter as soon as you notice that you can no longer adjust for zero ohms. If this is not done, the chemicals may leak out of the worn-out battery and corrode wire connections or fine-wire resistors in the multimeter. The battery case may also expand with age and jam the battery in its mounting.

Radio Troubleshooting with an Ohmmeter. A simple resistance measurement between the blades of the line-cord plug tells you a lot about an a-c/d-c radio. With the radio switch turned on, the ohmmeter reading should normally be around 200 ohms. This is the total cold resistance of the tube filaments in series. An appreciably lower reading means that some part is shorted or grounded. More measurements should then be made to locate the trouble, before the set is plugged into a power line. The chief suspect is a paper capacitor that is connected between one terminal of the on-off switch and the other side of the power line.

A reading of infinity usually means a burned-out tube, but could also mean an open line cord, a defective switch, or a break in filament circuit wiring.

Ohmmeter Precautions in Transistor Circuits. The lowest range of an ohmmeter can send enough current through a transistor to ruin it. Always use a higher resistance range, such as R $\times$ 10 or higher, when making resistance measurements in a circuit containing transistors.

Always remove the batteries from a transistor radio before making resistance measurements. The battery voltage may damage your ohmmeter, and the ohmmeter may provide a short-cut path from the batteries to a transistor that cannot withstand the full battery voltage.

The safest procedure is to make only d-c voltage measurements in transistor equipment when trying to locate the trouble. After you have found a stage in which the d-c voltages are off, you can quickly isolate the trouble with a few carefully planned resistance measurements, using the service manual as your guide.

When checking low-voltage electrolytic capacitors in transistor circuits with an ohmmeter, correct polarity must be observed to prevent damaging the capacitor. On some multimeters, the polarity markings do not represent the polarity of the internal battery for ohmmeter readings. Instructions for checking the battery polarity of your multimeter are given in the capacitor chapter.

How to Use a Multimeter 139

Measuring Direct Currents. Although practically all multimeters have several current ranges, these are seldom, if ever, used for testing or troubleshooting. Voltage or resistance measurements are made instead, since they are much easier to make and give equally useful information about a part or circuit.

To measure the amount of current flowing through a circuit, the meter must be inserted in the circuit so the current can flow through the meter. This involves turning off the set and disconnecting or even unsoldering or cutting a wire so you can connect the multimeter test leads to the two sides of the break in the circuit. The circuit must then be reconnected or resoldered after the measurement has been made. It sounds like a lot of work, and it is.

Detailed instructions for making current measurements are given in the instruction manual for the multimeter you buy. The d-c scales are read exactly the same as for d-c voltage ranges.

Care of Test Leads. Test leads are an important part of your multimeter, so handle them carefully. When using the slip-on alligator clips for receiver connections, wiggle each clip a few times after it is in place to cut through any oxide or dirt on the terminal or the chassis.

Alligator clips can be clipped onto the multimeter handle when not in use. Of course, if your multimeter has a storage compartment, this is the best place of all for the clips as well as the test leads. Your toolbox is another good place for test leads and clips.

When using the multimeter, keep the leads on the bench so they do not get pinched in a drawer or in a cabinet door. When you unplug the leads, grasp the prod handles instead of pulling on the wires. Pulling on test leads can eventually loosen or break the wire inside the prod handles.

If insulation is broken on a test lead so bare wire shows, wrap with No. 33 Scotch insulating tape to prevent the bare wire from touching the chassis while making tests.

Keep test leads free of kinks or knots. These put a strain on the wire and take up valuable lead length.

Keep the test leads away from your soldering gun. Keep them also away from hot tubes in the set. Heat can burn through or damage the insulation on the leads.

Multimeter test leads have a habit of catching on corners and knobs. Yanking on a test lead without looking first may pull a costly multimeter right off the bench. Test equipment does not bounce well!

Testing a Multimeter. To check a multimeter after an accidental overload or bump, measure the voltage of a flashlight cell. A new cell should read fairly close to 1.5 volts on the lowest d-c voltage range if the multimeter is undamaged. Measuring the a-c line voltage is another check; this should read within a few volts of 117 if the multimeter is good and the power-line voltage is normal in your locality.

Meter Zero Adjustment. Practically all meters have an adjusting screw that permits changing the at-rest position of the pointer a small amount. This is used to make the pointer rest exactly at the zero position on the d-c voltage scales when the meter is not in use. From time to time a slight adjustment of this screw may be needed, especially after an overload.

Using a small screwdriver, set the zero-adjuster screw on the front of the meter so that the pointer is on the scale line farthest to the left when nothing is connected to the test-prod leads. Tap the multimeter case lightly with your finger while doing this, to offset any slight friction in the bearings. If the multimeter is normally used in a vertical or a 45-degree position, make the zero adjustment in that position.

The pointer may move off zero a slight amount (generally less than one d-c scale division) when the meter is tilted from horizontal to vertical. To eliminate the nuisance of readjusting the zero position, use your multimeter in the same position all the time. If the pointer is bent as a result of severe overload, you can sometimes bring the pointer back to the zero mark with the zero-adjuster screw. Do this only as an emergency measure, however. For best performance, return the multimeter to the factory for repair if the pointer is noticeably bent from overload.

Replacing Ohmmeter Batteries. If the pointer cannot be brought all the way to the zero mark with the ohms-adjust control, the ohmmeter battery for that range is probably low and requires replacement. To replace the proper battery if more than one battery is used, check the operating instructions for your tester. Be sure you install the new battery in the same position as the old one, to obtain the correct polarity.

The center post of a 1.5-volt flashlight cell is positive. The corresponding terminal for this cell inside the multimeter will sometimes be marked + or painted red. If you get the battery in backward, the pointer will read off the left end of the scale when you connect the prods together for checking the ohmmeter zero. If two or more batteries are used on a particular range and one is reversed, the pointer will not come up to the zero-ohm mark.

Replace multimeter batteries with new ones having the same number and make as were in before, or use equivalent units recommended in the multimeter instruction book. Batteries may be obtained from the multimeter manufacturer or from your parts jobber. Some multimeters use standard-size flashlight batteries which may be obtained in drugstores.

Be sure the multimeter battery contacts are clean and have good spring tension against the battery. If a screw-in clamp is used to mount the battery, be sure to tighten the screws securely.

Common Troubles in Multimeters. Some of the more common troubles that develop in multimeters are:

1. Meter burned out because of overload. When this occurs, one or more of the resistors may also be damaged. Return the multimeter to the factory for repair.

2. Broken meter window glass due to dropping tools or other objects on it. Return to factory.

3. Smashed case or panel due to dropping the multimeter. This is often the most expensive repair. In addition to a new case or panel assembly, the meter will probably require new parts. Return to factory.

4. Bent pointer due to overload of meter. Try the meter zero adjustment first. If it will not bring the pointer to zero, return to factory.

5. Weak or loose plug or jack contact springs, usually caused by bumping a test prod while it is in the jack. Repair or replace jack or test leads yourself. New test leads and jacks can usually be obtained from jobbers.

6. Frayed or broken test-prod wire or broken test prods, due mostly to careless handling of test leads. Buy new test leads.

7. Low a-c voltage readings, due to defective rectifier in multimeter. Return to factory if correct replacement rectifier cannot be obtained from your jobber.

Getting a Multimeter Repaired. Return a damaged multimeter only to the manufacturer or to an instrument repair shop authorized by the multimeter manufacturer. Before shipping, write to the manufacturer and describe the damage. Be sure to give the model number and tell where you bought the instrument. The manufacturer will then send the necessary shipping instructions and will generally give you a rough estimate of the cost of repair.

Place the multimeter in a small cardboard box first, along with the test leads. Be careful not to jam the leads against the meter glass. Pack the cardboard box carefully in a wooden box. There should be at least a 1-inch thickness of crumpled newspaper or similar shock-absorbing material between the multimeter box and the outer box. Nail on the wood cover of the box with small nails.

Tie the box securely with heavy cord as an extra precaution. Letter the mailing address on it clearly with black ink, so the package will not go astray even if your pasted-on shipping label comes off.

Place a FRAGILE sticker on each side of the box. Generally these stickers and a shipping label will be supplied by the multimeter manufacturer along with the shipping instructions. Insure your package for its full value if shipping by parcel post.

7

Removing and Replacing Receiving Tubes

What Tubes Do. Vacuum tubes are the heart of the modern television or radio set. They have many different tasks. For example, rectifier tubes change a-c to d-c in the power supply. Amplifier tubes increase the intensity of the desired signal. Converter tubes change the incoming r-f signal to an intermediate frequency which can be amplified more easily. Other tubes have a variety of special jobs.

Tube Appearance. You will have little trouble spotting the tubes in a set because they look so different from other parts, and because tubes fit into sockets that permit easy removal and replacement. Examples of some of the different sizes and shapes of tubes you may have to replace are shown in Fig. 1.

Parts of a Tube. Though various types of television and radio tubes differ widely in appearance, all have an *envelope* or outer housing. This envelope holds a vacuum for the *elements* or *electrodes* inside.

All tubes have either a separate or self-contained *base* containing the *pins* that serve as terminals for the tube. A few tubes have an additional terminal on top of the tube, called a *top cap*. These and other external parts of tubes will be taken up one by one before considering what goes on inside the tubes.

Tube Envelopes. The glass housing or envelope that encloses the working elements of a radio or television tube comes in many different sizes and shapes.

A glass envelope often has a bright silvery coating on part of the inside surface. This chemical coating is called the *getter* because it is put on during manufacture to get or absorb the last trace of gas and air from inside the envelope after the air has been drawn out with vacuum pumps.

Some tubes may have a dull black coating of graphite over the entire inner surface. This is sprayed on during manufacture to provide an electrostatic shield for the working elements inside. Tubes like this do not ordinarily require separate metal shields.

Tubes are identified by a type number. This is generally etched or printed on the side or top of the envelope.

 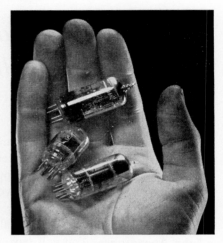

Fig. 1. Variety of tubes found in a television receiver. Tubes without bases, held in hand, are called miniature tubes. Picture-tube problems are covered in a later chapter. (Sylvania photos)

Tubes are also made with a metal envelope, called a shell, but these are used today chiefly as replacements in older sets. A *metal* envelope will not crack or break like glass when handled roughly, but the elements inside can be jarred out of place if the tube is dropped. The metal envelope acts as a self-contained electrostatic shield.

A metal envelope radiates heat better than glass. This keeps the tube elements cooler and increases tube life. Because it is getting rid of more heat, however, a metal tube will usually feel hotter than the equivalent glass tube. Metal tubes are made vacuumtight by welding, and their connecting leads are brought out through glass beads that are fused into holes in the base of the metal envelope.

Both metal and glass tubes can normally get hot enough to sizzle your fingers. It is a good idea to turn off the set and allow the tubes to cool for a few seconds before touching them.

Tube Bases. Construction features of various types of tube bases are shown in Fig. 2. A base may have anywhere from 4 to 21 terminal pins (also called prongs). Each pin connects to a thin wire lead that goes through the glass envelope to an element of the tube. These leads are made of special wire that seals to glass, so no air can leak into the tube around the leads.

When tubes have molded plastic bases, the pins that come through the

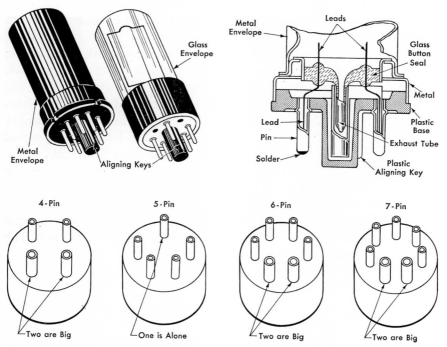

Fig. 2. Details of metal and glass octal tube bases and pin arrangements used on four older types of tube bases. In modern sets you will find chiefly miniature tubes and octal-base glass tubes

bases are tubular. At the outer end of each pin is a blob of solder that connects the inner lead securely to the pin. On some older bases, two pins are thicker than the others or the pins are spaced unequally. The tube can then be inserted only the one correct way in its socket.

Octal bases have provisions for eight equally spaced pins, with a molded plastic aligning key in the center of the base to ensure correct positioning in the socket. The socket has a corresponding notch in its central hole. Pins not needed in a particular tube are often omitted during manufacture, without changing the spacing of the remaining pins.

In metal-envelope octals, the leads come out of the envelope through

146   Television and Radio Repairing

glass beads that are sealed into holes in the envelope, in order to insulate the leads from the envelope.

Loktal bases have the same equally spaced arrangement of pins as octals but have a metal aligning key that has a lock-in groove which helps to keep the tube in its socket. Loktal tubes have solid-wire pins that go right through the glass to the elements inside. These tubes will be found only in older sets, chiefly auto radios. A sidewise pull is needed to release a loktal tube from its socket.

Miniature tubes have no separate bases. The pins come right out of the glass envelope, and the spacing of the pins controls insertion in a socket.

Subminiature tubes, like miniatures except for being much thinner, are rarely used in television and radio sets. One type found in some television sets is the type 5642 high-voltage rectifier. This has two flexible filament leads coming out of one end, and a flexible anode lead coming out of the other end. Such tubes are removed by unsoldering their leads.

Diodes. The simplest tube of all is one having only two working elements, a filament and a plate, mounted in a vacuum as shown in Fig. 3A. A tube having just two elements is called a diode. When the filament is cold, no current can be sent through the tube by the B battery because the gap between the filament and the plate is a break in the circuit. The plate is also known as the anode.

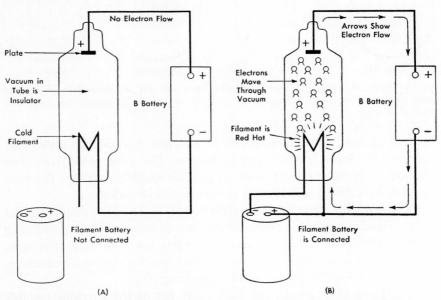

Fig. 3. Operation of a diode tube. Without a filament battery, no electrons can flow

Removing and Replacing Receiving Tubes 147

When the filament is heated by current obtained from a filament battery connected as in Fig. 3B, the filament gets red-hot. The resulting heat boils electrons right out of the wire, just as heat boils particles of steam out of water. These electrons are negative charges of electricity.

Once electrons are free of the filament, they are attracted by the positive charge that is applied to the plate by the B battery. There is thus a movement of electrons from the filament to the plate inside the tube. This stream of electrons moving through empty space is an electric current, just as are electrons moving through a wire.

If B-battery connections are reversed, the plate has a negative charge. Now the plate repels the electrons that are coming out of the hot filament, and there is no flow of electrons through the tube.

If an a-c voltage source is used in place of the B battery, the plate of the tube is alternately negative and positive. Electrons flow through the tube on the half-cycles when the plate is positive, but not on the alternate half-cycles when the plate is negative. Electron flow is thus always in the same direction, from filament to plate through the tube, even though the plate voltage is a-c. This action of a tube, called rectifying action, is widely used for changing a-c to d-c.

Indirectly Heated Cathodes. In any tube the electron-emitting element is called the *cathode*. The commonest type of cathode used in tubes is the indirectly heated cathode. The source of heat for this is an insulated filament of tungsten wire inside a metal sleeve. This filament is called the *heater* because its only job is to provide heat. The metal sleeve is covered with a chemical oxide coating that serves as the cathode. When heated by the heater element inside, this oxide coating gives off electrons much more efficiently than does a bare wire. The cathode has its own lead in these heater-type tubes. Symbols for filaments and cathodes are shown in Fig. 4, along with typical constructions.

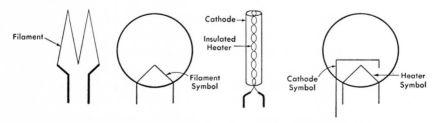

Fig. 4. Construction details and symbols for the two commonest electron sources used in tubes. Filaments and heaters use the same symbol, but a heater will always have a cathode symbol too

Triodes. Diodes are useful in rectifying alternating current, but do not boost the strength of weak signals. In 1906 Lee De Forest achieved amplifying action by placing another element between the cathode and the plate. Since this made three elements in all, the new tube was called a triode. Since the new element was a winding or grid of fine wire, it was called the grid. The grid closest to the cathode in a tube is called the *control grid*. A triode has only a control grid but other tubes have additional grids.

How a Grid Controls Current. The number of electrons that travel from the cathode to the plate in a tube can be controlled by changing the electrical charge or voltage on the grid. When the grid has no charge, electrons go through just as if the grid were not there.

When the grid is made more negative than the cathode, the grid repels electrons back toward the filament. The more negative the grid is, the fewer electrons get through the grid wires. A negative grid thus gives less plate current than is obtained without the grid. The grid can easily be made negative enough to block all the electrons, so that the plate current is zero.

In television and radio sets the grid of a triode is generally operated with a small negative d-c voltage called a bias voltage. This allows a certain amount of plate current to flow. When an a-c signal voltage is applied to the grid of a triode, the signal alternately adds and subtracts from the negative grid-bias voltage. As a result, the plate current of the tube alternately increases and decreases just like the a-c signal voltage on the grid. The signal current in the plate circuit can be several thousand times as strong as the signal current in the grid circuit. This means that the tube has amplified the signal several thousand times.

Triode Circuit. A typical triode amplifier circuit is shown in Fig. 5. A dotted-line symbol represents the grid in the tube circle. One battery, called the A battery, provides the current for the heater. Another, called the B battery, provides the plate voltage that places the positive charge on the plate. Note that the B battery connects between cathode and plate, not just to the plate. This is essential to make the plate positive with respect to the cathode.

The third battery, placed in the grid circuit to make the grid slightly negative with respect to the cathode, is called the C battery. With these three voltages, a triode is capable of boosting the strength of a weak input signal, so that a much stronger signal is sent to the load in the plate or output circuit.

Tetrodes. Adding a second grid to a tube gives four active elements in all. Such a tube is called a tetrode. The extra grid is a wire spiral or a wire screen placed between the control grid and the plate. The extra grid is operated at a positive voltage almost as high as the plate. It is called the *screen grid* because it serves as a screen that prevents the amplified signal on the plate from acting backward on the control grid. Such action in a backward direction, called feedback, would cause howling or other trouble in a set.

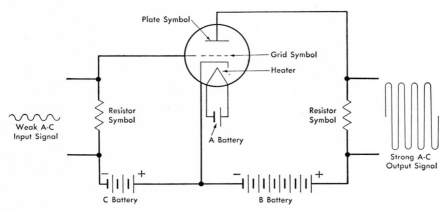

Fig. 5. The three voltages needed to make a triode tube amplify a weak signal can be obtained from separate batteries as shown here. In modern home sets the d-c operating voltages for tubes are obtained from an a-c power supply in a receiver

Pentodes. When still another grid is added, to give five active elements in all, the tube is called a pentode. This third grid is located between the screen grid and the plate and is connected to the cathode or ground. It serves to suppress secondary electrons that are splattered off the plate when the main stream of electrons hits the plate; hence this grid is called the *suppressor grid*.

Other Types of Tubes. Some tube types have still more grids for special applications. One example, widely used as the mixer–first detector in superheterodyne receivers, is the pentagrid converter which has five grids. Such a tube actually does the job of two tubes.

Two or more tubes are often combined in one envelope. Thus, the type 12AU7 tube has two triode sections, while the 6AL5 has two diode sections. Another common combination is the duodiode triode, which has two diodes and a triode all in the same glass or metal envelope. Symbols for some of these tubes are shown in Fig. 6.

An electron-ray or magic-eye tube is a special type of triode used as a tuning indicator. In one type, the plate is a cone-shaped electrode coated with fluorescent material that glows when electrons strike its surface. The amount of surface that glows depends on the grid voltage applied to the tube. When a station is tuned in properly, the grid voltage applied to the tuning-eye tube is most positive, and the fluorescent glow covers the maximum area for that station. On very strong signals the glow covers the entire area of the cone.

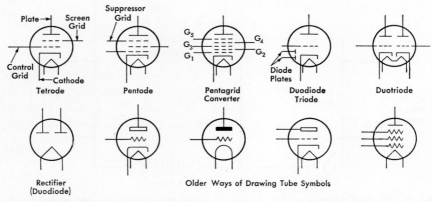

Fig. 6. Symbols for some of the tubes that are widely used in television and radio sets

Receiving-tube Type Numbers. An identifying type number is etched, stamped, or otherwise marked on the envelope or base of a tube. These type numbers are assigned to tubes by EIA (Electronic Industries Association).

In general, all octal-base tubes follow the EIA numbering system. Most of the newer tubes do also, but the type numbers of a few special types of new tubes and most old tubes have no meaning.

In the EIA tube-numbering system, the first number group usually indicates the filament voltage in 1-volt steps. The number 0 indicates a tube having no filament; the number 1 indicates a filament voltage between 0.1 and 2.0; the number 2 indicates a filament voltage between 2.1 and 2.9; the number 3 indicates a filament voltage between 3.0 and 3.9, and so on, to 117 which indicates a filament voltage between 117.0 and 117.9. Examples are 0Z4, 1A3, 5U8, 6AU8, and 35W4.

One exception to the numbering rule is 7, which is used to designate a 6.3-volt tube having a loktal base. No tubes have a filament voltage in the range of 7.0 to 7.9, so this causes no confusion. Another exception is 14, which designates a 12-volt tube having a loktal base.

The letters following the filament-describing number serve only to distinguish between tubes having the same set of numerals. Letters for rectifier tubes are issued backward starting from Z. Tubes like the 6Z5, 5Y3, and 5U4 are rectifiers. Letters for other tubes are issued starting from A and going through the whole alphabet, then continuing with BA, BB, BC, and so on. Thus, a tube with a letter near the end of the alphabet may or may not be a rectifier, but a tube with letters earlier in the alphabet will never be a rectifier.

The second number group in the EIA tube-numbering system indicates the number of useful elements connected by wire leads to the tube-base pins and the top cap. Examples of these elements are the filament, the cathode, each grid, each plate, and the shield or metal envelope. The filament is counted as one element even though it may come out to two or three pins. For example, a 12AT6 has six elements brought out by wire leads to the tube-base pins: filament, cathode, control grid, screen grid, plate, and shield.

A second letter group at the end of the tube type number is sometimes used to indicate a special tube characteristic. The meanings of these letters are given in Table 1.

Table 1. Meanings of Letters Used after Tube Type Numbers

G	glass envelope with a molded plastic base and locating lug.
GT	glass envelope smaller in size than type G envelope.
GT/G	tube which can be used interchangeably with either a G or a GT type.
A, B, C, D, E, or F	improvement over original type without the letter; for example, the 2X2-A is the same as a 2X2 but will withstand severe shock and vibration.
/	symbol meaning that type numbers before and after it are interchangeable.
S	external shield; thus, a 6F7S is different from a 6F7.

Tube Sockets. Sockets serve to connect the pins on a tube base to the receiver circuits. Sockets also provide a mounting for the tube and a guide for lining up the proper tube pins with the socket holes.

Each socket has spring contacts that make good electric connections to the tube pins. Each contact has a soldering lug underneath, usually punched with one or two holes and notches for wires.

Loktal tube sockets have a snap ring in the center hole of the socket.

152 *Television and Radio Repairing*

This ring holds the tube in position and serves as a ground connection between the lower metal shield on the tube and the receiver chassis.

Tube sockets are made from either molded plastic or punched-fiber insulating material, as in Fig. 7. The molded socket is generally used in applications where tubes are plugged in and out a large number of times, such as in tube testers. One type of molded socket is fastened with a split-type clamp ring. This snaps into a recess in the rim of the socket under the panel or chassis.

The punched-fiber wafer socket is most commonly used in television and radio receivers because of its low cost and small size. One style consists of

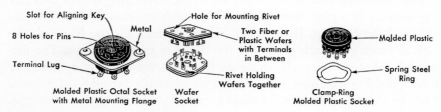

Fig. 7. Common types of octal-tube sockets

two $\frac{1}{16}$-inch-thick pieces of punched fiber with the spring contacts in between. Rivets hold the two fiber wafers together. Another style consists of one piece of $\frac{1}{16}$-inch-thick fiber with each spring contact individually riveted to the fiber. Wafer sockets are generally riveted to the receiver chassis.

The sockets used in printed circuits are covered in a later chapter, because special soldering techniques are required to remove them.

Pin Numbers. Tube pin numbers are assigned in a standard manner for each type of tube. These numbers are shown on almost all circuit diagrams and tube-base diagrams. When the numbers on a diagram are arranged in order around the circle of the tube base or socket, they are usually shown as viewed from the *bottom* of the base or socket. This helps trace the wiring on the underside of the chassis. The numbers then increase in a *clockwise* direction starting from the identifying guide, as shown in Fig. 8.

A guide is always provided in the tube socket to ensure placing the tube in the socket correctly and to furnish a reference point for numbering the socket contacts and tube pins. This guide may be the space between two large pins, a larger spacing between two or more unevenly spaced pins, or an aligning key, as shown in Fig. 8.

Removing and Replacing Receiving Tubes **153**

When looking at the bottom of a tube or the bottom of the socket, the first pin in a clockwise direction from the guide is pin No. 1. With octal tubes and sockets, omitted pins and terminals are always counted just as if they were there.

Tube-base Diagrams. Tube symbols are usually drawn as in the lower right corner of Fig. 8, with each lead numbered. Once you have learned how to find any pin number on any tube, you will find this type of symbol entirely adequate.

When some pins are omitted from octal tubes, the corresponding socket terminals are often used as convenient anchor points for the leads of resistors and capacitors.

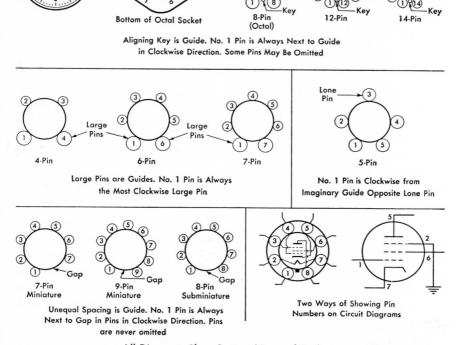

Fig. 8. Chart showing how to find the No. 1 terminal of any tube socket when looking at the bottom of the socket, and methods of showing pin numbers on circuit diagrams

Socket Troubles. When tube-socket contacts are corroded or dirty, they may be the cause of an intermittent trouble, noise, or an open circuit. To clean the contacts, remove the tubes and apply contact cleaner to each socket contact with a small brush. Now run a small piece of wire into each socket hole. Use a wire that is as thick as the tube pins. Roughen the wire a little by nicking it slightly all around with cutting pliers. The sharp edge of a straightened paper clip will often do nicely in miniature sockets. A steel bolt of the correct diameter, an inch or more long and having fairly sharp threads, will work even better as a cleaning tool.

Do not use too large a cleaning tool, as this will spread the socket contacts and do more harm than good. Twist the cleaning tool between your fingers as you move it up and down in the contact.

Insufficient contact pressure in a wafer socket can sometimes be corrected by squeezing the contacts together slightly under the socket with a pair of long-nose pliers. Do not exert too much pressure or bend the contacts too far, as you can easily break off the stiff contact material. An ice pick or other sharp-pointed tool can sometimes be used to tighten a loose socket contact, by working with it from the top of the chassis.

Bad Socket Terminals. A poor connection between a socket terminal and the set wiring is almost always due to a poorly soldered joint. If you suspect a particular socket, push each connection separately with a stick of wood or other insulating material while the set is turned on. Noise or intermittent operation when you press against a socket terminal is an almost certain indication of a bad joint there, even though the soldered connection looks good to the eye.

Before resoldering suspected socket-terminal connections, turn off the set and remove the tube from the socket. Unless the tube is removed, heat from the soldering gun may loosen the wire in a tube pin or even solder the pin to the socket.

Sockets for miniature tubes require still more care because here there is a great tendency for the solder to flow into the tiny contacts on the socket. If you have a dummy tube base with aluminum or steel pins, plug it into the socket while soldering. Dummy bases are made especially for this purpose and are sold by parts jobbers. The solder will not adhere as readily to the aluminum or steel pins as it does to the tinned or nickel-plated pins of the regular tube.

When resoldering a joint, use only a small amount of solder. Generally the solder already on the terminal is sufficient. All that you need do is apply heat and a little fresh solder to get its rosin flux.

Removing and Replacing Receiving Tubes 155

If too much solder gets on the contact, hold the set so excess solder can melt and run down the soldering gun, away from the contact. It can then be shaken off the gun easily.

Do not hold the gun on the contact any longer than it takes to melt the solder. Do not allow the gun to touch or get too close to the socket insulation. Heat can char or distort the socket insulation enough to move the contacts out of line or provide a high-resistance path between contacts.

Socket-insulation Troubles. A short-circuit between socket terminals, breakage of contacts, or breakage of socket insulation can often require complete replacement of the socket. This is a rather difficult and time-consuming job, so do it only when you cannot repair the socket. If the socket has arced over between contacts or if you accidentally char the insulation while soldering, first try to scrape off the charred surface with a sharp knife, or remove it by drilling.

Replacing a Defective Socket. Replacing a socket is difficult because of the large number of connections and because of the possibility of damaging other parts of the set. If only one contact is damaged, it is usually possible to replace it with a good contact taken from a similar old or new socket.

Most sockets are riveted to the chassis, so you will have to drill out the mounting rivets after first disconnecting all socket leads. Remove one wire at a time. Carefully tag each wire with the socket-terminal number written on a bit of cardboard that is pushed over the bare end of the wire. It is important that you keep track of each lead. If they once get mixed, you may have to put in several more hours of work tracing circuits.

An exact duplicate of the socket in the set is easiest to install. If it is not available, you may have to drill new mounting holes or enlarge the socket hole to accommodate the new socket. Mount the new socket with nuts and bolts rather than rivets. If the original socket is mounted with a snap ring or clamp, mount the new socket by forcing the lock ring over the socket.

Socket insulation may be broken by improperly plugging in a tube. In four-pin sockets this may enlarge some of the socket-pin holes. Such a socket need not be replaced if you mark the socket or chassis in some clear way to indicate the true large holes.

First Steps in Removing Tubes. If removing more than one tube at a time from a set for test, be sure that tube sockets are labeled or that there is a socket layout diagram fastened inside the cabinet, so you can get the tubes back correctly.

Layout diagrams are always *top views* of the chassis, as in Fig. 9, since they are intended as guides for removing and replacing tubes. If no layout is given, make one on a calling card before removing the tubes. Fasten it to the chassis or cabinet with speaker cement.

In sets having two or more tubes of the same type, it is best to mark each tube and socket with a different number or letter so you can get each tube back in its original socket. This will reduce the possibility of

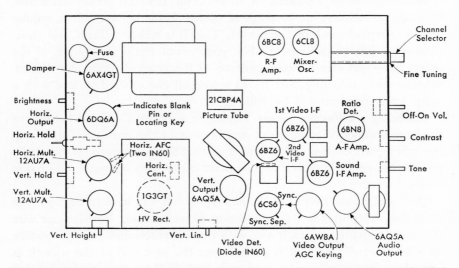

Fig. 9. Socket layout diagram for typical vhf television receiver. All controls are identified

circuit unbalance due to small variations in the characteristics of tubes having the same type number. A china-marking pencil works well for marking the glass tube envelope as well as the metal chassis. In television sets it is particularly important to get each tube back in its own socket.

Always turn off the set or unplug the line cord before replacing tubes. In an a-c set, removal of a tube reduces the load on the power transformer and other parts, making voltages go up on tubes remaining in the set. This rise in voltage can sometimes damage tubes still in the set or even affect other parts.

In an a-c/d-c set having tube filaments in series, a cold tube inserted in a string of hot tubes upsets voltages and may likewise cause premature failure of some tube or part.

Removing Tube Shields. The commonest type of tube shield is removed by twisting slightly as you pull up on it. Another type, constructed like bayonet-base auto lamps, must be pushed down and turned before it is pulled up.

Tubes requiring shields generally have metal clips or a raised metal section on the socket or chassis to make secure contact with the shield for grounding purposes.

Miniature tubes often have shields that are about three-fourths the length of the tube and are permanently soldered to the chassis. The tube projects just enough out of the shield to permit gripping it with fingers for removal or installation.

Tube shields are important. Be sure to replace each shield on the correct tube. Push the shield down firmly so it makes good contact with the chassis. If contact clips or grounding springs are loose, tighten them first to ensure good contact.

Removing Top-cap Connectors. The top cap used on some tubes is connected to the set with a flexible insulated wire having at its end a cup-shaped metal connector. This top-cap connector is removed by twisting it slightly while pulling up. The connector will normally come off easily, and you can then unplug the tube.

Do not twist the connector too much or pull on it with too much force, as you may break the cap off the tube itself. If the connector does not come off easily, try pushing down on it first. There will often be some play between the tube top cap and the connector, and downward pressure may break the bond that was causing the two parts to stick. If this does not work, pry between the cap and the connector carefully with a very small screwdriver.

To replace a top-cap connector, simply place it in position over the tube top cap and push it straight down into place.

Tube Clamps. Various types of holding devices are used to hold tubes securely in their sockets. These are found most often in portable sets, where vibration and shock otherwise might cause the tubes to creep out of their sockets.

Larger tubes, such as rectifier tubes, are often held by two semicircular pieces of spring steel that grip opposite sides of the tube base. These springs must be pushed down simultaneously with fingers or two screwdrivers, to release and pull out the tube.

Miniature tubes are sometimes held in their sockets by a spring-type clamp. The spring is easily pulled to the side when the tube is inserted into the socket. The spring is then placed over the tube to exert a constant pressure on the glass top and hold the tube in the socket. If the tube has a removable shield, the shield and tube can often be removed together and the tube then pushed out of the shield.

How to Pull Out a Tube. After removing the top-cap lead and shield, if present, grasp the tube firmly as close to the base as possible, and rock it very slightly from side to side as you pull up. Be sure the set is turned off. If you let the tubes cool after turning the set off, you will not burn your fingers on a heated tube envelope.

The rocking action is particularly essential on loktal tubes, to release the spring catch in the socket. Pulling at a slight angle instead of straight up will often help remove loktal tubes.

Tube-pulling Tools. If the tube does not come out with the above methods, try wedging a screwdriver between the chassis and the edge of the tube base. Before doing this, be sure the line cord is unplugged, so you do not receive a shock.

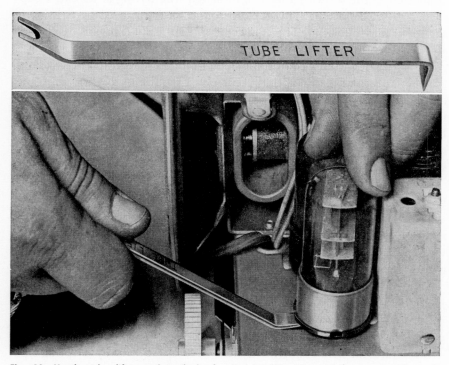

Fig. 10. Handy tube lifter and method of using it. Disconnect set from power line, tip tube slightly, insert tapered end of lifter under base so it goes around one pin, then press lifter handle down while guiding tube vertically with other hand. Use right-angled end when tubes are crowded together, as in television sets and in compact auto radios

Removing and Replacing Receiving Tubes 159

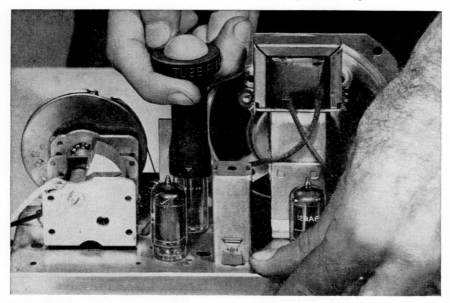

Fig. 11. Handy tube puller for miniature tubes, made from special rubber that resists tube heat and provides both suction and friction for gripping glass envelope

The screwdriver-type tube-pulling tool in Fig. 10 is particularly handy, as it was designed especially for getting out tight octal, loktal, and other large tubes. For miniature tubes the special rubber lifter in Fig. 11 is also available.

Replacing Tubes. When replacing a tube, plug it back into its proper socket by noting the position of the socket guide key and holding the tube to line up with the guide. Place the tube over the socket and rotate it slightly until the pins fall into the socket holes, then press the tube down into the socket. With octal and loktal tubes, just place the tube on the socket and rotate it until the aligning key drops into position, then push the tube firmly into place.

For miniature tubes in crowded locations, use a flashlight and a dental mirror or a small pocket mirror to help locate the tube-socket holes. Sometimes tube-location diagrams in service manuals show the key position for each tube.

One tube manufacturer has devised the magnetic pin locator shown in Fig. 12. After the old miniature tube is pulled out of its socket, a metal pin at the end of a wooden rod is set in the center hole of the socket. The locator is now placed over the wooden rod. Magnets molded into the

Fig. 12. Method of using GE pin locator for miniature tubes. (General Electric photos)

plastic base of the locator hold it in position on the steel chassis. The wooden rod is now removed, the new tube is inserted in the locator, and the tube is rotated until its pins fall into place.

Tube Failures. The vacuum tube is the most complicated part in a television or radio set. Many of the causes of faulty set operation can be traced to tube failure in one form or another. A tube can fail in many different ways and in many different parts of the set, making a receiver develop hum, noise, distortion, intermittent operation, and a host of other troubles.

Tube-filament Failure. A burned-out filament is one common defect in tubes. The filament, or heater, is a fine tungsten or alloy wire which is heated by current supplied by either a transformer or a battery. This heat is sufficient to make the filament glow bright red in some tubes. In other tubes, the glow may be only barely noticeable.

The filament, or heater, supplies the heat to boil off the electrons which enable the tube to operate. When the filament breaks, the tube stops working, and so does the set. In rare cases, some of the signal may pass around the burned-out tube and give weak or distorted operation.

The filament of a tube may open for any of the reasons that make an ordinary electric lamp burn out. Expansion of the filament wire with heat each time the set is turned on produces mechanical strains that can eventually break the filament. Sometimes this may occur as an intermittent

Removing and Replacing Receiving Tubes

break, where the filament is good when cold but opens as it heats up. When the filament cools off, it makes contact again, with the process repeating itself at regular intervals of a few seconds or minutes.

Overvoltage on the filament, caused by line-voltage surges or abnormally high line voltage, is a fairly common cause of tube trouble. If the power-line voltage in a locality is appreciably higher than 117 volts for long periods of time, explain to the customer that he must expect tube failures more often than normal unless the voltage is reduced.

Your jobber may have a tube saver or regulator for use when line voltage is high. The device plugs into a wall outlet, and contains an outlet into which the line cord of the set is plugged. Such a device reduces the line voltage a few volts and suppresses line-voltage surges.

Burned-out pilot lamps may overload the tapped section of the filament on rectifiers like the 35W4, making the tube burn out. For this reason, urge your customers to have pilot lamps replaced with correct replacement types as soon as possible after they burn out in universal a-c/d-c radio sets.

Battery-connecting Mistakes. Wrong connections to batteries or wrong polarity can cause almost instant burnout of tubes or transistors in battery sets. Make a connection diagram before removing batteries, unless the set already has unmistakable markings. Check twice before putting in a new battery, because you have to pay for your own mistakes.

Low Emission in Tubes. This is another common cause of tube trouble. It is due mostly to the normal aging of the electron-emitting surface of the filament or cathode. Nothing can be done to prevent this. The remedy is replacement of such weak tubes whenever a new tube gives noticeably improved performance or when a tube tester indicates that emission is low.

Loose Elements in Tubes. When an element in a tube gets loose for any reason, the set may give off a loud rasping noise, a rumble, or a growl when the tube is tapped. The tube is said to be noisy and should be replaced.

If the noise builds up gradually in volume, it is usually because of loudspeaker sound waves acting on the tube, and the tube is said to be microphonic. When this occurs, the howl or hum can generally be stopped by holding your hand on the tube. The tube should be replaced, as it is defective.

Open Elements in Tubes. Opening of the connection to a tube element can be due to a poorly soldered joint in the pin of a tube having a plastic base, to breakage of a welded connection within the tube, or to breakage of the cap on the top of the tube. These defects may block the signal completely or cause distorted reception. Tubes with poor connections should be replaced.

Loosening of the cement between the glass envelope and the tube base

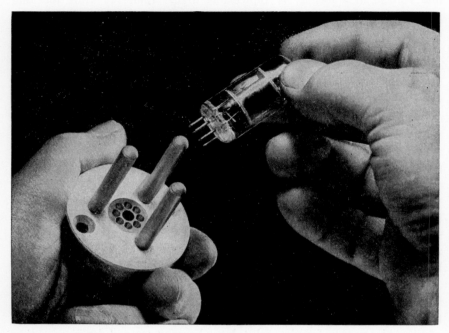

Fig. 13. Pin-straightening tool for 7-pin miniature tubes. Inserting tube in tapered holes does the straightening job automatically. Tube can be inserted in any position, since straightener has one extra hole. These pins often break off when straightened with pliers or with a screwdriver. (Sylvania photo)

can cause an open element by breaking the solder connection to a base pin, or can cause a shorted element by twisting the pin leads together inside the base. Such tubes should be replaced even if they test good, as they will eventually cause trouble.

Opens in tubes may also be due to rosin or dirt on tube pins. Scraping off the rosin fixes this trouble. Sandpaper is also good for cleaning tube pins.

Bent tube pins that do not touch socket contacts also cause opens. Miniature tubes give the most trouble, since their pins are simply the wire leads that come through the glass. Bent pins of miniature tubes can

Removing and Replacing Receiving Tubes

easily be straightened with a tool made for this purpose, shown in Fig. 13. Two tube-pin straighteners will be needed, one for seven-pin tubes and one for nine-pin tubes.

Tubes designed for high filament voltages, ranging from 25 to 117 volts, often have intermittent filament defects because their filaments are wound with finer wire. The tubes may open up because of excess heat during parts of the day when the a-c power-line voltage is high, yet operate satisfactorily at other times when the line voltage is normal or low. Such tubes are defective and should be replaced.

Open cathode leads inside rectifier tubes may be caused by a shorted filter capacitor in the power supply of the set. A filter capacitor may short because of age, long storage, or excessively high voltage. You will learn later how to check filter capacitors before replacing a rectifier tube.

Shorted Elements in Tubes. Short-circuits between the elements in a tube are caused by mechanical failure of a mounting support or insulator inside the tube. These failures in turn may be caused by excess vibration, excess heat, or continuous operation at higher than rated voltages.

Short-circuits between elements are often intermittent and show up as noise when the tube is tapped. Tubes with shorted elements should be replaced even if the set is working all right at the moment, because they can cause future trouble.

Insulation Leakage in Tubes. Electrical leakage between elements is caused by cracks or breaks in the insulation separating the various elements. The most common place for leakage to occur is between the filament and cathode of indirectly heated tubes. This will be shown up by a good tube tester.

In some circuits filament-cathode leakage does not affect receiver performance, so do not replace a tube unless the new tube actually works better in that socket. In other circuits, filament-cathode leakage may be too small to show up on a tube tester yet cause trouble. Here, replace the tube if a new tube works better.

Arcing may occur between the pins on the tube base itself. This may be due to excessive voltage between elements or more often to dust, moisture, or oil on the base. A mark or charred line in the base insulation between two of the base pins is a clue. If you run into trouble of this type, clean off the tube base with a cloth dipped in carbon tetrachloride or other cleaning fluid.

Gassy Tubes. Gas may develop in a tube because of overheating of one of the elements or actual leakage through tiny cracks in the tube envelope. Gas acts to produce excessive electron flow to the plate. It also reduces the effect of the control grid on the plate current. Two of the effects of gas are noise and distorted reception.

Gas gives the most trouble in audio output tubes. Try a new output tube whenever a radio or television set sounds distorted.

The presence of gas may often be noted by the appearance of blue haze near the plate and filament inside a tube. To find out whether the glow is gas haze or the normal fluorescent glow of some power tubes, hold a small permanent magnet near the tube. The harmless fluorescent glow will move as the magnet is moved, while the gas haze will not.

Ordering New Tubes. When replacing defective tubes, it is not necessary to get the same make. Tubes from any of the well-known and reliable manufacturers can generally be used interchangeably. Small variations do occur between tubes of the same type, but these are just as likely to occur in a batch all of one make.

Television receivers and radio sets with short-wave bands are most affected by variations in tube construction. If one new tube does not work right, try another one. Save the first tube, however; it will usually work perfectly in some other set.

If the old tubes are metal, try to replace with metal tubes. If the old ones are glass, try to replace with glass-envelope tubes. This is the safest procedure for avoiding trouble, though nine times out of ten you can interchange glass and metal tubes without affecting performance.

Noise may seem to develop in some sets when you have replaced the tubes with new ones. This is not necessarily an indication of defective tubes. It may be that the new tubes show up other troubles, such as a poor contact on the rotor of the tuning capacitor.

Interchanging Tube Types. Substituting other tube types for those in a set is an emergency practice forced on servicemen when correct tubes are not available. Tube manufacturers and radio jobbers have booklets giving lists of substitutions that are worth trying in such emergencies. Other booklets or charts give American equivalents for some types of foreign tubes.

Tubes with the same general type number but with a different letter group at the end are interchangeable for most applications. For example, 6C5, 6C5G, 6C5GT, 6C5GT/G, and 6C5MG are usually interchangeable with each other. Even tubes having entirely different type numbers can be interchanged under certain conditions, as explained in the tube-substitution booklets.

When a metal tube is replaced by its glass-tube equivalent, you may sometimes have to use a shield can over the glass tube. The shield stops squealing and howling in the loudspeaker due to feedback. Use the type of shield that grounds itself securely to the chassis through spring clips that you first bolt to the chassis.

Obsolete Tubes. Old sets using special tubes that are no longer available find their way into a service shop once in a while. In general it does not pay to repair these antique sets, and the customer should be so advised, tactfully. A modern set will generally cost less than the repair charges for the antique and will usually work much better than the old set ever did.

Foreign Tubes. Information on many foreign tubes can be found in charts available from some tube manufacturers. You may find it best to turn down repair jobs on foreign sets, however. Foreign tubes and parts are hard to get, as also are service manuals for foreign sets. The required parts will usually have to be ordered by mail from the American distributor for the foreign set.

Emergency Repairs on Tubes. A loose base or loose top cap on a tube can be repaired in a few minutes, allowing you to get a tube back in service quickly or do a repair job at minimum cost. A new tube is always better than a repair, though. Use the following makeshift repairs only when absolutely necessary.

To re-cement a loose top cap, first scrape the envelope under the cap as carefully as possible with a penknife. Take care not to exert pressure on the top-cap lead wire or you may break the glass seal where the wire lead comes through the tube. Apply a coating of Duco household cement, a good glass or china cement, or speaker cement to the tube surface under the top cap, and push the cap tightly against the tube. The cement should ooze out the side of the cap when it is pushed into place. Set the tube aside to dry for an hour or so. Always handle the top-cap connection carefully after it has been repaired.

To replace and resolder a top cap that has come off, heat the loose top cap by holding a soldering gun against it, then shake excess solder off the cap. Scrape the cap and the tube envelope, apply cement, bring the top-cap lead wire through the hole in the top cap, and push the can into position on the glass envelope. Now solder the top-cap lead wire to the cap. Do not leave the gun on too long, as the heat may crack the glass seal around the lead.

A broken top-cap connection can usually be repaired. Carefully scrape the short projecting lead. Wind a piece of bare tinned copper wire (about No. 22) around the lead a few times and squeeze the turns gently with pliers. Solder this joint quickly so as not to heat and crack the glass. Remove the cement from inside the top cap to make room for this joint. Complete the repair now just as for a top cap that has come off.

A loose tube base may be re-cemented with speaker cement. Use a stiff strip of paper to poke the cement in around the base, and place a few rubber bands around the tube and base lengthwise to apply pressure until the cement has hardened.

Repairing Tube Pins. An unsoldered tube pin may be resoldered. Scrape the end of the tube pin and as much of the copper wire lead inside as possible with a penknife. Heat the pin for about a minute with your soldering gun, then apply rosin-core solder to the open end of the pin. File or scrape off excess solder to make a neat, smooth pin.

Corroded or dirty tube pins may be cleaned by scraping, filing, or rubbing with fine sandpaper.

Reading Faint Tube Markings. One way to make indistinct tube markings visible is by holding them up to the light in different directions, to catch the light reflections. Putting the tube in an electric refrigerator for about 15 minutes to chill it is another way of making faint markings on glass visible.

Pilot Lamps. Pilot lamps or dial lights are special types of flashlight bulbs used in almost all radios for illuminating the tuning dial and indicating when the set is turned on. Pilot lamps are also used as on-off indicators on television sets, home audio systems, and other equipment.

Most modern sets use the T-3¼ miniature bayonet-base lamp shown in Fig. 14, as this stays tight under vibration. Older sets used miniature screw-base lamps.

The most popular pilot lamps are the types 44, 47, and 1847. Voltage and current ratings for these lamps are given in Table 2, along with ratings for older and less-used types. The 12-volt types are used in auto radios. Types 1847 and 1866 are designed to have an average life of over 5,000 hours, for use in hard-to-reach locations. If these long-life lamps are not readily available, the type 47 may be used in place of the 1847, and the type 44 may be used in place of the 1866.

When replacing a pilot lamp, use a new lamp with the same type number as the old one. The number is usually stamped on the base. Replacement lamps will have the correct volt and ampere ratings if the same number is used.

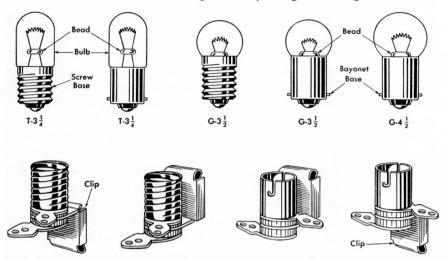

Fig. 14. Types of pilot lamps used in television and radio sets, shown actual size, and typical screw and bayonet sockets for these lamps. Numbers under lamps designate size and shape of glass envelope.

Table 2. Pilot Lamps Used in Television and Radio Sets

Type No.	Rated Volts	Rated Amperes	Bulb Style	Base
40	6.3	0.15	T-3¼	Screw
41	2.5	0.50	T-3¼	Screw
42	3.2	0.35	T-3¼	Screw
43	2.5	0.50	T-3¼	Bayonet
44	6.3	0.25	T-3¼	Bayonet
45	3.2	0.35	T-3¼	Bayonet
46	6.3	0.25	T-3¼	Screw
47	6.3	0.15	T-3¼	Bayonet
48	2.0	0.06	T-3¼	Screw
49	2.0	0.06	T-3¼	Bayonet
50	7.5	0.20	G-3½	Screw
51	6.0	0.20	G-3½	Bayonet
55	6.0	0.40	G-4½	Bayonet
57	12.0	0.24	G-4½	Bayonet
57X	12.0	0.24	G-4½	Bayonet
1490	3.2	0.16	T-3¼	Bayonet
1847	6.3	0.15	T-3¼	Bayonet
1866	6.3	0.25	T-3¼	Bayonet
1891	12.0	0.24	T-3¼	Bayonet
1892	12.0	0.12	T-3¼	Bayonet

Removing and Replacing Pilot Lamps. To remove a burned-out pilot lamp, first turn off the set and unplug the line cord. In most sets the pilot lamp can be removed by reaching in from the back of the chassis. If a clip-type socket mounting is used, slide the clip from the chassis and then remove the lamp from the socket. If you cannot get at the pilot lamp easily, remove the chassis from the cabinet first, then remove the pilot lamp.

To remove a screw-base pilot lamp, turn the bulb in a counterclockwise direction the same as you would an ordinary electric light bulb. To remove a bayonet-base lamp, press the bulb down into the socket, turn it a small amount counterclockwise, and release, the same as you would remove an automobile lamp.

If the lamp seems loose when you touch it, tighten it and try the set again before removing the lamp for replacement. If the lamp fits the socket loosely, remove the lamp and squeeze the sides of the socket slightly with the fingers or pliers to provide a little pressure on the side of the lamp base.

If a screw-base pilot lamp is cemented to the socket to prevent its working loose in shipment or in normal operation, try scraping the cement loose with a penknife. If this does not work, apply a few drops of acetone or nail-polish remover to soften the cement before unscrewing the lamp. After the lamp is out, scrape off excess cement on the inside of the socket. Speaker cement solvent is usually acetone.

When replacing the clip-on type of pilot-lamp socket, be sure you get the socket located in the right way so that the shade or reflector, if one is used, works properly. Also, make certain the clip-on bracket is seated firmly in place, as it usually provides the return path for lamp current.

Pilot-lamp Performance. The pilot lamp in an a-c/d-c set will glow brightly when the set is first turned on, then dim to a proper glow as the set warms up. This is normal and is due to the greater current which flows through the lamp when the tube filaments are cold.

A pilot lamp may flicker with loud signals or when tuning an a-c/d-c set. This is also normal and is due to the change in plate current caused by the loud signal. Part of this increased plate current flows through the lamp filament and causes it to change in brilliance.

Check both sections of the rectifier filament with an ohmmeter or tube tester if a new pilot lamp burns out quickly in an a-c/d-c set. The section of the rectifier filament that parallels the lamp may be burned out. In this case the full tube-filament current flows through the pilot lamp and causes it to burn out again.

If two or more pilot lamps are connected in series, all of them will go out if one burns out. To locate the burned-out lamp, remove the lamps one at a time and replace with good lamps. When you replace the burned-out lamp, all will light up.

If you have a low-range ohmmeter handy, it is sometimes faster to remove all the lamps and check them with an ohmmeter. A good lamp will read below 100 ohms, while a burned-out lamp will read infinity. Tube testers usually have provisions for testing pilot lamps, too.

Readjust Sets after Replacing Television Tubes. When a new tube is installed in a television receiver, it is often necessary to readjust one or more of the controls at the back of the set.

When the low-voltage rectifier tube is replaced, the vertical size, vertical linearity, and vertical centering controls may need readjusting to give an undistorted raster that overscans the screen by about 5 per cent. This keeps most of the picture in sight, while leaving a small margin of reserve sweep to compensate for aging of the new rectifier tube. Check the focus adjustment if the picture is slightly fuzzy. Even the ion-trap magnet may require a slight readjustment to give peak brightness and best focus.

When the vertical oscillator or vertical output tube is replaced, readjust the vertical size and linearity controls to fill the screen.

After replacing the horizontal oscillator tube, adjust the setting of the horizontal frequency control for proper lock-in at the center of the horizontal hold control adjustment range.

When you install a new horizontal output tube, adjust the width control to overscan the screen slightly, to allow for aging of the tube in its first hundred hours of operation. All other horizontal controls may need touching up also. New sync-system tubes usually call for small adjustments of associated controls.

8

Removing and Replacing Picture Tubes

How Picture Tubes Work. In ordinary radio tubes the electrons that are emitted from the cathode move outward in all directions to the surrounding cylindrical metal plate or anode. In a television picture tube, however, the electrons emitted by the cathode are concentrated in a narrow beam or ray that is aimed at a fluorescent screen on the inside of the tube face,

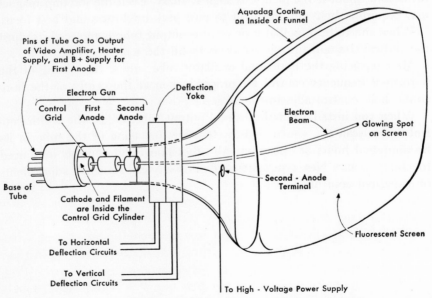

Fig. 1. Arrangement of electrodes in typical television picture tube. Second anode is connected to Aquadag coating inside funnel of tube, and coating is in turn connected to second-anode terminal recessed in wall of funnel

as indicated in Fig. 1. This screen glows wherever the electron beam hits it. A picture tube is technically known as a cathode-ray tube, and is sometimes also called a kinescope.

The filament in a picture tube provides the heat that makes the cathode emit electrons. The control grid, also called grid No. 1, determines the number of electrons that get into the beam. A high negative voltage on the control grid cuts off the electron beam entirely, as required for black portions of a television picture. A less negative control-grid voltage lets electrons go out into the beam to produce the bright spots of a picture.

The first anode, also called grid No. 2, works together with a small hole in the output end of the control grid to focus the electrons to a small point at the screen. The second anode requires a high voltage, generally somewhere between 10,000 and 20,000 volts, to give the electrons in the beam enough speed to produce the desired bright spot on the screen. The grid and anode electrodes together are called the electron gun.

Picture-tube Connections. The filament, cathode, control grid, and first anode have wire leads going to pins on the base of the picture tube. Most picture tubes in use today have a 12-pin base with an aligning key, but a few have a standard octal base. Sometimes only a half-socket is used with a 12-pin base, since the pins in use are grouped on one side of the base. The second anode connects to a conductive Aquadag coating on the inside of the tube, and this in turn connects to the second-anode terminal button on the funnel of the tube.

Deflection and Focusing. An electron beam can be deflected by a magnetic field or by an electrostatic field. Television picture tubes use electromagnetic fields to deflect the beam. These fields are produced by four coils in the deflection yoke that surrounds the neck of the tube. The deflection yoke makes the electron beam sweep across a rectangular area of the screen line by line from top to bottom, with the process repeated many times per second. At the same time, the picture signal is applied to the control grid in the electron gun to vary the beam strength, so the electron beam paints the picture on the screen.

Many older sets have an extra coil on the neck of the picture tube, used alone or in conjunction with a permanent magnet to aid the internal anode in focusing the electron beam. Newer picture tubes use only magnets for focusing, or have means for adjusting the voltage applied to a focusing anode.

Television Picture-tube Type Numbers. Cathode-ray-tube type numbers have useful information. The first numbers indicate the over-all diameter of a round screen or the diagonal dimension of the screen on a rectangular picture tube.

The first letter following these numbers distinguishes between tubes that would otherwise have the same type number.

The last letter and last numeral together specify the type of fluorescent screen. This combination is P4 for black-and-white television picture tubes, and is P22 for color picture tubes. A letter after the screen designation indicates a minor modification in tube design. Examples of picture-tube designations are 10BP4-A, 14ATP4, 17CP4, 21AWP4, and 21AXP22-A.

Aquadag Coating. The second anode of most picture tubes is a black conductive coating on the inside of the glass funnel. This coating is commonly known by the trade name Aquadag. The coating is connected to the high-voltage terminal button that goes through the funnel of the picture tube.

Most glass picture tubes also have an external Aquadag coating. This covers part or all of the funnel but does not come near the high-voltage terminal. A metal clip, band, or spring on the chassis presses against this external coating to ground it. The internal and external Aquadag coatings and the glass funnel form a filter capacitor for the high-voltage power-supply circuit. To avoid getting a shock from this capacitor when taking a picture tube out of a set, discharge it by shorting the high-voltage button to the Aquadag with a test lead.

Many troubles can occur if the external Aquadag coating is not properly grounded. In addition, an ungrounded coating builds up a voltage that can be high enough to give you a stinging shock. Always check the grounding contact when working on a picture tube.

High-voltage Connector. The high-voltage second-anode terminal on most picture tubes is a button-shaped metal cavity that is set into the glass funnel of the tube. A mating spring-contact plug on the end of the high-voltage lead, like that shown in Fig. 2, is pushed into this cavity to make connection with the second anode inside the tube.

The high-voltage lead is disconnected from the picture tube by grasping the plug with your fingers and pulling firmly straight out from the wall of the tube. Be sure to turn off the set first, then ground the high-voltage terminal to the chassis or to the Aquadag coating with an insulated test lead.

Some plugs have a rubber or plastic insulating cap. A rubber cap that has deteriorated through natural aging can cause arcing and corona in the region surrounding the high-voltage connector. If thorough cleaning

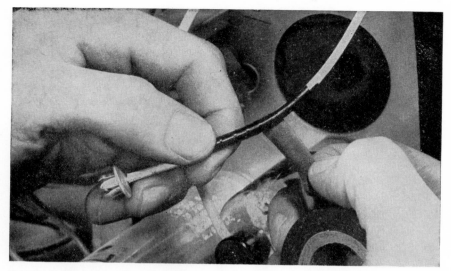

Fig. 2. Spring-contact plug used on high-voltage lead. No. 33 electrical tape is being wrapped around lead to cover insulation damage. (Minnesota Mining and Mfg. Co. photo)

of the cap does not clear up the trouble, an entire new high-voltage lead with cap should be installed.

Picture-tube Ion-spot Damage. Older types of television picture tubes require a single-magnet or double-magnet ion trap on their necks. When properly adjusted, this magnet prevents the heavy particles called ions from getting out of the electron gun and hitting the screen. Examples of ion traps are shown in Fig. 3.

Improper adjustment of the ion trap allows the ions to hit a small area near the center of the screen. The repeated impact of these heavy ions burns a brown circular or X-shaped spot in the coating. This spot is then permanently dead to any fluorescing effect. Once the damage is done, the only remedy is replacing the tube.

Ion-trap magnet adjustments are necessary to make the electron beam bend back into line exactly, so practically all the electrons pass through the hole in the end of the electron gun. The more electrons that hit the screen, the brighter is the picture on the screen. Ion traps are designed so adjusting for brightest picture automatically gives maximum trapping of ions.

The ion-trap magnet should be adjusted whenever a new picture tube is installed. The adjustment of the ion-trap magnet should also be checked after each adjustment of the focus coil.

174 *Television and Radio Repairing*

If the ion trap is so loose that it might shift in position when the set is bumped or moved during house cleaning, anchor it in the correct position with Scotch tape or service cement.

Adjusting Ion-trap Magnets. An ion trap should be placed on the neck of the tube in such a way that the arrow or other mark on the magnet points to the screen. If an ion trap has no marking, mark it yourself with

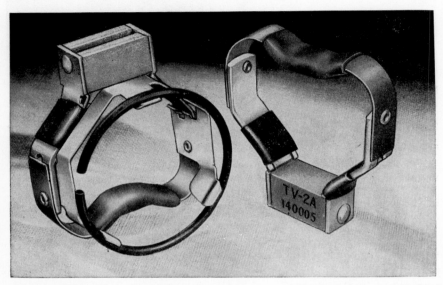

Fig. 3. Examples of ion traps. That at the left has two magnets, one being a block of Alnico and the other a circular piece of magnetized steel wire

a grease crayon before removing the trap. If the ion trap has two magnets, the stronger magnet should be at the base end of the tube.

With the picture tube operating and the brightness control adjusted for low intensity, the ion trap should be moved a short distance forward and backward and at the same time rotated to obtain the brightest raster (line pattern).

If two positions are found to give about equal brightness, the position nearest the base of the tube is the correct one. If some obstruction such as a focus or centering lever prevents the ion trap from being rotated to its best position, remove the ion trap, give it a half-turn to reverse the polarity of its magnetic field, and reinstall it. The new best-brightness position will now be on the opposite side of the tube neck.

If, in obtaining the brightest raster, the ion-trap magnet has to be

Removing and Replacing Picture Tubes 175

pushed against the focus coil, the magnet is probably weak. A new magnet should be tried.

As a final check, set the brightness control to obtain a raster of slightly above average brilliance. Adjust the focus for a clear line structure to simulate actual operating conditions with a picture. Now move the ion-trap magnet back and forth slightly to the exact position giving maximum brightness.

Never move the ion-trap magnet to remove a shadow from the raster if this decreases the intensity of the raster. The shadow should be eliminated by moving the focus or deflection coils. To avoid burning the screen, the ion-trap magnet should always be in the position that gives maximum raster brilliance.

If the picture tube has just been installed or the set has been moved, the brightness control should be kept low until after the initial adjustment of the magnet. Tubes have been ruined in 15 seconds of operation with the ion-trap magnet out of adjustment and the brightness control set too high.

Aluminum-backed Screens. Practically all modern picture tubes have aluminum-backed screens. A thin coating of aluminum is applied to the back of the fluorescent screen during manufacture. This coating passes electrons but blocks the heavier ions. This is why ion-trap magnets are not usually used on picture tubes having aluminum-backed or aluminized screens. If the original tube has a trap, transfer it to the new tube, however. Some manufacturers feel that an ion trap can prolong the life of an aluminized tube.

Deflection Yokes. The two pairs of coils that make up the deflection yoke of a picture tube are mounted on an insulating form. One set of coils, positioned vertically on opposite sides of the neck of the tube, makes the electron beam sweep horizontally back and forth. The other set of coils, mounted above and under the neck, makes the beam move slowly downward line by line and then sweep suddenly to the top of the picture. Under certain conditions, this return sweep can be seen as a diagonal white line on older sets or on newer sets that are out of adjustment.

Deflection yokes and picture tubes are rated according to the total angle through which they sweep the electron beam. In the early days of television, this angle was only about 50 degrees, which meant that the beam could be bent only 25 degrees in either direction from its undeflected position.

As engineers improved yoke and picture-tube design, the deflection angle

176   Television and Radio Repairing

has steadily increased. Many tubes use 110-degree or even larger deflection angles. This permits shortening the length of the picture tube while still providing a large-sized viewing screen.

Adjustment of Deflection Yoke. A deflection yoke should press against the funnel of the picture tube. If not far enough forward, dark areas will appear at the corners on the screen.

In older television sets the deflection yoke is usually supported by a bracket that is mounted on the chassis or cabinet of the set. This bracket has adjusting screws or wing nuts that permit changing the position of the yoke.

In sets having 110-degree picture tubes, the yoke is smaller and is supported entirely by the neck of the picture tube. A small clamp holds the deflection yoke and centering magnets in position. Loosening of the clamp permits removal or adjustment of the yoke.

Rotating a deflection yoke rotates the entire picture on the screen. If you have a set in which the picture is at an angle to the horizontal edges of the mask, loosen the yoke mounting clamp and rotate the yoke until the picture lines up perfectly.

Pincushion Correction Magnets. Deflection yokes for 110-degree picture tubes generally require one or two pairs of permanent magnets mounted on the yoke close to the funnel of the picture tube, as in Fig. 4. These magnets are set in the correct position at the factory to give a raster having straight edges, but may occasionally need resetting.

If the magnets are supported by easily bent brass straps, just bend the

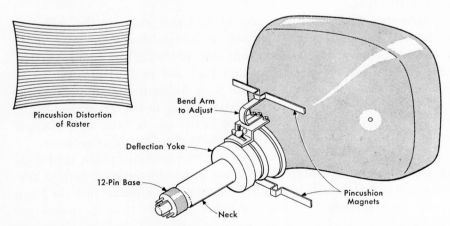

Fig. 4. Example of pincushion distortion affecting all four sides of picture, and one method of mounting pincushion magnets to eliminate this distortion

straps to make adjustments. If the magnets have locking screws, loosen the screws and move the magnet-supporting brackets.

Picture clues pointing to possible magnet misadjustment are nonlinearity (distortion that makes circles egg-shaped), accompanied by pincushioning at the sides of the picture or raster. This means that some sides of the picture bulge inward, as shown for all four sides in Fig. 4. It will usually be necessary to reduce the settings of the height and width controls or shift the picture off-center to see the bulge. Shaded corners on the picture are another clue.

Moving a pincushion magnet toward the funnel of the picture tube pulls the raster lines toward the magnet. Moving the magnet away reduces the effect of the magnet. To readjust a magnet, move it away from the picture tube, then slowly move it toward the tube until the edge of the picture is straight. If the magnet is brought too close to the tube, however, a corner of the picture or raster may become shaded.

Linearity and width controls should be adjusted for the best possible picture after adjusting the pincushion magnets.

Picture-centering Magnets. In newer types of picture tubes, the picture is usually centered on the screen by rotating two centering levers located just beyond the deflection-yoke assembly, as shown in Fig. 5. Each lever is a part of a washer-shaped ring magnet.

To center the picture after installing a new picture tube, first bring

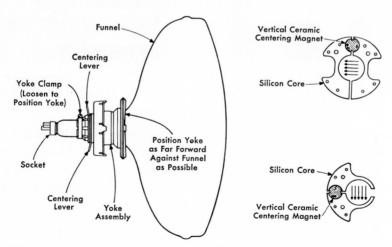

Fig. 5. Location of centering magnet levers on 21-inch picture tube using 110-degree deflection yoke, and example of ceramic centering magnet arrangement used behind deflection yoke on some picture tubes. Arrows indicate direction of magnetic field through neck of tube

the two levers together in a horizontal position on either side of the neck of the tube. Separate the levers until the picture is centered vertically, then rotate both levers together in either direction until the picture is centered horizontally. If vertical centering is now off, move either one of the levers slightly to correct it.

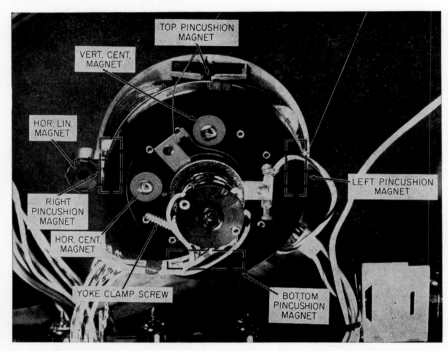

Fig. 6. Deflection yoke having total of seven permanent magnets for centering picture and correcting distortions. Yoke is held in position against funnel by a clamp having a single tightening screw. (Philco photo)

Another two-magnet centering arrangement, using rotatable disk-shaped ceramic permanent magnets set into silicon cores, is also shown in Fig. 5. The deflection-yoke assembly in Fig. 6 uses these centering magnets along with four ordinary pincushion magnets and an adjustable horizontal linearity magnet.

Mirror Aids Picture-tube Adjustments. When making adjustments on the picture tube in a set having a large cabinet, a mirror is almost a necessity for seeing the picture while working behind the set. A photographer's chromium-plated ferrotype plate is even better than a mirror. The highly polished plate provides excellent reflection, and will stand much more abuse than an ordinary mirror. Get a size that fits inside your tube caddy. It can be propped up on a chair in front of the set.

Adjusting Focus. When a new picture tube is installed in a black-and-white or color television receiver or when the picture tube ages, it is necessary to adjust the focus of the electron beam on the screen. This

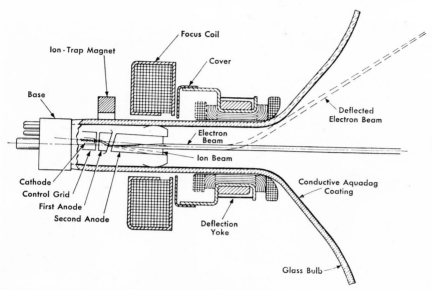

Fig. 7. Location of deflection yoke, focus coil, and ion trap on the neck of a picture tube having a bent gun

makes the beam produce the thinnest possible raster line on the screen, as required for a bright and clear picture. A fuzzy picture usually means that the focus is out of adjustment.

In many older television receivers, focus was adjusted by varying the direct current through a coil mounted behind the deflection yoke on the neck of the picture tube, as in Fig. 7. The rheostat (variable resistor) used to adjust the current was usually a back-of-set control. Centering was then achieved by tilting the focus coil.

The commonest type of focus control used today permits applying a choice of two or more different d-c voltages to an electrostatic focus anode in the picture tube. One version is a flat metal strap or spring that goes over two of the pins on the base of the picture tube, as shown in Fig. 8. The strap is used to connect the focus pin to whichever of the other two pins gives the best focus. Never try the focus strap on pins not specified in the service manual; other connections may damage the receiver.

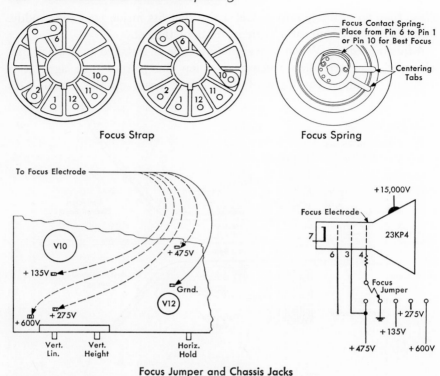

Fig. 8. Focus strap can be pushed over pins on base of picture tube to connect the focus anode (pin 6) to pin 2 on tube shown at left above. When strap is turned over, it can be put on pins 6 and 10. Use of spring in place of strap is shown at upper right. Focus jumper arrangement permits applying any of five different voltages to focus anode, which is pin 4 in the type 23KP4 picture tube shown

In other sets, three or more focus voltage jacks are provided on the chassis. An insulated wire coming from the picture-tube socket, with a plug at its free end, is inserted in the jack that gives the best focus. An example of this is shown in Fig. 8.

Some sets use a back-of-set focus-control potentiometer to provide continuous adjustment of the d-c voltage that is applied to the focus anode in the picture tube.

Tubes without Focus Control. You will also encounter sets in which there is no focus control. Here the picture tube is built to give a compromise electrostatic focus that manufacturers hoped would be acceptable for the life of the tube. Unfortunately, when voltages in the set change as a receiver ages, or when the picture tube itself ages, the focus gets worse.

Removing and Replacing Picture Tubes 181

When you encounter poor focus in a set having automatic electrostatic focus in its picture tube, try replacing the low-voltage rectifier tube first, then replace the high-voltage rectifier. If this does not cure the trouble, and the receiver operating voltages are essentially in agreement with those specified in the service manual, the picture tube will have to be replaced. Unfortunately, there is no guarantee that the new tube will give appreciably better focus.

In some new types of picture tubes having electrostatic focus, a beam aligner is placed on the neck of the picture tube to improve the focus. This aligner looks like an ion trap, but is simply adjusted for best possible focus. It may be needed on a replacement tube even if the original tube did not have it, and can be bought separately.

Testing Picture Tubes. With some tube testers you can purchase an adapter that permits testing the electron gun of a picture tube right in the set, as shown in Fig. 9. A socket at one end of the adapter cable fits the base of the picture tube. A plug at the other end of the cable is

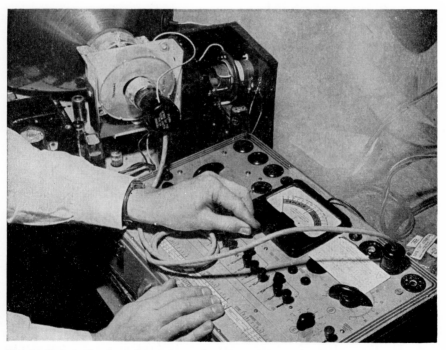

Fig. 9. Method of using adapter for testing picture tubes with a tube tester. The picture tube does not have to be removed from the receiver

inserted in a specified socket of the tube tester. Separate adapters are required for 8-pin and 12-pin bases. The adapter usually permits testing for emission, leakage, shorts between electrodes of the electron gun, and open filaments. An adapter is well worth purchasing, if available for the tube tester you plan to get. The profit on just one picture-tube replacement job will usually pay for the adapter.

Adapters do not permit complete testing of picture tubes. Substitution of a new tube is unsatisfactory except as a last resort, because of the time required and the cost of keeping spare picture tubes on hand. For these

Fig. 10. Example of special picture-tube tester (B & K model 440) and type 8XP4 8-inch check picture tube that can be used in practically any television set to check its picture tube

reasons, many shops use special picture-tube testers such as the B & K model 440 shown in Fig. 10. These testers come with the necessary cables for testing any picture tube while still in its set.

Picture-tube Troubleshooting. Trouble symptoms themselves are excellent clues to defective picture tubes, particularly when combined with simple tests that can be made with a multimeter.

The picture-tube troubleshooting chart in Table 1 combines symptoms, causes, and tests in an easy-to-use manner. In most cases these tests will enable you to determine quickly whether the trouble is in the picture tube or in the receiver itself. The customer is generally more willing to pay for a picture-tube replacement job when he knows that you have checked your decision in so many different ways. The test procedures are described in detail in later paragraphs.

Television Check Picture Tube. Inexpensive 8-inch picture tubes such as the type 8XP4 in Fig. 10 are available for test and substitution purposes. The check tube serves to tell whether the trouble is in the television receiver or in the old picture tube.

To use a check tube, insert it in the deflection yoke of the old tube as in Fig. 11, then place the high-voltage anode lead and the picture-tube

Table 1. Picture Tube Troubleshooting Guide

Symptom	Probable Cause	What to Do
No picture and no raster	Open filament in picture tube	Check with ohmmeter; if reading is infinity, check next cause
	Open in filament pin	Check between filament pins with ohmmeter while wiggling pins; if reading flickers, resolder one or both pins
	Open in filament wiring	Check wiring and filament contacts on picture-tube socket with ohmmeter or voltmeter; resolder where necessary
	Open cathode lead to picture tube	Check wire going to cathode terminal of socket, then resolder cathode pin of picture tube
Picture is too dim	Low cathode emission	Try tube brightener. If this works, it can be left on, but customer will eventually require new picture tube
	Ion trap misadjusted	Readjust ion trap
	Face of tube is dirty	Clean face of tube and safety-glass window
Intermittent picture and raster	Intermittent filament contact	Locate trouble with ohmmeter and resolder
	Intermittent short between electrodes inside picture tube	Try to get out shorting particle by tapping neck of picture tube while tube is in face-down position
	Intermittent open in cathode or control grid	Try resoldering pins
Picture is fuzzy	Focus needs adjustment	Readjust focus. Check focus electrode voltage. If tube has no focus adjustment and fuzziness is severe, replace picture tube
Picture blooms (enlarges and fades) when brightness control is advanced	Gassy tube or leakage between tube electrodes	Test for gassy tube. Try new high-voltage rectifier
Negative picture	Gassy tube	Test for gassy tube. If gassy, replace
	Weak emission	Test picture tube
Good picture but sound has crackle	Poor connection to Aquadag coating	Bend grounding spring to improve contact

Table 1. (Continued)

Symptom	Probable Cause	What to Do
Picture is too bright and cannot be reduced with brightness control	Leakage between control grid and another electrode	Replace picture tube, after making sure there are no defects in brightness-control circuit
Thin, dark vertical bar in picture and raster	Corona at Aquadag grounding spring	Bend grounding spring to improve contact
Broad, dark or light horizontal bars in picture and raster	Cathode-heater leakage	Check leakage with ohmmeter while tube is warm; if appreciably lower than for new tube, try isolation-type booster

socket on the check tube. No centering, ion-spot, or focusing adjustments are required on the check tube.

A check tube is designed to give a good picture under the wide variety of operating conditions encountered in different makes of television sets.

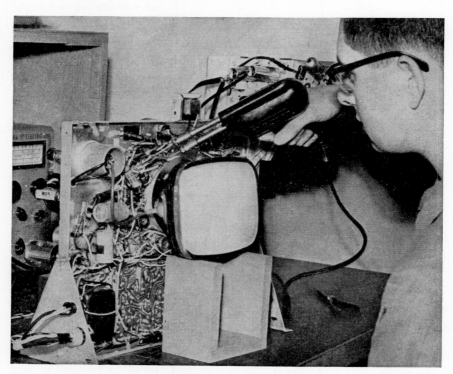

Fig. 11. Type 8XP4 check picture tube being used in vertical-chassis television set on service bench. U-shaped wood holder supports tube firmly against deflection yoke of set. (Sylvania photo)

It may not give a perfect picture on every set, but it will help you decide whether or not the set needs a new picture tube.

Picture-tube Filament Troubles. The simplest picture-tube trouble to recognize is an open filament. When this happens, there is no light in the electron gun at the base of the tube and no heat at the base. Look at picture tubes that are operating normally and feel their bases, to get an idea of how much light and heat there should be.

One common cause of an open filament is a poor contact inside one of the filament pins of the picture-tube base. This trouble can also be

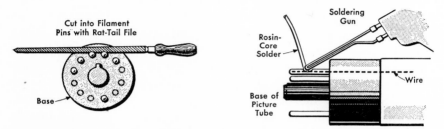

Fig. 12. Method of repairing picture tube having poor connections inside filament pins

intermittent, so that the picture tube operates normally for a while and then goes dark.

The filament of a picture tube can be checked easily with an ohmmeter. Take the socket off the picture tube and clip the ohmmeter leads to the filament pins. The reading should be around 2 ohms. With the ohmmeter leads clipped to the pins, tap the base of the tube lightly while watching the meter. Flickering of the meter pointer indicates a poor connection inside a filament pin. Resoldering the pins will often clear up the trouble. The filament pins are 1 and 12 for 12-pin bases and 1 and 8 for 8-pin bases.

To resolder a picture-tube pin while the tube is still in the set, disconnect the line cord and carefully wiggle the picture-tube socket off the tube. Hold the soldering-gun tip under one of the pins and heat it thoroughly while applying rosin-core solder inside the pin. Repeat for the other filament pin, then scrape off surplus solder outside the pins with a sharp knife.

A more positive repair can be made by filing partly through the pin carefully near the base as shown in Fig. 12, until the file just scrapes and cleans the filament wire inside. Now apply the soldering gun to the filed hole and feed in solder to fill up the hole. Scraping off surplus solder completes the job.

Shorted Picture Tubes. Another common trouble in picture tubes is a partial or complete short-circuit between the cathode and the control grid. When this occurs, the brightness control loses control, and the picture has maximum brightness at all times. Since a circuit defect can cause the same symptoms, always confirm your suspicions by making an ohmmeter test.

Measure a picture-tube short with the highest ohms range of your multimeter while the filament is still warm, so that conditions in the tube have not changed. To do this, operate the set for a while, then pull out the line cord, remove the picture-tube socket, and measure between the cathode and control-grid pins as quickly as possible. Note the reading, then quickly reverse the ohmmeter leads and read again. The *highest* reading is the one that counts. It should be over 10 megohms if the tube is good.

Weak Picture Tubes. Natural aging of a picture tube eventually results in weaker emission of electrons from the cathode. The result is a dim picture. Sometimes the dim picture looks like a photographic negative when the contrast or brightness controls are turned up. Circuit troubles in the receiver can also cause a weak or negative picture, so always check voltages at the picture-tube socket before blaming the tube.

Gassy Picture Tubes. When a small amount of air leaks into a picture tube, or when gas is released by overheated electrodes in the tube, the tube is said to be gassy. A colored glow seen between the electrodes of the electron gun while the set is operating in a darkened room is one indication of gas. Change of the getter coating on the inside of the neck from its normal silvery color to a milky color is another sign of gas. A blue glow just inside the neck of the tube is harmless, however.

Here is one multimeter test that points to a gassy tube in a set having a power transformer. Push a sharp straight pin through the cathode lead that goes to the cathode terminal of the picture-tube socket, and push another pin through the control-grid lead. Measure the d-c voltage between these two pins when the set is on. Turn off the set, carefully pull the socket off the picture tube, turn the set on, and read the voltage between the two pins again.

If there is an appreciable difference in the readings with the tube in and out of its socket, the tube may be gassy. A grid-to-cathode short inside the tube or a leaky coupling capacitor in these electrode circuits may cause this same change in readings, however, so check these two things next.

With the set off, a resistance reading of infinity between the control grid and cathode pins of the picture tube is normal, and a near-zero reading means the tube has a grid-to-cathode short and needs to be replaced for this reason.

Check for a leaky coupling capacitor in the control grid and cathode circuits of the picture tube next, using the ohmmeter section of your multimeter as described in the capacitor chapter. Replace the capacitor if leaky or if in doubt as to its quality, and try the set. If the trouble still exists, you can assume that the picture tube is gassy and needs replacing. Many tube testers will correctly indicate a gassy picture tube.

This test cannot be used on transformerless sets because the filament current to other tubes is interrupted when the socket of the picture tube is removed. Here all you can do is check for grid-to-cathode shorts and coupling-capacitor leakage, then try a new picture tube.

There are no remedies for a gassy tube. If your tests show that the old tube is gassy, and its performance is no longer acceptable to the customer, a new picture tube will be required. A gassy tube can eventually give a negative image like that produced by a weak tube.

Picture-tube Intermittents. If the picture on the screen of a television set is good for a while, then suddenly goes bad for a short time, look for an intermittent contact somewhere in the set. The picture tube itself is the logical place to start looking.

Tap the neck of the picture tube lightly with a pencil while watching the screen, to see if this causes the intermittent trouble. Pull and push on each lead going to the tube socket, to the deflection yoke, and to the high-voltage terminal of the picture tube while similarly watching the screen. (Use an insulated tool to move the high-voltage lead.) If there is a focus coil, pull on its leads also.

When you can duplicate the trouble by moving a particular lead, you will probably find a poor connection at the end of that lead. Resoldering should then cure the trouble.

Spot of Light after Set Is Turned Off. Television-set owners often complain that a bright spot of light occurs in the center of the screen for a few seconds after the set is turned off. The action is normal for many sets having aluminized-screen picture tubes and can do no harm.

The lingering-spot effect occurs because the cathode of the picture tube holds its heat for quite a few seconds after its heater current stops. At the same time, the capacitance between the inner and outer Aquadag coatings on the glass funnel of the picture tube holds its charge and attracts the electrons emitted by the cathode. Since all other circuits in the

188   Television and Radio Repairing

set stop when the set is turned off, the deflection yoke stops working and the electron beam goes straight to the center of the screen.

If the spot occurs on a set that did not have it when new, look for an open capacitor connected between the first anode (grid No. 2) and ground.

Using Picture-tube Brighteners. When lack of brightness is due to natural aging of the cathode of a picture tube, it is possible to increase

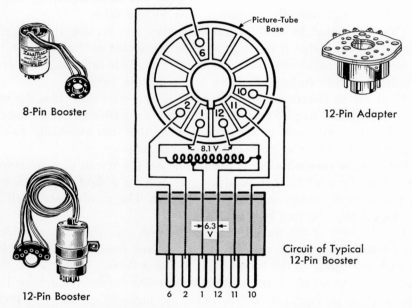

Fig. 13. Examples of picture-tube brighteners, also called boosters, that are used to increase the life of an aging picture tube by boosting its filament voltage. At upper right is an adapter that makes picture-tube socket leads conveniently accessible

the life of a tube by installing a picture-tube brightener or booster between the base of the tube and the tube socket. Essentially this brightener is nothing more than a small transformer that boosts the filament voltage. An example is shown in Fig. 13. Some brighteners work with either series-wired or parallel-wired filaments, while others work only for one type.

A higher filament voltage shortens the life of the filament, but there is no way to predict when the filament will burn out. With a brightener, a picture tube may burn out in a few weeks or may work fine for many months. It is worth trying, but do not guarantee anything. Be sure to order the correct type of booster for the picture tube, or keep one of each available type in stock. The instructions furnished with boosters tell you which one to get and how to put it on.

When you run into other picture-tube troubles, check with your jobber regarding picture-tube rejuvenators. He may be able to recommend one for the type of trouble you have. New equipment is being developed at all times. Some rejuvenators or restorers can take care of several different types of picture-tube troubles.

Measuring Picture-tube Voltages. Pull the picture-tube socket just far enough from the base so you can touch the base pins with your multimeter test prods. Measure between the first-anode terminal on the socket and the chassis with the multimeter set to a high d-c voltage range. This is the B+ voltage of the receiver, and should be somewhere between 200 and 400 volts d-c. The service manual for the set will give the exact value.

The adapter shown in Fig. 13 speeds picture-tube measurements. It is inserted between the base and socket of the picture tube, and has a convenient terminal lug for each pin.

Next, check the grid-to-cathode voltage. This should be zero when the brightness control is at maximum, and somewhere between 75 and 100 volts d-c when the brightness control is turned all the way in the other direction to its minimum setting.

High-voltage Check. The last check to make is the high-voltage supply. Servicemen rarely use a meter for this. Instead, they turn off the set, remove the second-anode lead from the picture-tube funnel and adjust the lead so its end is well away from other parts, then turn on the set and draw a spark from the lead with an insulated screwdriver. One hand is kept in the pocket, and the screwdriver is held only by its plastic handle while bringing its metal end near the high-voltage lead. If there is sufficient voltage, it will be possible to draw a spark that is $\frac{1}{8}$ to $\frac{1}{4}$ inch long on an average set. The exact amount of voltage is not critical. After making this test a few times, the serviceman learns to know when a spark indicates ample voltage.

Instead of using a screwdriver, some servicemen prefer to make this test with a grounded lead. Clip one end of a test lead firmly to the chassis, position the second-anode lead as above, then turn on the set and bring the other end of the grounded lead slowly toward the end of the high-voltage lead. If the high-voltage supply is working properly, the spark should jump from $\frac{1}{4}$ to $\frac{1}{2}$ inch. Do not touch the two leads together after getting a spark, however, as this may damage the high-voltage rectifier tube. The grounded lead thus gives about twice as long a spark as the screwdriver path does through the capacitance formed by the screwdriver handle and your body.

Never attempt to draw an arc between the second-anode button and the high-voltage lead. This usually burns an almost invisible hole in the anode button, through which air eventually enters and destroys the vacuum of the picture tube.

Corona and Arcing. When a sharp point exists on a high-voltage terminal, the air in the vicinity may become ionized. This ionization is called corona, and can be seen as a purple glow in a darkened room. The ozone gas that is produced by ionization of air will destroy rubber and many other types of insulation.

Arcing occurs when a grounded wire or plate is too close to a high-voltage terminal. Arcing corresponds to a discharge of lightning.

Both corona and arcing reduce the high voltage that is available for the second anode of the picture tube, thereby reducing picture brightness. These troubles also produce noise signals that are seen on the picture as snow, streaks, bright flashes, and loss of horizontal sync. Buzzes and crashes may be heard with the sound.

High humidity encourages corona. A humid atmosphere also makes dust particles adhere to insulating surfaces, weakening the insulation and causing both corona and arcing.

Locating Corona. If you turn on a television set in a darkened room, with the brightness-control setting reduced, you should be able to see any corona that exists. Be sure to remove the cover of the high-voltage power supply first, so you can see this region also while making the test.

The pungent odor of ozone is another indication of corona, but do not try to localize corona with your nose. High voltage will jump just as readily to your nose as to a metallic ground.

Locating Arcing. If arcing is suspected, darken the room and watch the high-voltage section of the receiver while the set is operating. It usually occurs in the vicinity of the high-voltage rectifier tube, inside the high-voltage power-supply compartment. Often you can hear the arcing directly as a hissing sound, particularly if the volume control of the set is turned down.

Arcing can occur in any type of picture tube, particularly when a new tube is first placed in service. Look for it in the region of the electron gun, but do not condemn the tube right away. Arcing inside a picture tube is generally due to particles that have loosened during shipment and lodged between electrodes. Operation of the set for a few days generally serves to burn out these particles. If the arcing persists, gentle tapping of the neck of the tube may help to clear it up. Try holding the set face down while tapping.

Handling Television Picture Tubes. A television picture tube is not dangerous when properly handled. The danger comes when persons handle or transport a picture tube carelessly while it still has a vacuum.

A television picture tube, like any other tube, operates under a high vacuum. The glass envelope is made with a large safety margin of strength, to withstand the pressure of air on the outside of the tube. The glass used for the face of some tubes is almost 1 inch thick.

When the glass envelope of a picture tube is scratched, particularly on the sides and corners of the faceplate, the glass is weakened. The picture tube can then implode (explode inward) if it is bumped or otherwise strained in the weakened region. When the glass breaks, the outside air rushes into the vacuum inside the tube, and the glass particles are thrown toward the center of the tube with great force. Some pieces of glass collide and bounce off each other. Others continue in a straight path through the tube and out the other side.

The following safety rules apply to the handling of television picture tubes:

1. Wear safety glasses when installing, handling, or working near live television picture tubes.

2. Keep to a minimum the time that you handle unprotected live television picture tubes.

3. Leave a new picture tube in the manufacturer's carton until needed, and put the old tube in this carton afterward.

4. Never transport an unprotected tube in a vehicle. In addition to bumps and flying stones, there are too many other ways in which the exposed glass can be weakened enough so immediate or delayed breakage can occur. The implosion of a tube in the confines of a car can scare the driver enough to cause a traffic accident, even if the occupants are lucky enough to escape injury from flying glass. When transporting a chassis on which the picture tube is mounted, place a double-folded blanket or furniture-mover's pad over the tube.

5. Never place meters, tools, or other heavy articles at levels higher than a picture tube on your workbench unless they are securely fastened.

192 *Television and Radio Repairing*

6. Never carry a picture tube by the neck. Grasp a tube at its widest part, around the faceplate.

7. Never leave television picture tubes where children can reach them, even if in cartons.

8. When it is necessary to rest a picture tube momentarily on a bench or table, place it face down on a clean, soft pad, as in Fig. 14.

Fig. 14. Place a picture tube face down on a clean, soft pad when it is necessary to set the tube aside for a few minutes. (Sylvania photo)

9. Never treat worn-out picture tubes as normal trash. Destroy the vacuum first by letting in air slowly at the base, as instructed later. The tube is then no more dangerous than a milk bottle.

10. Always discharge the high-voltage button to the Aquadag coating before removing a picture tube from a set.

There have been so few picture-tube implosion casualties among servicemen that unfortunately few today take the trouble to wear safety glasses. Nevertheless, it is definitely recommended that you wear safety goggles when handling live picture tubes.

Removing and Replacing Picture Tubes 193

Removing Television Picture Tubes. Picture tubes should be removed from television receivers only as a last resort. Make absolutely sure first that all the other tubes and parts in the set are good, because it takes a lot of careful work to change a picture tube.

With most television sets, the chassis must be removed from the cabinet in order to get out the picture tube. The procedure for removing the chassis was covered in an earlier chapter. The steps involved in removing the tube from a typical set will now be given. Be sure the cheater line cord is unplugged before you start.

1. Remove the second-anode connector from the terminal on the funnel of the tube, after first discharging it to the chassis with an insulated test lead. On some tubes this is a snap-on cap that requires a bit of force with your fingers to get it off. Note the position of the terminal so you will be sure to put in the new tube correctly.

2. Ground the exposed second-anode terminal on the funnel of the picture tube to the chassis for a few seconds with a heavily insulated test lead that is first clipped to the chassis. Do this several times, until no further spark is obtained. This is necessary because the second anode can store high-voltage energy for several hours after the picture tube is disconnected. The amount of energy is so small that there is no danger from electric shock, but it can definitely be felt. An unexpected sudden shock when picking up the picture tube might cause you to drop it.

3. Remove the picture-tube socket by grasping it with your hand and pulling straight back from the base of the picture tube while gently wiggling the socket. Be careful not to pull the base off the tube. Pry gently with a screwdriver inserted between socket and tube base, if necessary.

4. Remove the ion-trap magnet if one is present on the neck of the tube near the base. Usually the magnet is held in place only by a rubber-covered spring-steel clamp, and can easily be pulled off the neck of the tube.

5. Measure as accurately as possible the distance that the face of the picture tube projects ahead of the chassis, so you can get the new tube in this same position without time-consuming trial and error. This is necessary because picture tubes of the same type can vary considerably in length, yet the face of the new tube must fit against the mask on the screen.

6. Loosen the special hold-down strap that goes around the widest part of the picture tube to hold it on the chassis, after first studying its anchor screws carefully so you can get them back on correctly. If the picture tube

has a harness that goes completely around its thickest part as in Figs. 15 and 16, remove the screws from the harness brackets instead. Take the harness off the tube only if you have to replace the tube.

7. Withdraw the picture tube toward the front of the chassis by grasping the widest part of the tube. If the deflection yoke and focusing coil

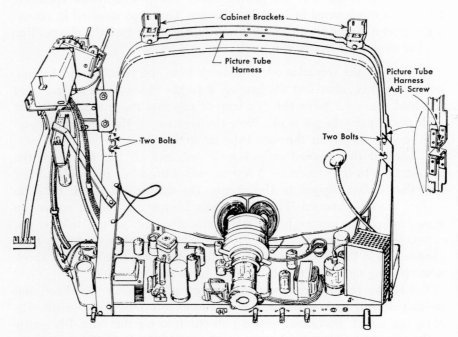

Fig. 15. Example of picture-tube mounting. Harness has an adjustment screw at right center as shown, and a similar screw at left center. Two brackets at top serve to fasten harness to cabinet, and two bolts on each side fasten harness to receiver chassis strap

are supported by brackets mounted on the chassis, these parts need not be disturbed when removing and replacing the picture tube.

8. If the centering magnets, deflection yoke, and focusing coil are supported only by the neck of the picture tube, as on some modern picture tubes having 110-degree deflection of the beam, remove these parts first. They can be slid off the tube neck after the holding clamp is loosened.

9. Place the defective picture tube in a carton for safe handling. If installing a new tube immediately, put the old tube in the carton in which the new tube was shipped. If setting a tube down temporarily, place it on its face on a soft cloth or pad.

Fig. 16. Mounting harness on new short-neck picture tube. Entire 110-degree yoke assembly is in position on the neck. (Philco photo)

Loosening a Frozen Yoke. After a picture tube has been in service for a number of years, the yoke may stick to the neck of the tube. If this occurs, disconnect the yoke and remove it with the picture tube. Place the tube face down on a soft pad. You can then work the yoke free with both hands, with minimum strain on the neck of the picture tube. Clean out any visible wax and dirt between the yoke and the neck first.

One way of loosening a completely frozen yoke is to heat the insulation between the yoke and the picture tube. This can be done by disconnecting the yoke from the receiver and applying about 50 volts a-c to the two horizontal deflection coils in series. Apply the voltage only long enough to loosen the yoke, because too much heat may damage the insulation.

The required a-c voltage can be obtained by placing the horizontal deflection coils of an old yoke or a 150-watt lamp in series with those of the yoke to be loosened, and applying line voltage to the combination. This cuts the line voltage about in half. Check with the a-c voltmeter section of your multimeter to be sure that the voltage across the pair of horizontal coils is somewhere between 50 and 60 volts.

Removing Cabinet-mounted Picture Tubes. Many of the newer and shorter picture tubes are mounted directly on the front of the cabinet rather than on the chassis. The necks of these tubes are made strong enough to support the deflection yoke and centering magnets, so no chassis brackets are required.

To remove one of these picture tubes, first remove the picture-tube socket, the high-voltage lead, and the deflection-yoke assembly from the tube. The chassis can now be removed from the cabinet.

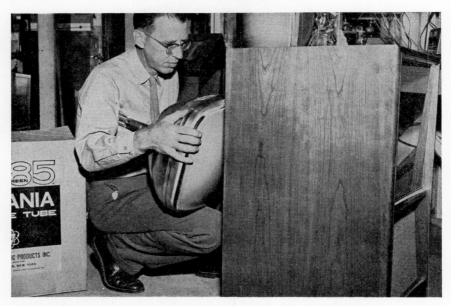

Fig. 17. Method of holding picture tube in preparation for mounting in cabinet. This tube has its own safety-glass faceplate, so no separate safety-glass window is needed. Safety goggles should be worn. (Sylvania photo)

A strap with brackets goes around the widest part of the picture tube. The brackets in turn are fastened to the cabinet. The lower bracket screws or nuts are removed first. Hold the picture tube with one hand to keep it from dropping while you remove the mounting screws or nuts from the brackets at the upper corners of the cabinet.

The picture tube and tube strap bracket are carefully removed from the cabinet as a unit. If replacing the tube, place the old tube face down on a clean cloth and remove the tube strap bracket after loosening its locking nut. The strap is then transferred to exactly the same position on the new tube.

Some servicemen prefer to leave the strap bracket on the cabinet. They loosen the locking nut while holding the picture tube, then take out only the tube, as shown in Fig. 17.

If the new tube does not line up properly with the mask in the cabinet, take the tube out and reposition the strap as required to get proper alignment.

Picture-tube Extension Cables. There will be times when it is desirable to try out a set while the picture tube is still mounted on the cabinet but the chassis is out of the cabinet. The various leads going to the picture tube are usually just a little too short to permit this.

Removing the picture tube from the cabinet and propping it up on top of the chassis is one solution, but there is great risk of damaging the picture tube. If you find yourself working on a number of television sets of this type, get a set of extension cables. This will include an extension for the high-voltage lead, an extension cable for each type of picture-tube socket, and an extension for deflection-yoke cables that have octal connectors.

Installing a New Picture Tube. When replacing a defective picture tube, the following steps will apply for most sets:

1. Remove the new tube from its carton, holding the tube only at its widest part. Do not grab the tube by its neck, as that is the weakest part of the tube. If the base has a cap, leave this on to protect the pins until you are ready to put on the socket.

2. Rotate the tube so its second-anode contact on the funnel is in approximately the same position as that for the old tube. With rectangular tubes you have only two choices. In the set shown in Fig. 18, the second-anode lead is on the opposite side of the set from the high-voltage compartment. The extra length of the lead would be your clue here.

3. Put the deflection yoke on the neck of the picture tube, as in Fig. 19, and fasten it in position with the yoke clamp. On older sets in which the yoke is mounted on the chassis, insert the neck of the picture tube through the deflection yoke while holding the tube by the edges of its face as shown in Fig. 20. Do not use force; if the yoke sticks, investigate and remove the cause.

4. Carefully adjust the position of the tube in its supporting cushion at the front of the chassis. Slide the tube forward or back on the chassis until its face has the same position as for the old tube. Place the protective cushioning strips of tape over the tube as they were before, put on the hold-down strap, and fasten the strap carefully by tightening the bolts until the strap has the same tension as before.

198 *Television and Radio Repairing*

5. Reinstall all parts that go on the neck of the tube, such as the deflection-yoke assembly if it is a separate unit, centering magnets if used, and an ion trap if used.

6. Place the socket on the picture-tube base.

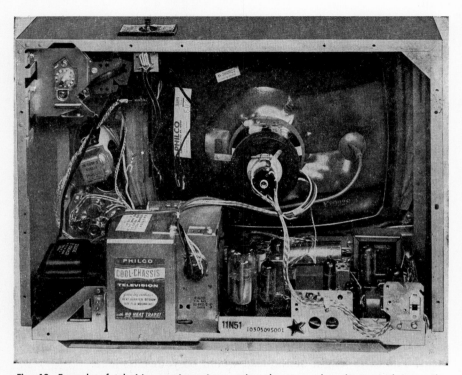

Fig. 18. Example of television receiver picture tube whose second-anode terminal is on the opposite side of the set from the high-voltage compartment. Picture tube is held against window of cabinet by metal straps that loop over the four corners of a rectangular wire frame. Deflection yoke and centering magnets are anchored on neck of picture tube by metal strap having single tightening bolt, just above socket of picture tube. (Philco photo)

7. Connect the high-voltage lead to the second-anode terminal on the funnel of the tube.

8. Wipe the face of the picture tube and the inside of the front-panel safety glass with a soft cloth to remove all dust and finger marks. For plastic safety windows, use only water for cleaning. For glass windows and the glass face of the tube, window cleaner can be used.

9. Slide the chassis back into the cabinet, insert the chassis bolts, but do not tighten them until you are sure the tube is in the right position.

Removing and Replacing Picture Tubes 199

The face of the picture tube should be against the mask, but should not be touching the safety-glass window. Sometimes the chassis can be slid forward or back enough to get the picture tube in the right position. If necessary, remove the chassis again and adjust the position of the picture tube.

10. Push the deflection yoke as far forward as it will go against the flare or funnel of the picture tube, then tighten the yoke-adjusting screws. This

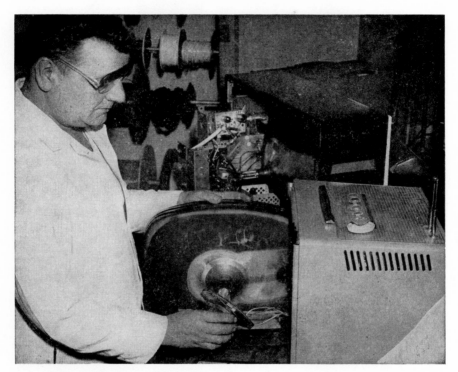

Fig. 19. Placing deflection yoke on neck of picture tube. On this portable set, the picture tube can be taken out from the front without removing the chassis. (General Electric photo)

adjustment prevents the electron beam from hitting the neck of the tube and causing shadows in the corners of the picture.

11. Replace the knobs on the front-panel controls.
12. Connect the loudspeaker cable and any other cables that were disconnected when removing the chassis.
13. Plug in the set and turn it on. If the picture tube requires an ion trap, turn down the brightness control before turning on the set, and adjust the ion-trap magnet immediately, as already described.

14. Adjust the deflection yoke to get a horizontal picture if it is tilted.

15. Adjust the focus if necessary, to eliminate corner shadows and get sharpest possible focus. Always recheck the ion-trap adjustment after adjusting focus.

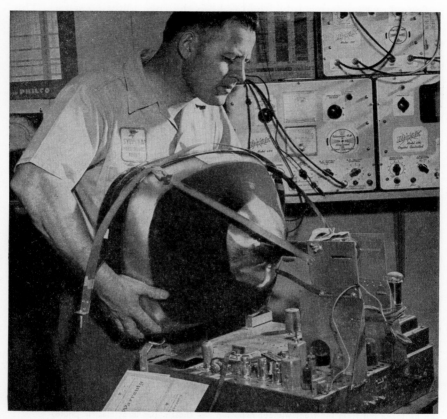

Fig. 20. Correct way to hold a picture tube when inserting it in the deflection yoke on a chassis. Safety goggles should be worn. (Sylvania photo)

16. Adjust horizontal and vertical size controls if necessary to make the picture fill the screen.

High-voltage Rectifier Precaution. After replacing a picture tube, it is usually desirable to put in a new high-voltage rectifier tube. The new picture tube draws much more current than the old one, and may cause a weak high-voltage rectifier to go bad in a short time.

Cleaning Picture Tubes. Both new and old picture tubes should be thoroughly clean around the neck and on the face. Use water or a good window cleaner for the face of the tube, and use alcohol to clean the neck and the area around the high-voltage terminal. Absolute cleanliness around the neck of the tube is especially important, since greasy smudges might carry leakage currents from the coils surrounding it.

Removing Safety Glass for Cleaning. Occasionally it is desirable to clean the inside of the safety-glass window and the face of the picture tube without removing the chassis. On most sets this can be done by taking out the safety-glass window.

Look for removable channels at the top or bottom edges of the safety glass. Sometimes these channels are held in position by screws. Some channels snap into position, and can be pried out by working carefully with a small screwdriver. There are many other methods of fastening the safety glass, but careful study will show the secret of each.

In one set, four sliding retainer bars are revealed when the channel strips are pried off. These bars come off when pushed to the right with a small screwdriver, permitting removal of the safety glass.

Tubes with Built-in Safety Glass. Some picture tubes have their own safety glass, permanently cemented to the face of the tube. This has the advantage that dust cannot collect between the safety glass and the face of the tube. The 23-inch rectangular picture tube was the first to be made in this manner.

Getting Credit for Old Picture Tubes. Many types of picture tubes can be turned in to your jobber for a credit allowance on a new picture tube. The glass or metal walls on these tubes can be re-used by the picture-tube manufacturer. The resulting rebuilt tubes are preferred by many customers because they cost less than tubes made entirely from new glass.

The glass face of an old tube can be re-used only if it is free from cracks and scratches. Defective tubes should be kept in cartons and handled just as carefully as new tubes if they are to be returned for credit.

Rebuilt Picture Tubes. The quality of a rebuilt tube depends greatly on the integrity and experience of the rebuilder. Rebuilt tubes having well-known brand names should give essentially the same performance as new tubes.

Reliable manufacturers use only the glass from an old tube. The entire phosphor screen and the Aquadag coating are cleaned out and replaced, and an entire new electron gun is installed. Less reliable manufacturers use the old screen and sometimes even the electron gun, and some may simply apply a rejuvenator to make the old tube test like a new tube for a few weeks.

Picture-tube Warranties. Most picture tubes are guaranteed against failure due to manufacturers' defects for a given period of time, usually one year. The picture tubes that come in new receivers are usually also covered by a warranty. It is a good idea to become familiar with the policies of your local jobbers and distributors, because definite procedures must be followed in order to get credit for a bad tube.

Picture-tube warranties provide only a new picture tube. There is no allowance for the labor involved in obtaining and installing the tube. If a customer asks you to replace a picture tube that is covered by a warranty, you are entitled to make your usual charges for labor. Be sure to add a reasonable extra amount to cover the time it takes to get the tube exchanged. Do not plunk the bad tube down on the parts counter and say, "I want a new tube free because this is still under warranty." You usually must show the customer's sales receipt for the set, to prove that the picture tube did not last out the warranty period. There will probably be forms to fill out.

You will somehow have to obtain an empty picture-tube carton for carrying the old tube. If the required new tube is out of stock, you will have to make an extra trip to the distributor later to get it. Explain all this to the customer beforehand, to avoid trouble when you present your bill.

Destroying Defective Picture Tubes. Bad picture tubes that have no credit value should never be discarded without first breaking the vacuum to let in air. This eliminates the danger of an implosion that may cause damage for which you might be held liable. Never give old picture tubes to children.

A simple and safe procedure used in many shops requires only a garbage can and an iron rod. Place the tube in the can, face down. Place the cover firmly on the can. Get an iron rod about ¼ inch thick, and insert it through a hole previously drilled in the side of the can near the top. Move the rod around vigorously inside the can until the rod taps and breaks the neck of the picture tube. There will be a popping sound as air rushes in through the broken neck, but as a rule the large part of the tube will remain intact.

With only an occasional picture-tube replacement, the cost of a garbage can is not justified. The corrugated cardboard shipping carton for the new picture tube can be used instead. Place the old tube face down in the carton on top of a wad of cloth that makes the tube base project above the top of the carton. Fold the carton flaps in position, bending or cutting them so the tube base projects. Now crush the plastic aligning key with side-cutting pliers. If a hiss of air is heard, the job is done. If the exhaust tip does not break at first, file off the glass tip as shown in Fig. 21, or reach in with side-cutting pliers or long-nose pliers to break it off. You can also drill through the bottom of the key with an electric drill to open the exhaust tip. Be sure to wear safety glasses.

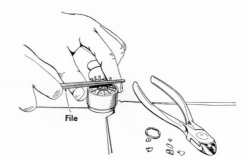

Fig. 21. Method of using file to break off glass tip inside base of picture tube while tube is in shipping carton. A hiss of air rushing in means the vacuum has been broken. The tube is then as safe to handle as any other glass object

Small Picture Tubes. In one all-transistor portable television set the screen of the picture tube is only 2 inches in diameter. Inside the set is a concave mirror that enlarges the image and projects it onto a slanted frosted-glass screen located under the viewing hood.

Because of the compactness of these portable television sets, special procedures are required for removing and replacing the picture tube. With the service manual at hand, the job is essentially no more difficult than replacing larger standard picture tubes.

Color Television Picture Tubes. Replacement of a color picture tube is a job requiring special test equipment and a great deal of experience in adjusting color television receiver circuits. It is therefore best to pass up color tube jobs until you have had the specialized training required. The tubes cost well over a hundred dollars each, so an accident during the installation can wipe out your profits from many other service jobs.

Halo Light. In a few television sets, an illuminated band called a halo light surrounds the picture tube. This is simply a rectangular fluorescent lamp that operates from its own power transformer in the set. There is sometimes a rheostat in series with the lamp to permit varying the brightness.

If the halo light goes out, check continuity between its terminals with an ohmmeter while the set is turned off, to make sure that the current-limiting resistors have not opened. Any reading above about 10,000 ohms

204 *Television and Radio Repairing*

means that the resistors are good. Since the power transformer rarely goes bad, you can now assume that the halo light needs replacement. This will have to be obtained from the manufacturer's distributor for that make of set.

Stocking Picture Tubes. Over a hundred different types of picture tubes are in common use today, and many more types are still being made for

Fig. 22. How technical improvements have shortened picture tubes. At left is 1950 tube designed for 70-degree deflection yoke. Next two use 90-degree yokes, and the two tubes at the right use 110-degree yokes. (Philco photo)

replacement in older sets. One reason for this is the continual improvement in design. By increasing the deflection angle and shortening the electron gun, tube manufacturers have been able to shorten picture tubes a little more each year without loss of picture quality, as illustrated in Fig. 22. Each such change means a new family of picture tubes, one for each desired size of screen.

The average cost of a picture tube to you as a serviceman will be somewhere around $25 per tube. The manufacturer's warranty usually starts the day you buy the picture tube, regardless of when you install it. For these reasons, picture tubes had best be left out of your tube stock. Buy them only as needed.

9

Testing Tubes without a Tube Tester

Three Methods of Testing Tubes. The most common cause of failure in television, radio, and home audio sets is a worn-out or defective tube. Therefore, it is never a waste of time to test the tubes first when a set is brought to you for fixing. Your customers expect you to do this, because they always hope the trouble is just a bad tube.

Tubes can be tested in three different ways: (1) by checking their performance in the set; (2) by substituting good tubes; (3) by using a tube tester. All three methods are used on radio and home audio sets, with the tube tester being most popular among servicemen. For television receivers the first two methods are used most, because tube testers do not reveal all the small differences that can affect television pictures.

In this chapter the first two methods of testing tubes are covered in detail. You can actually get started in television and radio servicing without having a tube tester, by following these instructions for testing tubes by *set performance* and by *substitution*.

No matter which way you test tubes, there are a number of things to be done first. These include verifying the customer's complaint, making sure tubes are well seated and in the correct sockets, looking over the top of the chassis for obvious defects, and seeing if any tubes fail to light up or warm up. Procedures for making these preliminary checks will be given first.

Verify Customer's Complaint. Always verify the customer's complaint of improper set performance before testing tubes. If you are making the check in the home, you may find that the cause of the trouble is something outside the set. There may be a poor antenna connection, a noisy motor in the vicinity, or a blown house fuse. If the customer has brought his set to your shop, ask him to turn it on and demonstrate the trouble to you.

Make Sure Each Tube Is in the Right Socket. If tubes are not in their correct sockets, the set may motorboat (make put-put sounds), squeal, or not operate at all. Tubes can get mixed up easily when they all have the same type of base. This often happens when the owner takes them all out for checking in a do-it-yourself tube tester in a drug or grocery store.

One way to check the location of each tube is by checking it with the tube-layout diagram. This diagram is generally fastened on the inside of the cabinet, pasted on the rear of the chassis, or pasted on the bottom of the cabinet. The diagram is usually in the service manual for the set too. If you cannot find the tube-layout diagram, look on the chassis alongside each tube socket for the tube type number. If it is not there, pull out a tube and see if the number is printed in the center of the socket.

If there are no socket markings or diagrams to check against, then assume the tubes are in their right sockets and proceed with the next step. Only once in a while will you find that the customer has removed the tubes and replaced them incorrectly. If you suspect something like this and there is nothing on the set to check against, hold up further work until you can get a service manual containing the tube-layout diagram.

Make Quick Inspection of Entire Set. Before removing the chassis from its cabinet and before even thinking of testing tubes, spend a few seconds looking for simple, obvious troubles that you can easily see or feel. After you have checked the following list of common external defects a few times, you will find that you can do it automatically, while listening to the customer describe how the set went bad.

1. Push each tube down into its socket. Oftentimes a tube creeps up and one of its pins no longer makes contact with its socket terminal.

2. Check the wall outlet. If in the customer's home, and the set is stone-cold-dead (no lights or heat from any of the tubes and no hum or other sounds when the set is turned on), make sure there is line voltage at the outlet into which the set is plugged. You can check the outlet quickly by plugging in your soldering gun or by measuring the a-c voltage at the outlet with your multimeter. If a floor lamp or table lamp is handy, try the lamp in that outlet.

3. Inspect the set's line-cord plug. Look for loose wiring and loose screws.

4. Inspect the line cord. Look for weak, frayed, pinched, or broken regions. Look especially for breaks at the point where the cord enters the plug and where it enters the cabinet and the chassis.

5. Look for a blown fuse in the set. Some television sets have their own fuses in the line-cord circuit. See that the fuse fits tightly in the fuse holder. Tap the fuse with a pencil to see if it is loose or makes intermittent contact. If the set is dead, pull out the line-cord plug and check the fuse with a low-range ohmmeter even though it may look good. A good fuse should have essentially zero resistance. Small cartridge fuses sometimes develop open circuits at the end of the fuse wires after they have been in use for a number of years.

6. Try the set on-off switch for good snap action. If there is any reason to suspect a defective switch, pull out the line-cord plug first and measure between its prongs with your ohmmeter. The reading should be infinity with the switch off and well below 500 ohms when the switch is turned on.

7. Look for running wax or tar on or near the power transformer if the set has one. This sometimes means that the transformer has been overheated or burned out. Blistered resistors are easily spotted. Note any unusual odors such as that of burned insulation, overheated enamel, or the rotten-egg odor of a defective selenium rectifier. Such symptoms are clues to serious troubles, indicating that you should take the chassis into your shop for thorough checking.

8. On battery sets, check battery voltages with the set turned on. If the voltage is more than about 15 per cent below the rated voltage, a new battery may be all that is needed. Be sure each battery terminal is clean and is making good contact. See that the battery is installed correctly. The center contact of the common flashlight-type cell is positive, and its corresponding connection to the set is generally colored red. Do not rely blindly on this, however, because some transistor radios use special mercury cells or other types in which the center contact is negative.

9. Look for toys and other metal objects that a child may have poked into the set through a ventilating hole.

Inspecting Tube Filaments. In a-c sets the tube filaments are connected in parallel, as in Fig. 1A. The rectifier tube may or may not have its own filament winding on the power transformer. The light from all filaments comes up to brightness gradually, taking as long as a minute in some a-c sets. Only a small amount of light is given off by modern tubes. In some, the only sign of light is a bright red glowing area about the size of the head of a pin.

Some tubes may not appear to light up when viewed from the top, because of the coating of silvery getter material on the inside of the glass.

208 *Television and Radio Repairing*

By looking at these tubes from the side, the filament glow can often be seen near the base of the tube. If the tube is covered by a metal shield, remove the shield first before checking for a filament glow, or touch the tube to see if it is warm.

If the pilot lamp does not light, remove it and examine it. A definite break in the filament shows the lamp is burned out and should be replaced. If you cannot see the open filament, check the lamp further with

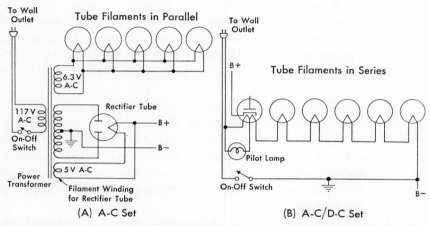

Fig. 1. Parallel arrangement of tube filaments in an a-c receiver, and series arrangement of filaments in a typical a-c/d-c radio set

an ohmmeter. A burned-out lamp will show infinite resistance, while a good lamp will read below 100 ohms.

Checking Tubes by Touch. If there is no obvious light coming from the center of a tube, leave the set on for a few minutes and then feel the tube to see if it is warm. Rectifier and power-outlet tubes generally run the hottest because heat is generated in the plate as well as in the filament. Touch each tube carefully. Some of them get hot enough in normal operation to give a bad burn. In particular, watch out for metal tubes like the 6L6, 5Z4, and 6AG7. The black metal shells give no indication that the tube is hot. Your fingers can be severely burned by innocently grabbing a tube after the set is on a while.

A tube which does not light up or warm up may have an open filament. This can be checked quickly with an ohmmeter, by following the instructions given later in this chapter.

Filament Warmup in A-C/D-C Sets. In a-c/d-c radios the filaments are connected in series, as shown in Fig. 1B. When such a set is first turned on, the power-amplifier tube (such as a 50C5 or 50L6), the rectifier tube (such as a 35Z5 or 35W4), and the pilot lamp may light up quite brightly. The remaining tubes will show just a faint light. After the set has been on a few seconds, the pilot lamp will dim to its normal brilliance.

This change in the amount of light as the set warms up is due to the series connection of the filaments. When the set is first turned on, the

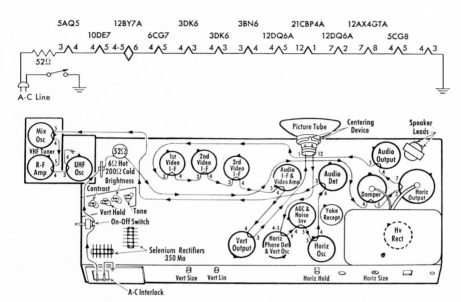

Fig. 2. Series-string arrangement of tube filaments in a typical television receiver, and pictorial diagram showing route taken by filament current in another series-string television receiver

filaments are cold and their resistance is low. This causes a relatively large amount of current to flow through all filaments. Because the rectifier and the power-amplifier tubes use the greatest amount of filament power and heat up faster, these tubes normally give off the greatest light.

After the set is on for a while, the slower-heating filaments gradually warm up and increase in resistance. This reduces the filament current flowing through all tubes, so the pilot lamp then has its normal brightness.

Television sets have more tubes but still use either series or parallel arrangements of tube filaments. In a-c/d-c television sets there may be one long string of tubes, as shown in Fig. 2.

Checking for Loose Tubes. The preliminary check of a receiver includes pushing the tubes down in their sockets. Rock each tube sideways gently

210 Television and Radio Repairing

in its socket when doing this, to clean the tube pins and the socket contacts. If any tube seems too loose, remove it and look for broken socket insulation around the keyway and at the tube-pin holes. These broken insulation defects often occur when the tube is forced improperly into the socket, with the key in the wrong position.

If the socket looks good, put the tube back in. Be sure it goes into position with some pressure, so that the tube pins fit snugly in the socket contact. If a tube slides loosely into place, the socket contacts may have lost their spring because of age or heat from the tube.

A loose contact on a socket can sometimes be fixed by removing the chassis and squeezing the bottom of the contact together a little with pliers. Bent pins on tubes should be straightened, as they may be the cause of poor contact in the socket.

Make certain each tube goes all the way into its socket. Sometimes a tube is held away from the socket contacts by solder in the socket or by the socket mounting screws.

Checking for Loose Top Caps. In older radio sets and in high-voltage sections of television sets you will often find tubes that have top caps for connections. Feel each tube top-cap clip and be sure it fits tightly on the top cap of the tube. If the clip is loose, squeeze the sides a little to increase the spring tension. *Warning:* In television sets, top caps may have dangerously high voltage. Do not touch them until the set has been off for several minutes.

Checking Filaments with an Ohmmeter in A-C Sets. In a-c sets, any tube which does not light or get warm after being on for a minute or so is very probably burned out. Remove it from the set, and check the filament for continuity with an ohmmeter by clipping the test leads to the

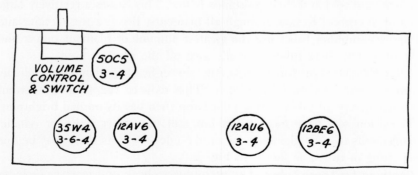

Fig. 3. Example of sketch that can be made to speed up the use of an ohmmeter or a-c voltmeter when testing tube filaments for continuity

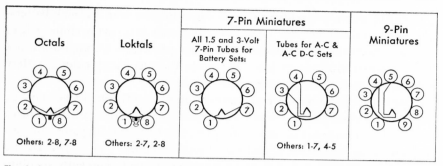

Fig. 4. Commonest filament pin numbers for receiving tubes. The pin numbers listed as exceptions under the symbols are found on only a few tube types, many of which are now obsolete

two filament pins. Any ohms range can be used. Replace any tube which reads infinity on the ohmmeter.

Generally, if a tube is burned out, the set will stop operating. In a few instances you may find an a-c set that will continue to operate, sometimes weakly, if one of the tubes is burned out.

Do not attempt to check for filament continuity while the tube is still in an a-c set. The resistance of the transformer filament winding is shunted across the defective tube, along with the filaments of other tubes in the set, as shown in Fig. 1B. The ohmmeter will therefore show continuity even though one tube is burned out.

Identifying Sockets from Underneath. When checking tube filaments, you have to know what tube is in a socket before you can find its filament pins. This means looking at the tube itself or the chassis marking alongside it, which is rather a nuisance when you have the chassis upside down for convenient testing. The solution is to make a rough sketch before you start testing, showing the sockets as they appear when looking at the bottom of the chassis. Mark each socket with the type number of the tube that is in it. Next, look up each tube in a tube manual and jot down the filament or heater pin numbers right on your diagram, somewhat as in Fig. 3. Making this diagram takes only a few minutes, but you will gain this time back through faster testing and you will minimize chances of making mistakes. After servicing for a while, you will find that you have memorized the filament pin numbers for the commonest tubes and will not have to look them up each time.

The filament pin numbers used most often on receiving tubes are given in Fig. 4. You need to look up the pin numbers only when you come to a tube that tests open at the commonest pin numbers.

Checking Filaments with an Ohmmeter in Battery Sets. In small battery-type portables and farm battery sets the filaments are usually connected in parallel. Follow the same procedure as for an a-c set in locating burned-out tubes. Examples of ohmmeter readings for the filaments of good battery tubes are 15 ohms for a 1S4 and 30 ohms for a 1R5.

Checking Filaments with an Ohmmeter in A-C/D-C Sets. In a-c/d-c sets, all the tubes go dark and get cold when one of them burns out, the same as for series-string Christmas-tree lights. You then have no obvious clues to the bad tube. The filaments of the tubes in a-c/d-c sets may be checked by removing each tube and measuring between filament pins with an ohmmeter, just as for a-c sets. For faster work, tubes can be left in the set and the following procedure used:

1. Unplug the line cord from the wall outlet.
2. Turn the set switch on.
3. Remove the chassis from the cabinet and turn the chassis over so you can see the wiring.
4. Trace the line cord to where it connects to the set wiring. One wire will generally go to the on-off switch. The other wire will go to one of the tube-socket terminals (generally the filament terminal of the rectifier).
5. Clip one test prod of the ohmmeter to the tube-socket terminal having the line-cord wire.
6. Touch the other prod of the ohmmeter to each of the other socket filament terminals in succession, working away from the first prod, until all filaments are checked. An example of how this is done is shown in Fig. 5.

Actual ohmmeter values do not matter here, as they can be different for every set. It is the infinity reading you are looking for, as this means you have just moved through an open circuit.

The last tube in the series string will generally have one of the filament terminals connected to the chassis, and the return path to the other side of the line will usually be through the chassis. The filament wires are all the same color in most sets, simplifying the tracing of wiring from one tube socket to the next. To find the rectifier filament tap for the pilot light, trace the wires from the pilot-lamp socket back to where they connect to the tube socket.

7. In making the continuity check, the first tube you come to which does not show a reading on the ohmmeter has an open filament. To confirm this, remove the tube from the set and connect the ohmmeter directly across the filament pins. On tubes like the 35Z5, test the whole filament across

pins 2 and 7, and test the tap section for the pilot lamp across pins 2 and 3. On the 35W4 miniature-tube equivalent of the 35Z5GT, pin 6 is the filament tap, and pins 3 and 4 are the ends of the filament. If either set of pins does not show continuity, the tube is defective.

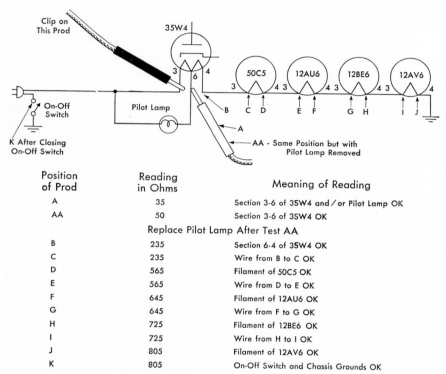

Position of Prod	Reading in Ohms	Meaning of Reading
A	35	Section 3-6 of 35W4 and / or Pilot Lamp OK
AA	50	Section 3-6 of 35W4 OK
	Replace Pilot Lamp After Test AA	
B	235	Section 6-4 of 35W4 OK
C	235	Wire from B to C OK
D	565	Filament of 50C5 OK
E	565	Wire from D to E OK
F	645	Filament of 12AU6 OK
G	645	Wire from F to G OK
H	725	Filament of 12BE6 OK
I	725	Wire from H to I OK
J	805	Filament of 12AV6 OK
K	805	On-Off Switch and Chassis Grounds OK

Fig. 5. Test-prod positions for checking filament circuit continuity of five-tube a-c/d-c radio with an ohmmeter to find a burned-out filament, with examples of readings that would be obtained when all filaments are good

Checking Filaments with an A-C Voltmeter in A-C/D-C Sets. In a-c/d-c sets the filament continuity of each tube may also be checked quickly with an a-c voltmeter. This method of testing has the advantage that it will show up tubes whose filaments are intermittently open. These tubes usually test good with an ohmmeter but open up as the filament heats and expands when the set is turned on. The test procedure is as follows:

1. Remove the chassis from the cabinet.
2. Set your multimeter to the 150-volt a-c range or to the next higher range above 130 volts, such as 200, 300, or 400 volts.

3. Trace the line cord to where it connects to the set wiring. One wire goes to a terminal on the on-off switch. Clip test prod to this terminal.

4. Plug the line cord into the wall outlet and turn the set on.

5. The second line-cord wire will generally go to one of the tube-socket terminals, such as the filament terminal of the rectifier. Hold the second voltmeter test prod on this terminal.

6. A voltmeter reading of about 115 volts indicates you are getting voltage to the set. Now move this second prod to the other filament terminal of the rectifier tube. If the voltmeter still reads about 115 volts, the rectifier filament is good.

7. Next, trace the filament wire from the rectifier to the next socket and move the second test prod to this socket terminal. Repeat this test for each filament terminal on each other socket, until you reach a tube socket on which you get no voltage reading at one of the filament terminals. The tube in this socket has an open filament and should be replaced.

The method is based on the fact that full line voltage appears across a break in a series-filament string. When the a-c voltmeter prods are both on one side of the break, no voltage is measured.

Caution: Remember that in some a-c/d-c sets the chassis is connected directly to the hot side of the a-c power line for one position of the line-cord plug when the set is on. This line voltage can be dangerous to you. The only way to work safely on one of these sets is with an isolation transformer between the set and the power line.

If you do not have an isolation transformer, keep your hands off the bare chassis whenever the set is plugged into a wall outlet. These a-c/d-c sets can give a dangerous shock *even with the switch off*, so respect them.

Checking Filaments with a Neon-type Tester. Many servicemen prefer to use a simple neon-type tester for locating an open filament. These testers consist simply of a small neon glow lamp, a tiny resistor, and two insulated leads with prods, costing around 50 cents at jobbers or hardware stores. The lamp glows red for a-c voltages above 60 volts.

The leads of a neon-type tester usually are not long enough to permit holding one test prod on the on-off switch terminal as for an a-c voltmeter. Instead, hold the test prods on the filament terminals of each tube socket in turn after plugging in the set and turning it on. Absence of a glow means the filament is good. A glow means the filament is open because full line voltage exists across an open filament in a series-string set. An example of this procedure is shown in Fig. 6.

Finding Intermittent Filaments in A-C/D-C Sets. An intermittent filament is one which will show continuity when the tube is cold but will open when the tube is hot.

If the tubes light up normally when an a-c/d-c set is turned on, start to go dim in a few minutes, and then, just before they go out entirely, start to get bright again, look for an intermittent filament. The problem in finding an intermittent tube is to catch the tube in the act of going out. This often calls for patient waiting at each tube in turn.

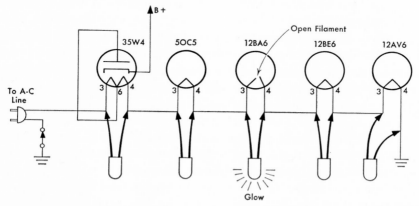

Fig. 6. Locating open filament in a-c/d-c radio with simple neon-type tester that glows for a-c voltages above about 60 volts

The a-c voltmeter test works the best for locating intermittents in a-c/d-c sets. When the voltmeter is across the filament of a tube with an intermittent filament, the meter will read the normal filament voltage for the tube under test when the tubes are bright. If the meter is across the bad tube, its reading will go up to the a-c line voltage, generally around 115 volts, when the tubes go out. If the meter is across a good tube, the meter reading will drop to almost zero when the intermittent occurs.

Remember—to find an intermittently open filament in an a-c/d-c set, connect the a-c voltmeter to the filament terminals of each tube in turn, not across the whole string of tubes. For each tube, wait until the set fades out.

Tapping a tube while it is being checked will sometimes hasten the intermittent condition. Rest the chassis on its side so you can hold the prod under the chassis with one hand while tapping the side of the tube with a pencil. Be sure to use an a-c voltage range above 115 volts, so you will not burn out the multimeter when you find the intermittent.

Finding Intermittents in A-C and Battery Sets. When tube filaments are in parallel as in these sets, the multimeter will not readily reveal the location of an intermittent tube. Your eyes and fingers will find the fault because only the filament glow in the bad tube will go out and only this tube will get cold.

Precautions When Installing a New Tube. After replacing a tube having a burned-out filament, watch the set carefully as you turn it on. Be ready to turn it off instantly if the new tube gets too bright or anything else unusual happens, such as a red-hot plate in a tube or a blue haze between the electrodes of the tube. The cause of the open filament may still be in the set, ready to burn out the new tubes too. If the new tube does burn out, a shorted capacitor is the most likely cause of the trouble.

When the filament of a rectifier tube burns out, there is always the possibility that the burnout was due to a short-circuit. If this condition exists, a new tube will burn out similarly.

To save the price of a new rectifier tube, keep on hand a supply of older rectifier tubes of various types that were taken out because of low emission. Insert a weak tube of the correct type in the rectifier socket first. If it operates normally, you can safely install a new tube. If the weak tube burns out in a minute or two, locate the short-circuit before trying any more tubes.

Finding Noisy Tubes. If the customer complains of noisy operation, turn the set on, tune between stations, and turn the volume up. Tap on the top and side of each tube or tap the tube shields so as to get motion in all directions. Use a finger or the end of a pencil. Do not hit the tube too hard.

If the noise occurs for all tubes, you are tapping too hard and jarring the chassis. This jars the defective part and causes the noise without revealing the troublemakers. Too sharp a rap may also break the glass or jar the tube elements out of shape.

If a tube produces a loud noise or howl in the speaker when tapped, replace it with a new tube. If this corrects the trouble, discard the old tube as noisy.

Sometimes the tube elements may be loose and vibrate without actually becoming open-circuited. Tubes of this type will set up a howl in the speaker, and are called microphonic. The howl may stop if you touch the tube, but the best remedy is to replace with a new tube.

On shielded tubes, look also for loose shields, for loose top-cap connections, and for shields that touch the top cap lightly enough to produce noise when vibrated by sound waves from the speaker.

A review of the past history of the set is helpful in checking for noise. Noise is more apt to show up in the tubes if the set has just been moved in a truck, if the set is subjected to street vibrations produced by subway trains or heavy trucks, or if the customer is in the habit of running the set at high volume. Some common radio troubles due to tubes are listed in Table 1.

Table 1. Common Radio Troubles Caused by Bad Tubes

Observed Effect	Probable Cause	Remedy
No sound, no filament glow on one tube in a-c set	Open filament	Replace that tube
No sound, no filament glow on all tubes in a-c/d-c set	Open filament on one or more tubes	Check filament continuity of all tubes
Low output volume	Rectifier or output tube weak	Replace rectifier or output tube
Pilot lamp out, but set plays	Open pilot lamp	Replace lamp
No signal, but tubes light	Dead oscillator	Replace oscillator or mixer tube
Noise when tube is tapped	Loose element in tube	Replace tube that makes most noise
Programs sound distorted	Gassy output tube	Replace output tube

Advantages and Drawbacks of Substitution Test for Tubes. Trying a new tube is one of the most accurate ways of locating faulty tubes. This method is widely used by television servicemen and is often used by radio servicemen for checking critical tubes such as oscillators and mixers.

The disadvantages of the tube-substitution method are the high initial cost of the tube stock, the space required for storing tubes, the inability to check tubes in the customer's home unless the types needed are determined beforehand and brought in the toolbox, the inability to check tubes that are brought in by the customer, and possible loss of tube sales by not detecting tubes that work but are weak and likely to fail in a few months. Tube testers usually show up these weak tubes.

Testing Radio-set Tubes by Substitution. To make a tube-substitution test, proceed as follows:

1. Turn off the radio set.
2. Examine the set to see if it has a socket-layout diagram or other socket identification; if not, mark each socket yourself or make a rough layout diagram.
3. Take all tubes out of set carefully and set them aside.
4. Put in an entire set of new tubes from your tube stock.

218 *Television and Radio Repairing*

5. Turn on the set. If it still does not work with all new tubes in place, leave the new tubes in and proceed to check individual parts until the trouble is located elsewhere in the set. When the trouble is found and fixed, put back the old tubes one at a time as described below, and watch for any difference in set performance. It is quite possible to have a number of bad tubes in a set in addition to other defects.

6. If the new tubes correct the trouble and the set starts working again, the fault was in one of the old tubes. To find exactly which tube of the group was defective, put the old tubes back in the set one at a time. Work from left to right or in some other definite order so no tubes are skipped. Turn off the set each time you change a tube. When you come to the tube which makes the set misbehave again, you have located the defective tube. Replace this defective tube with a good tube and continue to put back the old tubes. Be on the lookout for a second tube which might cause the set to misbehave again. A new tube is left in only when it makes the set work better than the old tube.

An alternate to the above method of substitution is to replace each tube one at a time with a new tube. The tube which causes the set to perform again is generally replacing the defective tube. Begin with the rectifier and the audio output tubes as these are the ones most often found defective.

Testing Television-set Tubes by Substitution. It is fairly easy to substitute new tubes for old in a radio receiver having only 4 to 8 tubes. Television sets have many more tubes, however, so that replacing them all is quite a chore.

Replacing all television tubes blindly is also highly undesirable for technical reasons. A television set may have two or more of one type of tube, differing enough in their characteristics so the set works properly only when these tubes are in their original sockets. Interchanging them just introduces a new trouble to confuse the servicing problem.

Fortunately a television set usually gives clues to the section of the receiver having a bad tube. Some of these clues are listed in Table 2, along with the receiver sections that should be suspected of having bad tubes. For most troubles, this narrows your troublehunting search to half a dozen tubes or less. You then need to find the tubes that are in one particular section. Table 3 is a convenient guide for this at the start.

Tube-identification data are generally given on a label pasted inside the cabinet of a television set. The same data are in the manufacturer's service manual for a television set, as well as in collections of service manuals. In addition, television tube placement data are given in handy toolbox-size

Testing Tubes without a Tube Tester 219

Table 2. Television-receiver Tube Troubleshooting Chart

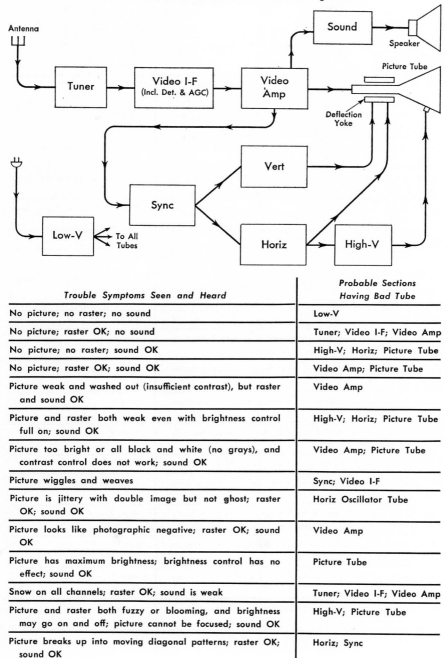

Trouble Symptoms Seen and Heard	Probable Sections Having Bad Tube
No picture; no raster; no sound	Low-V
No picture; raster OK; no sound	Tuner; Video I-F; Video Amp
No picture; no raster; sound OK	High-V; Horiz; Picture Tube
No picture; raster OK; sound OK	Video Amp; Picture Tube
Picture weak and washed out (insufficient contrast), but raster and sound OK	Video Amp
Picture and raster both weak even with brightness control full on; sound OK	High-V; Horiz; Picture Tube
Picture too bright or all black and white (no grays), and contrast control does not work; sound OK	Video Amp; Picture Tube
Picture wiggles and weaves	Sync; Video I-F
Picture is jittery with double image but not ghost; raster OK; sound OK	Horiz Oscillator Tube
Picture looks like photographic negative; raster OK; sound OK	Video Amp
Picture has maximum brightness; brightness control has no effect; sound OK	Picture Tube
Snow on all channels; raster OK; sound is weak	Tuner; Video I-F; Video Amp
Picture and raster both fuzzy or blooming, and brightness may go on and off; picture cannot be focused; sound OK	High-V; Picture Tube
Picture breaks up into moving diagonal patterns; raster OK; sound OK	Horiz; Sync

Table 2. (Continued)

Trouble Symptoms Seen and Heard	Probable Sections Having Bad Tube
Picture drifts up or down but not sideways; raster OK; sound OK	Vert; Sync
Picture breaks up diagonally and drifts vertically; raster OK; sound OK	Sync
Picture breaks up diagonally and drifts vertically; picture and raster do not fill screen; sound OK or weak	Low-V
Picture good but intermittently on and off; raster OK; sound OK	Video Amp; Picture Tube
Picture and raster both good but intermittently on and off; sound OK	High-V; Horiz; Picture Tube
Picture and sound both good but intermittently on and off; raster OK	Tuner; Video I-F; Video Amp
Picture, raster, and sound intermittently on and off	Low-V
Picture and raster both are compressed at top or at bottom; sound OK	Vert
Picture is nonlinear horizontally (compressed at right or left side); raster OK; sound OK	Horiz
Picture folds over on itself at one side; raster OK; sound OK	Horiz
Picture and raster both have alternate dark and light horizontal hum bars; sound may have hum	Low-V
Picture has alternate dark and light horizontal bars; raster OK; sound OK	Video Amp; Tuner; Video I-F
Picture and raster both have white horizontal bar at bottom or top; sound OK	Vert
Picture has white horizontal lines; flashes of white lines in raster; sound OK	High-V
Only thin horizontal line on screen; no raster; sound OK	Vert
Picture and raster fill screen vertically but not horizontally at sides; sound OK	Horiz; Low-V
Picture and raster fill screen horizontally but not vertically at top and bottom; sound OK	Vert; Low-V
Picture and raster do not fill screen either horizontally or vertically; sound OK	Low-V
Picture and raster both enlarged, so too big for screen; sound OK	High-V; Horiz
Picture and raster enlarge excessively when brightness is increased; sound OK	High-V
Picture and raster OK; sound dead, weak, or distorted	Audio

Table 3. Typical Names of Stages Used in Each Section of Television Receivers

TUNER Also called head end and front end. Generally a separate unit directly back of station-selector switch
Stage Names: R-F Amp; Converter; Mixer; Oscillator

VIDEO I-F Includes three or four video i-f amplifier stages, the video detector (often a crystal), and one or more automatic gain control (AGC) stages
Stage Names: 1st Video I-F; 2nd Video I-F; 3rd Video I-F; Video Detector; AGC; Keyed AGC; AGC Amp

VIDEO AMP Includes the video amplifier stages between the video detector and the picture tube
Stage Names: Video Amp; Video Output

SOUND Includes one or more sound i-f amplifier stages, the sound detector, and audio amplifier stages up to loudspeaker
Stage Names: 1st Sound I-F; 2nd Sound I-F; Limiter; Ratio Detector; Discriminator; A-F Amp; Audio Output

SYNC Includes all stages between the point in the video amplifier where sync signals are taken out and the point where the vertical sync signal is separated from the horizontal sync signals
Stage Names: Sync Amp; Sync Clipper; Sync Limiter; Sync Separator

HORIZ Includes stages that handle only horizontal sync signals, located between the sync separator and the horizontal output transformer
Stage Names: Horizontal Sweep Generator; Horizontal Oscillator; Horizontal Output; Damper; Horizontal Multivibrator; Horizontal AFC; Horizontal Phase Detector

VERT Generally has only two stages, located between the sync section and the vertical deflection yoke of the picture tube
Stage Names: Vertical Amplifier; Vertical Oscillator; Vertical Sweep Generator; Vertical Multivibrator; Vertical Sweep Output; Vertical Output

HIGH-V The high-voltage power supply, which generates 10,000 to 15,000 volts for the second anode of the picture tube, is generally inside a protective metal compartment on top of the chassis. The power required for both the filament and anode of the high-voltage rectifier tube is taken from the horizontal output transformer, so tube failure in the horizontal section affects the high-voltage section
Stage Names: High-voltage Rectifier

LOW-V A single low-voltage rectifier tube is generally used, or there may be one or two selenium or silicon rectifiers
Stage Name: Low-voltage Rectifier

booklets that may be available from your jobber. Newest sets are not likely to be in these booklets, so either Photofact Folders or the manufacturer's own manuals are needed for recent models.

The trouble symptoms listed in Table 2 can be caused by defects in many other parts besides tubes. Generally the defect will be in the sections specified for possible bad tubes. If no bad tube is found, look for trouble in other parts.

Television Tube-substitution Procedure. To test tubes by substituting new tubes in a modern television receiver having intercarrier sound, proceed as follows:

1. From the symptoms heard and seen, determine from Table 2 which sections of the receiver could have a bad tube. Remember that a raster is the pattern of horizontal lines seen when a good set is tuned to a channel on which there is no station. Use Table 3 as a guide for recognizing the names of the stages in each section, until you become familiar with the variety of names used for the same stage or tube by various manufacturers. Make a list of the suspected tubes.

2. Turn off the set, remove one of the suspected tubes, and insert a new tube of the same type.

3. Turn on the set, wait for tubes to warm up, readjust tuning controls if necessary, and note performance. After changing a horizontal oscillator tube, the horizontal hold control and other horizontal controls may need readjustment. Adjustments may also be needed after replacing a tube in the vertical circuits.

4. If the trouble is still there, repeat steps 2 and 3 for each other tube in the suspected sections in turn. Replace only one tube at a time. Leave the new tubes in for the time being, but keep a record of where they are.

5. When you come to the socket where a new tube restores performance, mark the old tube as defective and set it aside. Now remove your other new tubes and put the good old tubes back, each in the exact socket from which it came.

Television Tube-substitution Tips. When two or more of the suspected tubes are the same type, be sure to mark them and their sockets with a crayon in some way so you can get each tube back in its own socket.

When putting in a new tube for a substitution test, it is a good idea to place the empty tube carton over or near the tube in the set. This will

Testing Tubes without a Tube Tester   223

speed up locating the unneeded new tubes after you have found the bad tube, so you can put your new tubes back in your stock.

Leaving the new tubes in the set until the trouble is fixed takes care of the rare cases where two or more tubes are bad. Replacing one of the bad tubes then does not clear up the trouble. With a double-trouble set, the trouble will come back when you put the old tubes back in. The old tube that brings back the trouble is, of course, also bad, and should be replaced.

If there are two or more tubes of a given type in a set and you do not have a new tube of that type, try interchanging the tubes. If the set then works, leave the tubes that way; if not, put them back in their original sockets.

Tube substitution can be speeded up by wearing a lightweight leather glove on your tube-removing hand. With a glove, you do not have to wait until the tube cools enough to touch it.

High-voltage Precautions. Although the safest procedure is to turn off the television set each time you change tubes, experienced servicemen may change some tubes without turning off the set. From experience, sometimes quite shocking, they have learned how far they must keep their fingers from the high-voltage power supply and the second-anode terminal of the picture tube.

For tubes close to the high-voltage points or inside the high-voltage power-pack shield it is always necessary to turn off the set. The real danger is in the high voltage on the picture tube. This voltage is technically harmless in most modern sets, as the current it can send through you is low, but your involuntary muscular reaction to something like 15,000 volts can do a lot of damage.

Touch the spark-plug terminal of an automobile while the engine is running, and you will understand what this means. Entire television sets have crashed to the floor as the victim jerked back; elbows have cracked against pipes or posts as a hand jerked back. Play safe and turn off the television set for each tube swap, until you have become thoroughly familiar with the locations of the high-voltage points in each make of set.

Substituting Tubes in Transformerless Television Sets. In a-c/d-c television sets there is no power transformer, and the tube filaments are connected together in one or more series strings much like series-type Christmas-tree lights. When the filament of one tube in a string burns out, all the tubes in that string will get cold. The only way to locate the bad tube without instruments is to replace the tubes in the string one by one until the entire string of tubes lights up and gets warm again. The tubes in a given string may be scattered all over the chassis, so do not overlook any. Transformerless television sets generally have a hot chassis, so use the same precautions as for a universal a-c/d-c radio set.

Setting Up a Tube Stock. The first step in testing tubes by substitution is establishing a tube stock. Go slowly at first. Buy only what you know you need. The best way to know what you need is to service sets for a

Fig. 7. Example of tube inventory guide sheet available from RCA distributors

few months first. You probably will not give the fastest service at the start, but you will learn a lot that will save you money later on. Loading up with stock that may never be needed is one of the best ways to go out of business before you get halfway started!

The perfect tube stock would consist of a complete substitution set of tubes for each type of set in your neighborhood. In addition, it would have at least two extras of each type to help you give fast service on tube replacements. This could mean something like two thousand tubes, costing over $3,000 and taking valuable space in your shop, so some compromise must be made.

The practical tube stock is an estimate based on the types most often needed. Jobbers and distributors can be of real help in selecting the tubes most needed in your locality. Some furnish actual figures on tube sales.

Such sales lists change each year, so get the latest one from your jobber or distributor. For your first tube order, pick out only tubes having the highest sales.

You may decide to build up your tube library in several steps. Thus, you might buy 50 tubes the first time, 75 the second time as your cash on hand increases, and 275 the third time when you go into full-time servicing. For the first order you would get one of each of the 50 most needed types. On the second order, however, get one each of 30 types you do not have, plus extras of the most-needed types. For the third order you could choose about 120 new types that you do not have, and get still more extras of tubes that are already in your stock. This would give you about 400 tubes in all, of 200 different types. This is more than enough variety for a full-time servicing business. From then on, build up your stock of the tubes that you use most often, until you have at least a week's supply of each of these.

The important thing to remember is to get expert advice from a tube distributor when you start to build up a tube stock. Ask also for their inventory guides, like that in Fig. 7, to help keep your stock up to date. When tubes are sold, put the empty carton on the shelf backward or set aside the cover of the carton as a record. Replace tubes every time you buy parts, or at least once a week.

Taking Tubes on Service Calls. As a rule it is impractical to take along on a service call one tube of each type in your entire service library. Just take an assortment of the most-needed tubes. If you can find out from your customer the make and model number of the set, look up its tubes in the service manual for the set and be sure to take these tubes.

Tube Caddy. Tubes for home calls should be carried in a tube caddy. This is a special carrying case designed so you can see your entire tube stock when the case is open, as in Fig. 8. These caddies are available at reasonable prices from jobbers and from some tube manufacturers.

A half-stocked, carelessly arranged tube caddy looks unprofessional to the set owner. Form the habit, right from the start, of keeping tubes in order. Place empty tube boxes backward so you can see the missing tubes at a glance.

It pays to use only well-advertised, standard brands of tubes in clean cartons. When a customer says, "I never heard of that brand," she is really wondering if you are trying to slip into her set some cut-price tube for which you will charge her full price. Stay away from bargain sales on tubes.

There are certain tube types that should be in every tube caddy. For

226 *Television and Radio Repairing*

television receivers, you need at least one of each of the following: 1B3GT, 1X2B, 3BC5, 3BZ6, 3CB6, 4BQ7A, 5U4GB, 6AL5, 6AQ5A, 6AU6A, 6AV6, 6AX4GT, 6BC5, 6BQ6GTB, 6BQ7A, 6BZ7, 6CB6A, 6CD6, 6CG7, 6DQ6A, 6J6, 6K6GT, 6SN7GTB, 6T8A, 6U8A, 6V6GT, 6W4GT, 12AT7, 12AU7A, and 12BH7A.

Fig. 8. Example of good tube caddy being used on television service call. (Sylvania photo)

The tube types you stock will depend primarily on the makes of television sets that are most popular in your area. Certain tube types are used by only one or two receiver manufacturers, so you may never need them.

Portable radios are usually brought to your shop for repair. Oftentimes, when you are on a house call for some other set, the customer will mention that her portable is dead. For situations like this, you may want to include in your tube caddy the following more popular tubes used in portables: 1L6, 1R5, 1S4, 1S5, 1T4, 1U4, 1U5, 3S4, 3V4.

Testing Tubes without a Tube Tester 227

For home radios, the following tubes in your caddy will take care of a good percentage of the sets: 5Y3GT, 6AT6, 6BA6, 6BE6, 6SA7, 6SK7, 6SQ7, 6X4, 12AT6, 12AV6, 12BA6, 12BE6, 12SA7, 12SK7, 12SQ7, 35B5, 35C5, 35L6GT, 35W4, 35Z5GT, 50B5, 50C5, 50L6GT.

Tube Storage Shelves. Your stock of new tubes may be stored in drawers under the workbench, on shelves alongside the bench, or on shelves at

Fig. 9. Example of good arrangement of shelves for tubes. (Electrical Merchandising Week photo)

the rear above the bench. Regardless of the method used, arrange the tubes in order of type numbers. Start with the lowest type number (0Z4, 1B3, etc.) at the top left section of your storage space. Leave empty spaces on each shelf for tubes which you plan to get later.

If the tubes are stored on book-size shelves, place the tube cartons near the front edge. If you need space for several tubes of one type, stack the extras behind.

If you are building shelves for tube storage, space the shelves about 6 inches apart and make them about 6 inches deep. Build shelves in sections so you can easily move them around later as your business and your work warrant. A convenient width is 3 feet. Shelves used by one well-established service business are shown in Fig. 9. One coat of aluminum paint on tube shelves costs very little and gives the units a more finished appearance.

Using Tube Stock. Keep your stock of tubes fresh by taking the oldest when you have more than one in stock. Old, dirty, or dusty cartons give the customer the impression that you are selling him an inferior tube.

After making a substitution check, put all unneeded tubes back in their cartons and return these to the correct positions in the shelves. Be careful also not to mix new tubes with the old tubes you just took out of the set. Remember that time used in keeping your tube stock in order is chargeable to each customer requiring new tubes, as a part of your minimum charge or overhead expense.

Whenever you use or sell a tube, make some record of it as a reminder to order a new tube of that type. Some tube cartons have a tear-off tab with the type number printed on it for this purpose. With other tube cartons, the cover itself can be torn off for this purpose. Keep a box on your tube storage shelf to hold these tabs, with a blank card in it for jotting down type numbers when you do not have a carton tab. Once a week, or whenever you order replacement parts, you can sort out the tabs and order all the tubes needed to replenish your tube stock.

Buying Tubes. Build up your tube stock from well-known makes. The tube brand names most commonly known are GE, CBS, Raytheon, RCA, Sylvania, and Tung-Sol. Tubes with the brand name of the receiver manufacturer on them are generally equally good, since they are made by one of the large tube manufacturers.

Once you pick a brand of tube and a supplier, give him most of your business. It pays to know your supplier and have him know you. Not all tube manufacturers make all brands, so you will have to make exceptions once in a while. If you are a regular customer, a jobber will usually bill you once a month for all your tubes and parts.

Most tube suppliers allow a discount of 40 to 50 per cent off list price, with an additional 2 per cent if your bill is paid promptly.

Selling Tubes. Selling new tubes is a big part of a servicing business. Charging list price for a replacement tube is the fair and only way you can cover expenses of getting, storing, and selling tubes, including your time.

Testing tubes free is a profitable way to get to know your potential customers and to advertise. It is not necessary to advertise the free service because the public has come to expect this type of service courtesy as part of a serviceman's job. Give the customer's tubes an accurate and careful check. Do not rush the job. Above all, be honest. Remember that the man getting his tubes checked is a potential customer for a new tube now and for a repair job later.

Guaranteeing Tubes. When you sell a tube, you can safely guarantee it for 90 days against every trouble except glass breakage. Tube manufacturers in turn generally guarantee you a tube life of about 1,000 hours, which is about 11 hours of operation a day for the 90-day guarantee period. Few sets are operated this much. Most tubes actually last much longer than 1,000 hours.

Test new tubes before you sell them. Some tube cartons are made so that this can be done without taking the tube from the package. It is possible for a new tube to develop a short or an open filament due to shock during shipment or handling.

When you find a new tube that is defective, keep it for your tube salesman. He will replace it with a new tube.

Occasionally a new tube works all right for a while and then burns out. It is quite possible for a new tube to work for a few hours and then go out. Before you give the customer a second tube, however, try to get permission to check his set. Ask him to bring it in or stop by to see it. This will take a little time, but is advisable because the set may have a defect that will make the second new tube burn out also.

When the tube is at fault, explain that once in a while even the best-quality tubes will fail prematurely. Replace the tube free with a smile. Return the bad tube to your jobber, as it may be covered by the manufacturer's guarantee.

10

Using a Tube Tester

Reasons for Using a Tube Tester. A tube tester is needed in a servicing business for the following purposes:

1. To make a quick check of all tubes in a radio set.
2. To test tubes brought in by a customer.
3. To make reasonably certain the new tubes you sell are good. Sometimes new tubes are damaged in shipment or otherwise become defective after leaving the tube manufacturer's plant.
4. To check radio tubes in the customer's house.
5. To make a tube check in the presence of the customer so as to prove to him that the tubes are in good condition and that your charge for replacing some other part is legitimate.
6. To make a more complete check of a radio set even after you have located and repaired a fault. In this way you may uncover a weak tube which could cause trouble after the set has been used by the customer for a couple of months.

How a Tube Tester Works. The switching system of a tube tester applies the proper test voltages to the tube under test. This system consists of a group of sockets (generally one for each type of tube base) plus lever, circular, or pushbutton switches for connecting the socket terminals to the various power-supply voltages and the quality-indicating meter in the tester.

Lever switching generally consists of eight or more switches arranged in a row, with one switch for each tube pin. Toggle switches or slide switches are sometimes used in place of levers.

Circular switching consists of two or more multicontact rotary switches.

Using a Tube Tester 231

An example is given in Fig. 1. This type of switching is widely used in tube testers.

Punched-card switching is used in several makes of the more expensive tube testers. Several hundred prepunched plastic cards, one for each tube

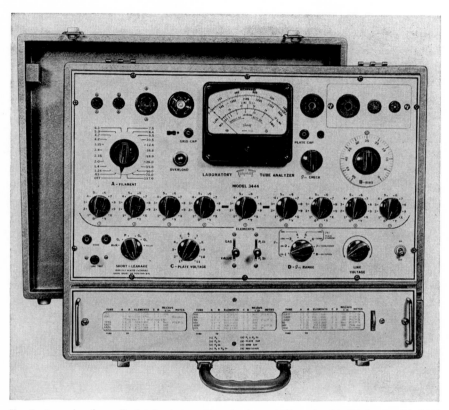

Fig. 1. Example of excellent tube tester using circular switching. In this Triplett model 3444 tester, the roll chart is visible through windows at the bottom of the tester panel and is moved by spinning the knurled wheel projecting up through the panel at the right. An adapter that permits testing picture tubes while in a set or carton is available as an accessory. (Triplett photo)

type to be tested, come with the tester. Insertion of a card activates switch contacts that produce the correct circuits and settings automatically for the tube to be tested.

Pushbutton switching, illustrated in Fig. 2, involves a number of pushbuttons, generally arranged in one or more rows, that are pushed down in various combinations, depending on the tube being tested. A master release button is provided at the end of each row to release all buttons at the end of a test.

The tube-quality indicator on most tube testers is a meter. The indicator for shorted-electrode and continuity tests is generally a neon lamp.

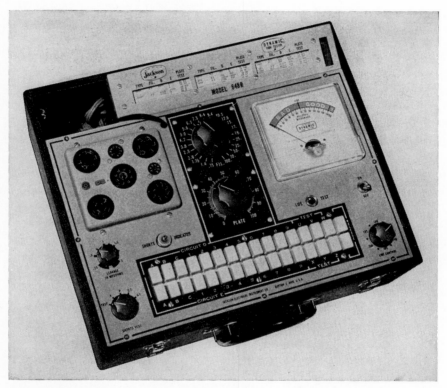

Fig. 2. Good tube tester having pushbutton switching and three-column roll chart. In this Jackson model 648R tester the ? section of the meter scale is above GOOD, where it means the tube is probably gassy, particularly if it is an old tube. The manufacturer furnishes instructions for making up a picture-tube adapter. (Jackson photo)

The quality-indicating meter is generally a d-c microammeter or milliammeter. The meter scale is divided into three parts to indicate the overall condition of the tube under test as GOOD—?—BAD.

Some testers also have a secondary scale marked 0–100 or 0–1,000 to indicate the condition of tubes more precisely than does a GOOD—?—BAD scale. The numerical scale also serves for diodes that cannot make the meter read GOOD even when new. A few testers also use a numerical scale to indicate the equivalent of the tube quality in micromhos. This is a technical unit for grid-to-plate transconductance, but is rarely needed in service work.

Choosing a Tube Tester. For the beginning serviceman, a tube tester is next in importance to the multimeter. A tube tester should meet the following requirements:

1. Give a reasonably accurate test of all the common receiving tubes. This means it should have a shorts-test circuit, a line-voltage adjustment, and a tube-quality meter. False assurance that a tube is good can make you waste hours looking in other parts for trouble that is in the tube.

2. Have a useful life of at least 5 years, as evidenced by the promise of the manufacturer to make test settings for new tubes available or furnish new charts at a nominal price each year.

3. Be easy to operate, permit rapid setup for each tube, and minimize the possibility of a mistake in setup. This calls for uniformly arranged controls and large, easily read panel markings.

4. Have a built-in roll chart, punched cards, or other equally convenient means of determining the correct settings for each tube. A separate booklet of tester settings is not good. It takes time to thumb through the pages, and such a booklet is easily misplaced on a crowded bench.

5. Have a rugged, neat-looking combination portable and counter-type case, with some color for customer eye appeal so that it can be used equally well in the shop, on the counter, or in the customer's home.

Operating-instructions Booklet. Clear, well-printed, and well-illustrated operating instructions are important to understanding your tube tester. Too many tube-tester manufacturers try to cut costs by putting out poorly written and poorly printed instruction books. Look for a tube tester with a good instruction book, both as a help to you and as an indication of a careful manufacturer.

Read the operating instructions carefully before you try to use the tester. This is the best practice to follow with all test equipment. Hold back the urge to plug it in and try it out as soon as you get the tube tester. Too many pieces of test equipment find their way back to the manufacturer because of such overeagerness. Go through the procedures for setting up controls for various types of tubes while you read the instructions, but do not connect the tester to the line.

After you have used the tester for several tubes without the power line connected, read through all the instructions again. You will find them

234 *Television and Radio Repairing*

clearer now since you are familiar with the controls and some of the settings.

Keep the operating instructions with the tester at all times, even when you think you have them memorized. If they are on a card, place it in a protective celluloid or cellophane cover to keep it from being torn or spotted. If they are in a booklet, keep it in a large envelope fastened inside the tester cover.

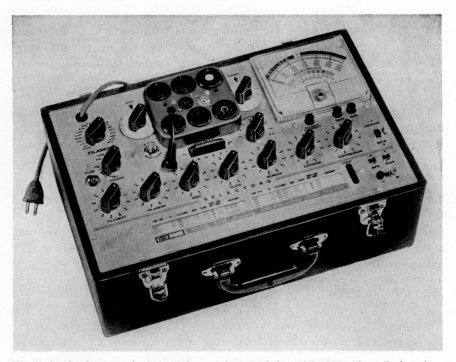

Fig. 3. Good tube tester having circular switching (Hickok model 6000). The roll chart here is printed in two sections, so that it need be turned through only half the distance of a one-column chart to find a particular tube. Picture-tube adapter is available. (Hickok photo)

Roll Chart. A smooth-operating roll chart with large numbers and rapid roll movement is a sign of a good tube checker. The chart should list all the tubes in common use in television and radio receivers. A separate chart or card may contain the settings for obsolete and less-used tubes.

The roll chart should be mounted to permit regular replacement with a new chart giving settings for the latest tubes. Some tester manufacturers make up new charts for this purpose, generally at least once a year. Other manufacturers issue lists of new settings from time to time. These lists may be published in a company magazine or sent directly to every registered owner of a manufacturer's tube tester. Sometimes a nominal charge is made for up-to-date charts or lists of tester settings.

The roll chart should not catch or bind when used. It should easily index against a reference line on a heavy glass or plastic cover.

The roll chart should be located so that you can run your finger along the columns as you set the controls. The controls should be located along the space above the chart in approximately the same sequence as the columns on the chart. In this way you can read the chart and set the controls from left to right for ease of setup.

If the tube chart has more than one column of data, as in Fig. 3, the first tube type number in the second column should be shown clearly on the panel. This saves time in spinning to the start of the second column for a tube type which is actually at the end of the first column. When testers do not have this indication, letter it on a small tab of paper and fasten it to the tester panel with transparent adhesive tape.

Line-voltage Adjustment. Almost all tube testers have an adjustment to compensate for variations in line voltage and tube load. It is generally either a potentiometer or a rotary selector switch marked LINE ADJ. on the tester panel.

The line voltage may be different at various locations in your neighborhood, and may change from hour to hour at your service bench. For this reason, make a rough adjustment of the line-voltage control when the tester is first turned on. This compensates for line-voltage variations due to your location and to the load on the line.

Make a final line-voltage adjustment after the tester is set up with the tube in its socket and warmed up ready for test. This final adjustment compensates for the load placed on the transformer in the tube tester by the heater and plate circuits of the tube under test.

The correct line-voltage setting is generally indicated by a mark on the tube-tester meter scale. The meter is converted to an a-c voltmeter for this test by a rectifier inside.

Tube-quality Tests. Several types of quality tests are used in tube testers. The emission test is the simplest. It indicates the ability of the heated filament or cathode to supply electrons. Emission drops off with the life of the tube; hence low emission means that a tube is near or at the end of its useful life.

Other types of quality tests have various names, such as transconductance test, mutual-conductance test, and power-output test. All serve to check the ability of the cathode to emit electrons and the ability of the tube to amplify signals. Because these tests combine an emission test with a test of some of the other properties of a tube, they usually come closer to measuring the tube's actual performance.

Numerical Scale for Diode Tests. With some tube testers, the GOOD—?—BAD scale of the tube tester cannot be used for some of the smaller diodes because the tube plate current is much lower than the lowest current range of the tube tester. Diodes such as the 6H6 and diode sections of multipurpose tubes such as the 6SF7 are examples.

For these, a tube-tester chart may have a special notation such as GOOD ABOVE 10. This means the tube is good if the pointer goes above 10 on the 0–100 scale, even though 10 is in the BAD region of the GOOD—?—BAD scale.

Tube-life Test. A filament-activity test switch is provided on some testers. This switch reduces the filament voltage 10 per cent while keeping all the other voltages constant. The purpose of the switch is to give a rough indication of the probable life of the tube being tested. If the reading on the numerical scale drops to three-fourths of its normal reading when the filament-activity test is made, the tube is presumed to be near the end of its useful life. The accuracy and usefulness of this type of life test are questioned by some engineers, however.

Testing Pilot Lamps and Ballast Resistors. These parts can be given a simple continuity test with a tube tester to determine whether they are open or not. Since a ballast resistor is simply a wirewound resistor mounted in a tube housing, the test is usually easier and quicker to make with an ohmmeter. The tube tester is used for this purpose only if its chart provides test settings for ballasts.

To test a pilot lamp, first find the lamp voltage from a pilot-lamp table or from markings on the lamp. Set the filament switch of the tester to this voltage. Hold the pilot lamp in the socket provided on the tester, generally in the center of the large seven-prong socket, and see whether the lamp lights.

Magic-eye tubes are checked in a tube tester only for brilliance of glow and for open and closed eye. Voltage-regulator tubes likewise get only a visual check for glow in most tube testers. No meter reading is used on these tubes. Visual checks are only a rough indication of quality, as a tube may have an apparently normal glow and yet be unsatisfactory in its circuit. Trying a new tube is the best test for some of these special types.

Using a Tube Tester 237

Setting Controls for Quality Test. Tube-tester controls are marked with letters or numbers. Considerable care must be used to set these controls exactly as specified on the chart. Pay particular attention to controls with fine markings from 0 to 50 or from 0 to 100. You may often find that a change in setting of only one small division will change the meter reading considerably. An error of a few divisions in a setting can change a GOOD reading to a BAD reading or can make a bad tube read GOOD.

Meaning of Quality Test. After the tube has warmed up and has been checked for shorts, the quality test is made according to the instructions supplied by the tube-tester manufacturer. You will note that some tubes, such as the filament-type 1R5, heat up within a few seconds. Other tubes with indirectly heated cathodes, such as the 35Z5 and the 50L6, may require a full minute to reach a steady indication.

A steady pointer indication on the BAD section of the scale indicates the tube is defective and should be replaced. An indication in the ? section indicates the tube will probably go bad in a short time. Such a tube should be replaced if the customer is willing, even though the tube may still work all right in the set.

Tube manufacturers permit a variation of 20 to 30 per cent in the ratings of a given tube type, so that you can expect differences in the readings for brand-new tubes even though all are the same type. New tubes should all test GOOD, however.

Tube testers are generally made so a tube rejected by the tester is definitely bad. Occasionally a tube testing good on a tube checker may not work in unusually critical circuits of the set.

Unusual Pointer Motion. The way the pointer moves across the scale can tell you something about the tube you are testing.

If the pointer moves rapidly across the scale and hits the full-scale stop with considerable force, pull out the tube quickly or turn off the tester. If the pointer is allowed to stay off scale under these conditions, the tester may be damaged. This behavior generally indicates an improper setup or a defective tube.

If the pointer tip seems to be blurred on a particular tube yet not on all tubes, it is generally an indication that a-c is getting into the meter circuit and causing the pointer to vibrate. This may be due to an improper setup, so turn off the tester and check the settings of the controls. If the tube setup is correct and the pointer still vibrates, it is a pretty good indication of a shorted tube, even though the defect was not revealed by the preliminary shorts test.

If the pointer tip vibrates excessively on all tubes (more than $\frac{1}{32}$ inch) while you make the line-voltage check, the line-check rectifier in the tube tester may be faulty and require replacement. This is a job for the manufacturer of the tester.

If the pointer jumps around considerably and does not settle at a reading, it indicates a bad connection within the tube or at the tube pins in the socket. To check for this, remove the tube, clean off the pins, and straighten them if necessary. If the pointer continues to jump around and appears to move more violently when the tube is tapped, it indicates a defective tube which probably has an internal intermittent short or open due to a loose element.

If the tube has not completely warmed up during the shorts test, the indicating pointer will gradually come up the scale to some steady reading which is the quality indication on that tube tester. If you hold the switch closed, the reading may drop slightly because of heating of the tube elements and tester components. Disregard this lower reading and use the first steady indication as the quality reading.

Slow and small erratic movements of the pointer generally indicate fluctuations in line voltage and not anything peculiar about the tube.

Pointer motion slowly up and down the scale repeatedly indicates a defective tube with an intermittent open or short due to the heat of the filament. For example, when the tube is cold, the filament makes contact. When voltage is applied, the tube starts to warm up, and the pointer begins to move up the scale. Before the tube is completely warm, the expansion of the filament section due to the heat causes the filament wire to open. This removes the source of heat, and the tube starts to cool off. The cathode emission decreases, and the pointer moves downscale. When the tube has cooled sufficiently for the filament again to make contact, the process is repeated. The pointer continues to move slowly up and down the scale as long as the tube is in the circuit. This tube is definitely bad.

Deflection past the GOOD section of the scale does not ordinarily mean that the tube is exceptionally good. Rather, it may also indicate that the tube is defective because of gas or a short, that the tester is set up incorrectly, or that the tube is outside the limits for proper performance in some critical circuits.

If the tube has high emission, the pointer of an emission tester will generally move slowly upscale and may even hit lightly against the full-scale bumper stop. Most emission testers have the GOOD—?—BAD scale calibrated so that a 100 per cent good tube indicates around the ¾ scale mark. An extra-good tube could then indicate near the end scale mark.

A tube for which the tester controls are set up incorrectly, on the other hand, may cause the pointer on an emission tester to move rapidly across the scale and hit the end-of-scale stop rather sharply. A tube that behaves this way should be unplugged immediately. A gassy tube may sometimes also make the pointer hit the stop.

Requirements for a Shorts Test. The most-used indicator for shorts between tube electrodes is a neon lamp. This lamp should give a bright glow, visible in normal room light, to indicate a short between the elements of a tube. If the lamp projects from the panel a small amount, it can be seen without looking directly down on it. A plastic protective cover is good protection against accidental lamp damage.

In some testers, the tube-quality meter is connected to an extra switching circuit that makes the meter serve also to indicate shorts. The circuit is calibrated to read BAD for interelectrode leakage below a certain resistance value.

Making a Shorts Test. A check for shorts between tube electrodes is always made before the actual GOOD—?—BAD test. Do not skip the shorts test to save time, as a shorted tube may give false readings or damage the tube tester itself.

If the tube shows a short, it is defective. Do not make any other tests on this tube.

To make a shorts test, rotary or lever switches are moved one at a time as called for in the tube-chart instructions for the tube under test. This connects each element in turn to the shorts-test circuit. With well-designed lever switching, when a certain lever makes the shorts-test lamp glow, the number of that lever is the number of the tube pin that is shorted to some other electrode.

Make the shorts test after the tube has warmed up in the tester. This may take only a few seconds for filament-type tubes like a 1R5 but a full minute for slow filaments like a 35Z5 or a 50L6.

The shorts test on multisection tubes, like the double-diode 6SQ7 which requires three quality tests, is generally made only before the first test. Testers differ in this respect, however, so study the operating instructions for your own tester carefully.

Some tubes may cause the neon shorts test to blink for a moment as the shorts check switch is moved from one position to the next. This is a normal behavior which may be caused by the discharge of the series capacitor used in some shorts-test circuits. Do not confuse this momentary glow during switching with the intermittent glow (indicating a defective tube) which you may notice on some tubes while they are being tapped.

For all shorts tests, tap the sides of the tube lightly with the fingers or a pencil while watching the shorts-test lamp. The tapping often causes loose tube elements or poor connections within the tube to change position, with a resulting intermittent glow on the shorts-test lamp. Tubes with intermittently shorting elements such as this often cause noisy radio reception or intermittent performance; hence they should be replaced.

Meaning of Shorts Test. The commonest type of trouble shown by the shorts test is heater-to-cathode leakage. In i-f amplifier tubes and audio output tubes this leakage can cause hum. In other tubes it may have no effect on performance. Considerable heater-to-cathode leakage is permissible in most auto radio tubes. Oftentimes a tube that is causing hum in one stage because of leakage will work perfectly well in another stage.

Most a-c/d-c sets are not critical as to heater-to-cathode leakage, even though considerable voltage may exist between the heater and cathode. This is due to the fact that the speakers used in small sets are generally not responsive to the 60- or 120-cycle hum which might be picked up by a leaky tube.

Taps and internal jumpers that should show intentional shorts in tubes are designated on the tube chart by a note such as SHOWS SHORT ON 1 AND 5.

Noise Test. Provisions for a noise test are not essential on a tube tester, as most of the noise-test circuits available at present do not tell much. They consist of nothing more than a panel jack connected in series with the plate of the tube under test. When earphones are plugged into this jack and the tube is tapped, noise will be heard only if the tube is extremely bad. A noisy tube will reveal itself anyway by making noise when in the set.

Gas Test. A sensitive gas test is not essential for your first tube tester. The high-quality tester you eventually get for shop use should have a gas test, however, to give the best possible check of r-f and i-f amplifier tubes.

Most of the present tube testers provide a compromise gas test. Here an off-scale indication beyond the GOOD region or appreciable upscale drifting is an indication of a gassy tube.

If you suspect a tube of being gassy, allow it to heat up first by leaving it in the tube tester for a few minutes before making the gas test. This is especially important on power-output tubes. On these tubes, call them gassy when in doubt, and put in new ones.

The number of tubes which are defective because of excessive gas is relatively small with modern tube-manufacturing techniques, except for output tubes where the heat of normal operation may drive gas from the metal tube parts.

In general, a tube which checks gassy in a tube tester is quite certain to perform poorly in a set. Remember—any tube which reads better than GOOD is probably gassy.

Easy-to-use Tube-tester Controls. For ease of operation the controls for a tube tester should be simple to operate. They should also be well spaced in convenient locations so as to not interfere with the tube or with reading the meter. The ideal tube tester is so clearly marked that it can be used properly without even reading the instruction book.

Ease of operation is important, to keep down the time for running a test. It also helps reduce the possibilities of mistakes in setting controls.

Special buttons or switches with markings like NORMAL PLATE, SECOND PLATE, or RECTIFIER AND DIODES are not recommended, as they are often overlooked and lead to error in reading and to possible tube damage.

The control knobs and buttons should be large enough to be operated easily without fumbling. The controls should be spaced far enough apart to reduce the possibilities of accidental moving while setting an adjacent control. There should be plenty of space for the panel markings associated with each control.

Control knobs should be securely fastened to their shafts, preferably with one or two setscrews in metal inserts, to prevent the knobs from loosening and changing the control settings. Lever switch knobs should be tightly fitted to the switch levers.

All switches should have a good snap action to index them into position, yet the motion between positions should be smooth and uniform without any signs of sticking.

The average time required for a tube test on a well-laid-out tube tester should be about 10 seconds to set up the controls, up to 60 seconds for the tube to come up to temperature, and 30 seconds to check for shorts and read the meter.

Filament-voltage Controls. Filament supply voltages should be provided in a tube tester for all tube filament voltages from 0.63 to 117 volts. The filament panel switch should be clearly marked with each of the filament voltages, to minimize chances of putting the wrong filament voltage on a tube.

Tube-tester Sockets. Individual sockets should be provided for each of the popular tube bases, in a logical arrangement on the panel. One of each of the following sockets should be provided: 4, 5, 6, combination small and large 7, miniature 7, octal 8, loktal or lock-in 8, noval 9. One of the sockets should have a center hole for testing pilot lamps.

Space may be provided for a spare socket, to be used if a new type tube base comes out. The tube-tester manufacturer will then provide instructions for installing and connecting the new socket.

Sockets should make good firm connections to the tube pins and should be securely fastened to the panel. Try a few tubes in the sockets before you buy. See that the tube goes in smoothly with a good snug fit, yet not so tightly that you have to pull the tester off the table to get the tube back out.

Two identical octal sockets are used on some testers because of inadequate switching circuits. These are confusing, so choose a tester that has only one socket of each type.

Testing Unlisted Tubes. Tubes not listed in the tube chart may often be tested by following the chart listing for similar type designations. Thus, tubes with a G, GT, or other tail-end designation are tested the same as corresponding tubes without these letters. For example, use the 6SA7 settings for testing 6SA7G or 6SA7GT tubes.

New tube types which are not listed in numerical order on the tube chart may be found in supplementary charts supplied with the tester or available from the manufacturer. If you have a new type of tube that is not on the chart or supplement sheet yet is in active tube production, write to the tube-tester manufacturer requesting a copy of the latest tube data for your tester. Give the type number of the tube you have to test.

Do not write for tube data just because you see the tube type listed in magazines or on some tube manufacturer's technical sheets. These technical listings often include many experimental types which do not become commercially available until a year or so later.

Tube-tester Socket Trouble. If a tube seems to require excessive force when plugging in, you may be making one of the following errors:

1. Trying to plug the tube into the wrong socket.
2. Trying to plug the tube into the socket incorrectly. For example, you may not have the guide on an octal tube lined up correctly with the guide in the octal socket, or you may not have the large pins on a four-pin tube lined up with the large socket holes.
3. There may be too much solder on one of the tube pins or the pins may be bent out of shape.
4. There may be something in one of the socket holes of the tube

tester. For example, a piece of solder or wire insulation can easily fall into one of the socket holes if you have the tester located near where you are doing set repairs. To remove this obstruction, turn the tester upside down and rap sharply on the back and sides. If the piece will not fall out, take the tester out of the case and push the obstruction out of the socket from the back.

If a tube drops loosely into the socket or seems loose after it is plugged in, the socket connections may be worn, or they may be out of shape. You may also find the tube pins themselves bent out of shape. Look particularly for bent tube pins on tubes with bases like the miniature seven-pin button base.

Tube-tester Precautions. When you are testing multisection tubes like the 6SQ7, it is usually safe to change the settings for the second and third tests without unplugging the tube. Read your tester instruction booklet carefully beforehand, however, because some testers may require special precautions. Some tubes may be damaged if settings are changed while the tubes are in the tester.

When you are through testing tubes, remove the last tube from the tester, turn it off, and reset all the controls to off or normal. Get the habit of doing this, to minimize chances of burning out tubes or damaging the tester. For example, if you leave the filament switch set at 6.3 volts and absentmindedly plug in a 1.5-volt tube the next time before checking the settings, the tube will burn out with a bright flash, and you will be out the price of one tube.

Always set the controls to the specified settings before plugging a tube into the tester, to avoid burning out the tube.

Most testers have an indicator to show that power is on. A pilot lamp is the best. It may be by itself or doing some other job also, such as lighting up the roll chart, lighting up the meter scale, or serving as a fuse.

It is not necessary to unplug the tester when you are through making a test. Be sure, however, to turn it off and set the controls to neutral as specified in the individual tester instructions.

Tube-tester Overload Indicators. Most tube testers have some means of protection against overloads. It may be a high-resistance lamp, a fuse, or a fine-wire fixed resistor designed to burn out before any other parts in the tester are damaged.

In normal tube-tester operation, the overload indicator lamp will not glow at all or will be very dim on most tubes. Only rectifier and output tubes normally make the lamp glow brightly.

If the tube-tester switches are not set correctly or if the tube under test has a low-resistance short between some of the elements, such as across the two filament leads, the tube draws high current. This current also flows through the overload lamp, making it glow brightly. When this happens, remove the tube at once and recheck the tester settings. If they are correct, try another tube of the same type. If it tests normally, the first tube is defective. In case of an extremely severe overload, the lamp will burn out. By thus acting as a fuse, it protects the tester from damage.

Do not leave the tube in the tester if the overload lamp glows brightly, as this may seriously damage the tube tester. Only on large rectifier and output tubes is a bright glow normal.

If the tube tester is provided with an overload fuse, this should be located on the panel so it can be replaced without having to remove the panel.

One tube tester uses a 50-ohm resistor in its test circuit as a fuse. In case of severe overload, this resistor burns out. This will prevent serious damage to the tester. Replacing this resistor is a relatively simple and inexpensive operation. Tube-tester instruction manuals tell how to replace these overload resistors, fuses, and lamps.

Even if the tester does not have an overload indicator, you can often tell when excessive current is being drawn. The transformer may hum more than normal when the tube is plugged in, the roll-chart lights may dim, and the line-voltage indicator may drop down the scale.

Do not leave tubes in a tester any longer than is needed to make the test. This particularly applies to rectifier and output tubes.

Tube-tester Case. A rugged, neat-looking combination portable and counter case will provide the best all-round use for your tube tester. If the case is wood or fabric, be sure it has leather or metal corner braces. Rubber or metal feet should be on the back for supporting the tester on a counter. There should also be feet on one side, for keeping the tester upright when you set it down.

A good-fitting cover is important for protecting the panel markings, meter, and controls while you are carrying the tester and for preventing rain from getting in when you are on calls. The cover should have a pocket or straps for the line cord. Slip hinges should be provided, so the cover can be removed when using the tester for benchwork.

Why a Tube Tester May Read Wrong. The voltages applied by a tube tester are rarely the same as those applied to the tube in the set. A tube may read good at the low plate and screen-grid voltages and the low currents of a tester yet distort or otherwise operate improperly at the higher voltages in the set. This variation occurs mostly on power-output tubes, where heat in the plate and screen grid at high voltages may release gas or cause secondary emission.

The bias and signal voltages applied to the control grid by the tube tester may differ from those in the set. This can cause a good tube to read bad in a tester because of overload in the grid circuit.

The filament voltage in your tester may be correct, but that in the customer's set may be too high or too low. If too low in the set, a tube that tests good can work poorly in the set because of low emission. If too high in the set, a tube that tests good can short out in the set because of higher filament voltage and greater heat.

Because of all these things, your final decision on any tube should be based on how it works in the set. When in doubt, try a new tube.

Tubes That Rattle. Occasionally a tube will rattle when it is shaken and yet test good in a tube tester. If the rattle is caused by loose material inside the glass envelope, the tube structure has very likely been damaged as a result of mechanical shock or excessive vibration. This tube should be rejected, as it probably will not last its normal life.

If the rattle seems to be inside the base, it is probably just a piece of loose cement between the tube envelope and the base. This will generally have little or no effect on the tube performance. If the tube tests good in the tester and does not behave suspiciously in the set, you can consider it a good tube.

Tube-tester Manufacturers. Good quality in components, in their assembly, and in calibration is important to satisfactory performance over the 5-year life of a tube tester. The best over-all criterion of quality is the name of the manufacturer and his experience in the tester business. Experience is important, as quality alone is no assurance that the manufacturer understands the serviceman's angle in tube-tester requirements.

Although manufacturers of tube testers bring out new models that may be better or worse than previous models, the following manufacturers have established reputations for consistent quality and reliability in their tube testers: Hickok; Jackson; Precision; Simpson; Triplett; Weston. The tube testers shown in this chapter are all good and are highly recommended for you.

There are other good tube-tester manufacturers, generally less known. If you decide on one, be sure to look the tester over carefully in reference to the many features mentioned in this chapter. There are also some very bad tube tester manufacturers, selling poor and makeshift testers at enticingly cheap prices yet making absurdly fantastic claims for them. Stick to the well-known makes and you will be safe.

Price. The cost of your first tube tester should be between $125 and $250. You may be able to get along for a while with a cheaper tester, such as the Triplett model 3414, but do not go below $75. Get the best tester you can afford.

Do Not Buy a Secondhand Tube Tester. Secondhand tube testers are a bad investment, especially if the particular model is no longer in production. Old secondhand tube testers will not check all the new types of tubes. The socket contacts may be worn, the line-voltage calibration may be out of adjustment, the meter may be damaged, and there may be a host of other faults.

Avoid Bargain Sales. Always buy the latest model of the tube tester you choose, as old models get out of date faster. Buying one that has been on the shelf for a year or more is false economy, even though you get it at a reduced price. Your testers are the tools of your trade. Modern tools promote efficiency in operation and increased profits.

Guarantee on Tube Testers. A definite user's warranty and service guarantee are important factors in the purchase of a tube tester. Before you buy, study the manufacturer's printed literature to see just what you can expect in the way of a warranty. Study also the procedures provided for getting data on new tubes and for repair service. Availability of an authorized service station in your locality for a particular make of test equipment is a definite advantage, since even the best of equipment can have an accident or go bad unpredictably.

Most tube-tester manufacturers agree to provide tube-chart data on new radio receiving tubes for 5 years after the particular tube-tester model is discontinued from production. Sometimes a nominal charge is made for these data. The important thing is to be sure the data will be available, for otherwise new tubes would make your tester out of date.

The tube-tester manufacturer generally warrants to the purchaser that any part defective in material or workmanship within 90 days from the purchase of the new tester will be repaired or replaced free of charge.

Tube-tester Kits. A number of manufacturers sell tube testers in kit form, with instructions for putting them together and making the hundreds of soldered connections that are required. These are definitely not recommended for anyone going into the servicing business on a professional basis, for the following reasons:

1. It is too easy to make a mistake in wiring or make just one poor soldered connection out of the hundreds that are required. If this mistake is not discovered, you end up with a tube tester that gives erroneous readings. This means many hours of valuable time wasted looking for circuit troubles each time your tester passes a bad tube, and many dollars spent replacing good tubes that are called bad by your unreliable tester.

2. Even if you do succeed in putting the kit together properly, there is no assurance that your tube tester is correctly calibrated. You will not be able to rely on readings that are near the mid-point or ? region of the meter scale.

3. Your customers will generally have more confidence in a factory-assembled professional tester than in one that you put together from a do-it-yourself kit. If they see that you have cheap equipment, they may expect lower charges or argue about any charge.

11

How to Solder

Importance of Soldering. In general, a television or radio set is fixed by removing the defective part and connecting the new part in its place. Except for tubes and pilot lamps, this involves unsoldering wires from the old part and soldering wires for the new part. Learning how to solder is therefore a highly important part of your training.

Soldering involves melting a lead-tin mixture in such a way that it flows around and adheres to the two or more parts or wires being joined together. The purpose of soldering is to provide a good electrical path that will not loosen or deteriorate with age or rough handling.

It is easy to make good soldered joints. All you need are your regular tools, a good soldering gun for ordinary wiring, a pencil-type soldering iron for printed circuits, the right kind of solder, and the know-how. By following the few simple instructions given here and in the printed-circuit chapter, you can make smoothly rounded, glistening joints of which any serviceman would be proud.

Advantages of Soldering Guns. For money-making servicing, most servicemen today use a soldering gun, so called because it looks a little like a gun and has a trigger-type switch. An excellent soldering gun for service work is the dual-heat model shown in Fig. 1. This is rated at 90 watts for low heat and 125 watts for high heat. There is a built-in spotlight for illuminating the joint being soldered. Pulling the trigger switch to the first click turns on the spotlight and gives low heat for routine soldering of joints. Pulling the trigger all the way gives high heat for heavier work, such as for soldering a wire to the chassis.

The main advantage of a soldering gun is that it gets hot enough for use in only about 5 seconds after the trigger is pulled, whereas ordinary electric soldering irons take from 3 to 5 minutes. Such quick heating makes it unnecessary to keep a soldering gun on all day long just so it will be ready for use when needed. Another advantage of the gun type is the long narrow tip, which is ideal for work in a compact, crowded chassis.

A soldering gun is simply a power transformer mounted in a pistol-shaped plastic housing. The primary winding is connected to the a-c power

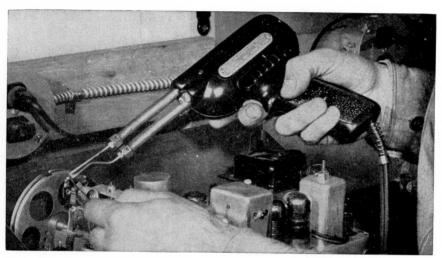

Fig. 1. Using a Weller soldering gun for unsoldering the leads of a defective part in a radio receiver. Pull on the lead with one hand while applying heat to its soldered joint with the soldering gun. Solder is held in left hand. Spotlight bulb on gun throws needed light directly on joint. (General Electric photo)

line through the trigger switch. The secondary winding has only a few turns of heavy wire, to send a large heating current through the loop of wire that forms the soldering tip. The voltage at the tip is less than 1 volt, so there is no danger of shock.

Using a Soldering Gun. To solder a joint with a soldering gun, pull the trigger and wait a few seconds for the gun to heat up. Now hold the tip of the gun against one side of the joint, and apply a dab of rosin-core solder between the tip and the joint to aid heat transfer. Next, apply solder to the other side of the joint until it melts and flows freely over the area of the joint. Take the tip off the joint, release the trigger, and the job is done.

Use the high-heat position only for soldering heavy pieces that require it. This high-heat position is for intermittent duty only. Continuous use on high heat will shorten the life of the tip and may overheat the transformer in the gun.

To unsolder a joint, apply the tip of the gun to the joint and pull the trigger. Apply rosin-core solder between the tip and the joint to aid heat transfer. When the solder on the joint melts, use long-nose pliers to pull the joint apart.

Care of Soldering Gun. The tip of a soldering gun should be kept clean and well tinned. It should be tinned the first time it is heated, by rubbing rosin-core solder over the entire tip. If the tip is not tinned, an oxide will form that is hard to remove and is a poor conductor of heat. The oxide also forms during use, making the tip look dirty and slowing up the transfer of heat to the work. When this occurs, scrape or sand the tip until shiny, then apply fresh rosin-core solder to retin the tip.

The tip of a soldering gun will eventually corrode. Replace the tip long before it actually breaks. A corroded tip takes longer to heat up and takes longer to transfer heat to the work. A new tip is inexpensive and can be installed in a few minutes. Simply loosen the nuts on the two studs and slide out the old tip. Place the nuts on the new tip, bend the ends of the new tip to fit in the holes in the studs, then tighten the nuts securely with a wrench to finish the job. Check these nuts for tightness occasionally, as loose nuts can cause improper heat or no heat.

When a tip takes too long to heat up but is not yet worn out, remove the tip and clean its leads with sandpaper, then replace.

Choice of Solder. There is only one correct solder for television and radio repair work—rosin-core solder. The smaller thicknesses, hardly larger than hookup wire, are most convenient for servicing.

Solder is made with several different ratios of lead to tin. The easiest to use and the fastest to cool is 60-40 solder (60 per cent tin, 40 per cent lead), but this is a little more expensive and not always available. Next best is 50-50 solder, which takes just a little longer to melt and cool. Even 40-60 solder is satisfactory, though much slower to melt and cool. Recommendation: Get 50-50 if you can.

Rosin flux is essential for cleaning off the thin film of metal oxide that forms on metals. Solder will not adhere to a dirty or oxidized surface.

Although faster-acting acid fluxes are available, they should never be used for servicing. The acid flux will stay on the joint and eat into the copper under the solder, causing corrosion that soon destroys the electrical connection at the joint. Play safe—use only rosin-core solder for electrical joints, regardless of what anyone may tell you. You want your soldered joints to stand up for a lifetime, not for just a few months.

When using solder to fasten metal parts together or to solder a bracket to the chassis during repair work, it is permissible to use a special noncorrosive paste soldering flux. This will clean dirty or oxidized metal surfaces so that solder can adhere, thus speeding up your work.

Sloppy soldered joints will be a discredit to your reputation among servicemen. Learn to make professional-quality joints right from the start. Practice making each type of joint as soon as you get the necessary tools and materials.

Unsoldering. The first use you will have for your soldering gun is unsoldering a joint so you can pull out the lead belonging to the defective part. To do this, simply hold the hot tip of the gun against the joint while pulling on the lead with long-nose pliers. If the joint heats too slowly, apply a small amount of solder between the tip and the joint to aid in the transfer of heat. It is just as important to use solder properly for unsoldering as for soldering.

As soon as the solder on a joint has softened, wiggle and pull the lead gently until it comes loose. Use long-nose pliers for bare leads, to avoid burning your fingers. If the joint is too tight, pry open the hook in the end of the lead with an ice pick before you pull out the wire. Never use force on a socket terminal lug, as these lugs often break off easily.

If the joint seems hopelessly tight, as many do, remove the soldering gun and wiggle the lead while the solder on the joint is cooling. Now you have a better chance to pry open the lead or cut off its end before repeating the unsoldering process.

As a last resort, cut off the old lead close to the joint with side-cutting pliers. If the loop of wire remaining is in the way of the new lead, you can easily remove it now with long-nose pliers while holding the gun against the joint.

During unsoldering, the hot solder will often flow down from the joint to a location where it may make trouble. For this reason, always remove surplus solder from the joint after unsoldering. You can easily pick up this solder with a freshly wiped soldering-gun tip. Shake or wipe the surplus solder from the hot tip and repeat the picking-off process as often as necessary to get all lumps of solder off. The cleaner the tip, the more solder it will pick up when doing this.

Unsoldering can often be made easier by removing the excess solder first, before wiggling and pulling on the lead. Practice unsoldering a dozen joints or so in an old set, and you will soon develop your own technique.

When unsoldering or soldering a joint deep down in a crowded chassis, a piece of asbestos paper about 4 inches square can be used to protect adjacent parts and insulation against being burned by the hot tip of the soldering gun.

Importance of Good Soldering. There is no more difficult trouble to locate in a set than a poorly soldered joint which is causing intermittent reception. The joint may look perfectly good yet have a bad contact underneath the solder. A poor joint can cause many other kinds of troubles also.

Too much solder can cause trouble by flowing to adjacent terminals or wires and producing short-circuits. When fixing a set, you certainly do not want to introduce new troubles, because they will be back eventually to plague you.

Preparation for Soldering. For good soldering, all parts of the joint should be clean. In most cases, the wires and terminals will be clean enough or will have been tinned (coated with solder) during manufacture so that cleaning is unnecessary.

Whenever any part being soldered is corroded or dirty, it should be scraped with a knife, filed, or sanded with a small piece of sandpaper until bright. This cleaning is essential for two reasons: so the hot solder will adhere to the metal and so there will be a good electrical path between the solder and the various parts of the joint.

Removing Insulation from Wire. Ordinary cotton or plastic insulation on hookup wire can almost always be split lengthwise by squeezing and crushing with pliers over the length of insulation to be removed. This can be done with the working jaws of long-nose pliers or ordinary gas pliers. Some side-cutting pliers have flat jaws for this purpose between the handles, spaced exactly far enough to squeeze the insulation but not the wire.

Squeezing the wire in a vise is another way of crushing the insulation. The strips of insulation can then be pulled off easily with pliers or fingers, and any loose threads can be cut off with side-cutting pliers, scissors, or a pocketknife. For an ordinary joint, it is usually sufficient to remove ½ to ¾ inch of insulation from the end of the wire.

Solid insulated hookup wire is easier to use than stranded wire. Both types come tinned, which means that the copper wire has been given a coating of solder during manufacture. Solid wire has one weakness, though; it is more likely to break after being bent back and forth a few times. This bending occurs during normal moving of wires when remov-

How to Solder **253**

ing and replacing parts in a crowded chassis. Stranded wire takes a little more time to connect but can withstand this rough handling. For your repair work, therefore, it is preferable to get stranded tinned wire. Either the No. 20 or No. 18 size (No. 18 is a little bigger) will do, with whatever type of insulation you prefer.

Cutting Off Insulation. When insulation cannot be removed by squeezing, use your pocketknife. This calls for a sharp knife and extreme care to avoid cutting too far. Even the slightest nick in the copper wire will weaken the wire enough to cause a break eventually at that point. Slice off the insulation, much as you would sharpen a pencil with a knife. It calls for an extremely sharp knife and a delicate sense of touch to avoid cutting the wire.

Cleaning Enameled Wire. A coating of hard-baked reddish-brown enamel is widely used as insulation on fine copper wire for coils and transformers. In the smaller sizes, this enameled wire is known as magnet wire because it was originally used in winding coils for electromagnets. The enamel should be removed from the end of such wire before soldering.

Scraping off enamel carefully with a sharp knife is the simplest and most convenient way. Be careful not to nick the copper wire, and be sure to get off all the enamel. If a small piece of fine sandpaper is handy, a faster and better job can be done by folding this over the wire so the grit bites into the enamel on opposite sides. Slide the sandpaper up and down the wire over and around the length to be cleaned while rotating the wire slowly.

Precautions for Plastic Wire. Plastic insulation in bright colors is widely used on wiring in new sets. This wire is thinner than cotton-insulated wire because the plastic coating has excellent insulating qualities. Most plastic insulation on wires has two weaknesses, however.

First of all, the plastic covering softens and melts away whenever touched accidentally by a hot soldering gun. The resulting damage to the plastic insulation may go unnoticed, and may not cause trouble until some time later when the wire has shifted so that the bare spot touches a bare terminal.

Second, if a plastic-covered wire presses against a sharp edge of a part or against a sharp corner of a terminal, the plastic will usually flow. The sharp corner may then eventually cut through the plastic and touch the wire inside. This can cause a short-circuit which is extremely difficult to locate because it does not show until the wire is moved.

When working with plastic-insulated wires, do your soldering quickly.

Do not hold the soldering gun on the joint any longer than necessary. Be careful not to touch any of the insulation with the gun. Excess heat at a joint can melt and weaken the plastic insulation for as much as an inch away from the joint.

Plastic-covered wire is sometimes called cellulose acetate hookup wire or thermoplastic hookup wire. It is excellent for connections if its two weaknesses are recognized.

Tinning Solid Wire. A wire should be tinned as soon as it is cleaned, before oxides form. Even factory-tinned solid or stranded hookup wire should be given a fresh coat of solder before making a joint.

To tin solid wire, support it over the edge of your workbench so you can hold the tip of the soldering gun under the wire with one hand and apply solder to the top of the wire with the other hand.

Another way is to unroll a few inches of solder from its spool and bend it to project over the edge of the bench. Now hold the wire under the solder with one hand. Bring the hot tip of the gun under the wire with the other hand, so the heated wire melts the solder.

A pair of pliers with a strong rubber band around the handles makes a handy miniature vise for holding a wire being tinned. It can also be used for holding a small resistor or capacitor while tinning its lead.

If the solder does not flow smoothly over the area to be tinned, scrape the wire some more. Rubber-covered wire will give the most trouble because the sulphur in the rubber causes a type of corrosion that has to be scraped off completely before solder will stick.

Tinning Stranded Wire. New tinned stranded wire is retinned the same way as solid wire. Old stranded wire usually gives trouble, however. You will have to clean the strands individually if they are enameled or dirty. Untwist the strands for about half an inch from the end and spread them out fanwise. Scrape them all at once with a knife or rub with sandpaper, first on one side of the fan and then on the other. Now untwist a bit more to rotate each strand, and scrape or sand again. Repeat until each individual strand is clean all around. Now tin the strands by heating them from one side with the soldering gun while applying solder to the other side. Shake off surplus solder while it is still molten, or tap the strands quickly with the heated soldering gun so that any molten solder between the strands drops out. After the wire has cooled, twist the strands together again. You can then proceed just as if working with solid tinned wire.

Untinned stranded copper wire is occasionally used in television and radio sets, especially for power-line cords. If the wire is new and fairly clean, it can usually be tinned without untwisting. If it is only slightly dirty, you can untwist and do a quick scraping or sanding job, then twist the cleaned wire together before tinning, to speed things up.

Burning Off Enamel on Wires. There is one easier way of cleaning enameled wire, if you have an alcohol burner handy. With stranded enameled wire, untwist the strands for about one and a half inches and spread them out so no wires touch. Hold the spread-out strands just below the tip of the inner cone of flame from the alcohol lamp until the wires are red-hot for almost an inch from the end. Now, immerse the heated wires quickly in a little can of alcohol. The enamel will dissolve immediately, leaving a bright and clean copper surface.

Enameled wire can be heated with a gas burner or any other hot flame just as well, then immersed in a pan of alcohol. Solid enameled wire can be cleaned the same way.

Tinning Soldering Lugs. Almost all the soldered joints made in servicing involve attaching a wire to a small, flat tab of metal called a soldering lug. Fortunately, these lugs are usually tinned during manufacture and do not require cleaning. The lugs are used extensively as terminals for tube sockets and other parts, as anchor points for wires on insulating strips, and for making connections to the metal chassis.

If a soldering lug is dirty, it must be cleaned by scraping, sanding, or filing, then tinned. When the lug is being tinned, the hole in the soldering lug usually fills up with solder. This solder can be lifted out with a freshly wiped soldering gun, so the wire can be looped through the hole.

Another way of getting a wire into a hole that has filled up with solder is to heat the lug until the solder melts, then poke the wire through. The wire is bent to shape after the solder has hardened, and the joint is then completed in a normal manner. A handy tool for reaming out solder-filled holes is an ice pick.

It is possible to make a soldered connection directly between a wire and the metal chassis, but this is not easy, and the resulting joint is usually messy in appearance and lacking in mechanical strength. Always try to make ground (chassis) connections to soldering lugs that are bolted, riveted, or otherwise connected to the chassis.

Joints for Soldering Lugs. The permanent and temporary joints used in connecting a wire to a soldering lug are shown in Fig. 2. Each has its own advantages for particular purposes. All four require that both the wire and the lug be cleaned and tinned.

Permanent Joints. The most widely used joint in service work is the permanent hook joint. To make it, the cleaned and tinned end of the wire is pushed into the hole in the soldering lug. The end of the wire is bent back to form a hook, the hook is squeezed together with long-nose pliers, and the joint is soldered by holding the soldering-gun tip against one side while applying solder to the other side. This provides a permanent soldered connection having much-needed mechanical strength as well as good electrical contact.

Another good permanent joint is the wrap-around joint. This is generally used on soldering lugs that have side notches instead of holes, but

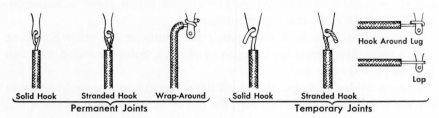

Fig. 2. Types of joints used for connecting a wire to a soldering lug

will work well on any lugs. A wrap-around joint is often the only way you can make a good connection to a terminal that already has a number of wires on it. The wire is wrapped tightly around the lug 1½ times with long-nose pliers, surplus wire is cut off, and the joint is soldered.

To unsolder a wrap-around joint, grasp the end of the wire with long-nose pliers and pull while holding the soldering-gun tip against the joint. With a little practice, you will find it easy to keep the solder molten with the soldering gun in one hand, while loosening the wire with long-nose pliers in the other hand. It takes practice to work with both hands in this way, just as it takes practice to play a piano with two hands.

Some prefer to push the end of an ice pick through a wrap-around joint and pry the wire apart while keeping the solder melted. On really tight joints, it is sometimes faster to cut the wire going to the defective part, and wrap the wire lead for the new part right over the old joint.

Temporary Joints. A temporary hook joint is used between a wire and a soldering lug when it may be necessary to remove the wire later, as in making test connections. The procedure is exactly the same as for making a permanent hook joint, except that the hook is not squeezed with pliers before soldering. The hook should be bent together just enough so the wire will not fall out of the hole in the lug, as shown in Fig. 2. Enough solder should be used to get a solid bond between the lug and the hooked wire, but the entire open hook need not be filled with solder.

With temporary joints, be especially careful not to move the wire until the solder has hardened. If you watch closely, you can actually see this hardening process. Molten solder has a shiny, mirror-like gray color, while hardened solder is duller and just a shade whiter in color. You will have to be a living statue for about 30 seconds after removing the gun from a temporary joint, because the slightest bump will move the wire and spoil the joint.

Whenever a joint is accidentally moved during the hardening process, reapply the soldering gun to melt the solder, then try again to hold your breath while it cools. Movement crystallizes the hardening solder, weakening it mechanically and spoiling the electrical connection.

Lap Joint. This is the most temporary of temporary joints. It is made by holding the straight end of the tinned wire against the soldering lug, holding the soldering-gun tip underneath the joint, and applying solder to the top of the joint. Sure, this requires three hands, which is why it is not used much. Actually, you can often bend the wire so that it will stay in the correct position against the soldering lug. You can then solder the joint conventionally with the gun in one hand and solder in the other. Sometimes you can hold both wire and solder in one hand.

Here is another way that often works, even though contrary to good soldering practice. Hold the wire in position on the terminal lug with one hand, press the heated soldering-gun tip over the solder so some will adhere to the lower face of the tip, then apply the soldering gun to the wire so that the solder transfers itself to the wire and the lug. Wire and lug both must be previously tinned, and the wire must be held steady while the solder is cooling. It helps to rest the hand against the chassis while holding the wire.

Dressing a Joint. When making a joint, the wire should be positioned so that the insulation is at least $\frac{1}{8}$ inch away from the lug, to avoid burning the insulation.

In all cases, the joint is dressed after being soldered, by arranging the wires neatly, straightening out kinks with the fingers, and snipping off any surplus wire projecting from the joint. When working with bare leads of parts, dressing also involves moving this bare lead away from other bare leads and away from adjacent terminals.

Good Joints. In a soldered joint, the solder should flow smoothly between the wire and the lug. Such joints are simple and easy to do once you

258 *Television and Radio Repairing*

learn them. The important thing is that your joint must be clean and hot enough to melt the solder itself.

Always apply a dab of solder between the gun tip and the joint at the start, so heat will be transferred faster to the joint. After this, apply solder only to the joint. Use just enough solder to get a smooth joint.

Bad Joints. One common mistake in soldering is applying solder directly to the gun tip and allowing it to flow off the tip onto the cold joint like wax dripping from a candle. This joint is messy, requires a lot more solder, and is not good electrically or mechanically. The joint can usually be

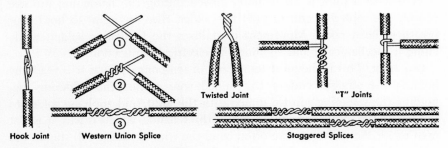

Fig. 3. Types of joints used for connecting two wires together

wiggled loose with your fingers, because hot solder does not stick well to a cold surface.

There are many servicemen in business today who make messy, puddled joints because they do not know any better. These men are continually in trouble with their customers because their repairs do not stand up. You have the edge on these servicemen just as soon as you have mastered the simple trick of making good soldered joints.

Soldering Two Wires Together. The diagrams in Fig. 3 illustrate the different types of joints that can be used for connecting one wire to another, as is required when there is no convenient soldering lug to serve as anchor for the wires. In all cases the wires must be previously cleaned and tinned. The soldering procedure is exactly the same as for soldering wires to lugs: hold the gun on one side of the joint and apply solder to the other side.

A hook joint is used where no mechanical strength is required. Here it is extremely important not to move the wires while the solder is hardening. Squeezing the hook together with pliers before soldering makes a semi-rigid and more or less permanent joint.

The Western Union splice is used where mechanical strength is required, as in antenna systems. To make it, clean each wire for at least 1½ inches from the end, grasp one wire in each hand and cross the bared ends, then twist each wire in turn around the other in opposite directions with your fingers. Keep the turns of wire far enough apart so solder can flow between the turns. You can either cut off the ends of the wire with side-cutting pliers or do the final bending with pliers, since the ends will invariably be too stiff to bend with fingers. Straighten the finished joint neatly with your fingers, then solder it.

The twisted joint is frequently used because it is easy to make in crowded locations, but has little mechanical strength. Simply twist the bared ends of the wires together with your fingers, solder the joint, then snip off the projecting ends with side-cutting pliers.

The T joints in Fig. 3 are used for connecting one wire to some point other than the end of another wire. In all cases, remove the insulation at the desired point by cutting or crushing, being careful not to nick the wire. The squeezed hook joint is of course more permanent than the open hook. The twisted-around T joint provides maximum permanence along with mechanical strength.

Insulating Wire Joints. Joints between wires should be taped. The best tape for service work is No. 33 Scotch electrical tape or an equivalent plastic tape. This tape should be wound tightly around the joint and well over adjacent insulation, so that there are at least two thicknesses of tape over all exposed wire. Where the joint is out in the open, work with the entire roll. When working in cramped quarters, however, tear off approximately the required length of tape. This tape has sufficiently good insulating qualities for use on line cords or other wires carrying dangerously high voltages.

To wrap the tape, start at one end of the joint. Hold the end of the tape in position against the insulation on the wire with one hand and stretch the tape while wrapping it around the joint with the other hand. Keep your fingers off the freshly exposed surface of the tape as much as possible, because fingerprints will reduce its sticking qualities. Overlap the tape at least half. When the entire joint has been covered with two thicknesses of tape, the job is done.

Always stagger the splices as in Fig. 3 when splicing line cords or other lines or cables having more than one wire. Then, even if the tape on the joints gets loose, the bare joints cannot touch each other and cause trouble.

Soldering Connections for New Parts. Although detailed instructions for installing and connecting new parts are given later in this book, there are a few general rules worth learning now. The actual joints are always made exactly as described in this chapter.

The secrets of fast and good soldering are summarized in Table 1. They are well worth extra study right now so you will never forget them.

Table 1. Rules for Good Soldering

1. Clean and tin the tip of the soldering gun or soldering iron frequently.
2. File, scrape, or sand the work if dirty.
3. Tin each part of joint separately first if untinned.
4. Make joint mechanically strong first whenever possible.
5. Use only rosin-core solder.
6. Be sure the soldering gun or iron is hot enough.
7. Apply solder between the heating tip and the work first, to speed up heat transfer.
8. Heat the work until it melts solder itself; then apply solder only to the work in completing the joint.
9. Be sure solder flows smoothly over entire joint.
10. Do not hold the soldering gun on the joint any longer than is necessary to get a good joint.
11. Do not use too much solder.
12. Do not move joint until solder hardens.
13. Avoid burning adjacent parts.
14. Do not spatter molten solder on other parts.

New transformers and electrolytic filter capacitors often have insulated leads, and these will usually be longer than are needed. Do not be afraid to shorten leads that are too long; cut them just long enough to leave a reasonable amount of slack for adjusting the position of the wire later, then remove insulation from the ends and proceed exactly as you would when connecting a plain wire to a soldering lug.

Resistors, paper capacitors, and mica capacitors will invariably have tinned bare copper leads, and these likewise will usually be too long. The lengths of the leads on the old parts can often serve as a guide for shortening the new leads. In general, however, hold the new part approximately in its final position to determine how long the leads need to be. Allow $\frac{1}{4}$ inch or so for bending back the hook joint, and allow extra length for safety.

If fairly long bare leads are required and there is danger of their touching other wires or terminals, slip a length of insulating tubing, called spaghetti, over each lead before making the joints. You can get this spaghetti in assorted sizes from any parts jobber or distributor. A bundle of assorted sizes costs little and will last a long time.

Avoid cutting the leads of paper capacitors too short, because then the heat from the soldering gun at the joint will travel along the wire and melt the wax in the capacitor. It is safer to have the leads too long rather than too short.

Do not make sharp-cornered bends in the leads with pliers, as this may weaken the wire. When bending leads with fingers, start at least ¼ inch away from the body of the resistor or capacitor.

Do not hold a soldering gun on a joint any longer than is necessary to get solder flowing smoothly over the joint, because heat may travel to delicate adjacent parts and cause damage. This is particularly important when working on printed circuits, transistors, crystal diodes, and crystal pickups for phonographs.

Wire-wrap Connections. In some receivers you will find solderless wire-wrap connections like that shown in Fig. 4. These are made with an auto-

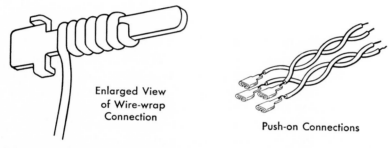

Fig. 4. Two types of solderless connections used in television and radio sets

matic tool that wraps the bare wire around the terminal post tightly at high speed. The wire actually fuses to the terminal metal at the sharp corners, giving a joint even better than a soldered connection.

A wire-wrap connection can be removed easily by unwinding the wire with long-nose pliers. Once the wire is removed, however, an ordinary soldered connection should be used for replacement. Wire-wrap connections cannot be made properly by hand.

Push-on Connections. Spring clips like those shown in Fig. 4 are often used on leads for loudspeakers, phonograph pickups, magnetic-tape recorder leads, and other parts that may be frequently disconnected. One clip is soldered or crimped on the exposed end of the wire, and pushed over a mating terminal lug or pin on the component.

A push-on connection can be removed and replaced with your fingers, as many times as desired. Never pull on the wire lead, however; grasp the clip itself.

12

Soldering and Repairing Printed Circuits

How Printed Circuits Are Made. Many television and radio sets have one or more printed circuits. These generally consist of a pattern of copper foil strips on one side of an insulating board. Holes are punched through the board to take the leads or terminals of the parts that are dropped into position from the unwired side of the board. Sometimes the projecting end of each lead is bent against the printed wiring after a part is installed. This gives a good mechanical joint before soldering.

In another form of printed circuit, the copper is plated on the board to form the desired wiring pattern. The copper may be plated on both sides of the board as well as inside the prepunched holes. This eliminates the need for the separate crossover wires that are required when wiring is on only one side of the board.

After parts have been installed, a printed-circuit board is lowered into a large pot of molten solder, wiring side down, to make all soldered connections at once.

How Printed Circuits Are Used. All the wiring for a radio set or phonograph can be placed on a single printed-circuit board that serves also as the chassis for the set. Line-cord, loudspeaker, and antenna connections are made to soldering lugs or terminals on the board.

Special plug-in terminals are often used on tube sockets, r-f and i-f transformers, and other multiterminal components used in printed circuits. These terminals are pushed through punched holes in the board, so they get soldered along with all the wire leads when the board is dipped.

Television sets may have as many as five separate printed-circuit boards.

These are usually bolted to a metal frame or chassis. Ordinary soldered connections, special connecting links, spring clips, or solderless wire-wrapped connections are used between the boards.

Replacing Tubes in Printed Circuits. Although tube sockets are mounted differently on printed-circuit boards than on an ordinary metal chassis, the tubes themselves are plugged in and removed in the usual way. The only thing you need to remember is that the laminated plastic board is far more brittle than a metal chassis. Never use force when pushing in or removing a tube.

When pulling out a tube that fits tightly in its socket, try to push down on the board near the socket with one hand while pulling on the tube with the other hand. Rock the tube gently, but not too far, while pulling it out. If the tube sticks, use a flat wood or plastic rod to pry it out.

When inserting a tube in a tight socket, rock it a small amount in all directions as you push it in. If you can conveniently get your hand on the opposite side of the printed-circuit board, support the board from that side with one hand while pushing in the tube.

Printed Circuits Are Easy to Fix. It is often easier to replace a bad component (part) in a printed circuit than on an ordinary chassis, because no untwisting of leads is necessary. On many jobs it is not even necessary to remove the chassis from the cabinet to test and replace a part.

Some manufacturers have made printed-circuit repairs still easier by stamping part numbers, values, and voltages right on the printed-circuit board. Sometimes the tube-socket terminals are also identified. In addition, service manuals are often printed in two shades of black or two colors to show the component on one side of the board and the printed wiring on the other side.

Service manuals often have numbers or letters to identify test points on printed-circuit boards and on circuit diagrams. This is particularly useful when measuring voltages.

Soldering on Printed Circuits. The copper foil of printed wiring is anchored to the laminated plastic board with a cement that cannot withstand much heat. Excess heat will loosen the foil from the board, after which it is easily broken. This is why a small soldering iron is safest for repair work on printed circuits.

Printed wiring is usually closely spaced. This means that special care is needed to avoid shorting adjacent wires with excessive solder.

Remember—heat, pressure, and excessive solder are the main enemies of printed circuits. If you recognize this and carefully follow the correct repair procedures given in this chapter, work on printed circuits will be just as easy as on individually wired circuits.

Choosing a Soldering Iron for Printed Circuits. The most popular soldering tool for printed circuits is a pencil-type soldering iron rated at about 40 watts. It can have a slotted tip, a wedge-shaped tip, or a variety of other tip shapes, as shown in Fig. 1. Usually the tips are interchangeable.

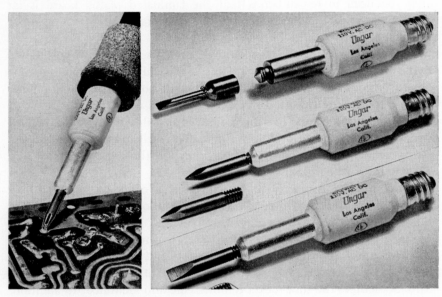

Fig. 1. Method of holding slotted tip of pencil-type soldering iron for unsoldering resistor lead on printed circuit, and examples of other tip shapes. (Ungar photos)

A soldering gun or a heavier soldering iron can be used on printed circuits if you recognize the danger of overheating the printed wiring. The less time you leave an iron on a printed circuit, the less is the risk that heat will flow through the printed wiring and loosen it from the board.

How to Use a Pencil-type Soldering Iron. With printed circuits, the tip of the iron should be up to solder-melting temperature and completely tinned before you touch the wiring board with it. Pencil-type irons heat more slowly than a soldering gun, so be patient. Allow several minutes for the iron to get hot, then test by tinning it with fresh 60-40 rosin-core solder.

The faster you can solder or unsolder a joint on a printed-circuit board, the better. Apply the heated iron to the board with firm pressure, wiggling or moving it slightly to break through the oxide film more rapidly and thereby improve heat transfer. As soon as the solder melts and flows, remove the iron. Do not leave the iron on the joint while brushing away solder, or while prying up a bent-over lead or terminal. Reheat the joint to complete the job, if you have to. These precautions are necessary to avoid loosening the adhesive that holds the thin copper-foil wiring on the board.

A heated soldering iron will usually melt through the protective coating that is sprayed over the wiring side of a printed-circuit board by the manufacturer. The only time you need to remove this coating first is when working on a damaged section of printed wiring. Wipe off the coating over the affected area with a cloth dipped in alcohol or some other solvent.

Surplus solder can be lifted away from a joint with an iron that has been freshly tinned, then wiped with a cloth. Wherever possible, take advantage of gravity when removing solder. On a vertically mounted board, apply the iron to the lower side of the joint so the molten solder can flow down to the iron by gravity. Even better results are obtained by working on the bottom of the board while it is flat. Here the board has to be supported high enough above the workbench so that you can see what you are doing.

Unsoldering a Clinched Lead. When a lead or terminal of a part is bent over on the wiring side, it must be straightened before the part can be removed. Apply a small soldering iron to the joint to melt the solder, then remove the iron and immediately pry up the bent-over lead with a knife or other tool. Repeat if necessary until the lead is straight. Do not overheat the connection.

Resoldering Joints. When resoldering a joint after installing a new part in a printed circuit, hold the heated iron against the lead or terminal of the component until it is hot enough to melt solder. Apply a small dab of 60-40 rosin-core solder to the terminal, then move the iron down quickly to the printed wiring. Hold the iron there only long enough for the solder to bridge smoothly from the terminal to the printed wiring.

Care of Pencil-type Soldering Tips. The tips of soldering irons corrode rapidly when heated for long periods of time. It is best to plug in the iron only when needed for a particular job. If you leave a pencil-type soldering iron plugged in all day long, you will usually have to clean and retin it at regular intervals. Tinning protects the tip of the iron against corrosion due to heat, and also aids in the transfer of heat from the tip to the joint being soldered.

Cleaning the Tip of an Iron. Cleaning is most easily done while an iron is hot. Wiping the hot copper tip with a cloth or a wad of steel wool will usually give a bright, clean surface.

If marks of corrosion remain after wiping, rest the hot soldering iron against a vise and clean one face of the tip until smooth. Use fine sandpaper first. Just loop a strip of sandpaper over a flat stick and use like a file.

Only if there are actual holes in the copper should you use a file for cleaning, because each filing removes some copper from the tip. Excessive filing will shorten the tip so much that it will have to be replaced. New tips cost only a few cents, however, so use a file when in doubt; it is far better to file your iron clean than to struggle with sandpaper.

When filing the tip of your soldering iron, hold the file flat against the entire face so as not to change the angle of the tip. Too blunt a tip is awkward to use in crowded locations, and blocks your view of the joint while soldering. A tip that is too slim, on the other hand, cools too rapidly at the sharp point and therefore cannot heat the work adequately.

After filing or sanding the tip, apply rosin-core solder immediately to the cleaned face. Wipe off surplus solder with a cloth, then inspect to make sure there is a coating of solder over the entire face. If spots show, repeat the filing and tinning. Clean and tin each other face in turn.

Precautions for Plated Tips. Some tips for pencil-type irons have a special iron plating over a copper core to minimize corrosion. When new, the iron will have a silver plating to prevent rusting until used. Plated tips should be retinned each time they are heated. Flow solder over the entire tip, then wipe it clean with a cloth. Filing a plated tip will of course destroy the plating, so do not do this until the tip can no longer be cleaned by wiping. Once you have filed a plated tip, treat it the same as a plain copper tip.

Solid iron tips also come with a rust-resisting silver coating. These should be tinned immediately after heating each time, the same as for iron-plated tips. Wipe and retin all iron tips before putting them away for the night.

When the coating of solder has burned off a solid iron tip, let the iron cool to room temperature, then file the tip down to fresh iron. Reheat the iron and retin, using solder paste or sal ammoniac. After tinning, flow fresh rosin-core solder over the entire surface to remove all traces of the paste flux and complete the tinning. An iron tip cannot be tinned directly with rosin-core solder.

Special Soldering-iron Tips. A number of special shapes of tips are available for use with pencil-type soldering irons. The pyramid tip has four flat faces coming to a point. The chisel tip has two flat faces coming to an edge. The choice between these two tips is largely one of personal preference. Try the pyramid tip first.

Special shapes of tips are available for heating two or more terminals at a time. Circular and triangular tips are used on terminals arranged in a circle, as for a tube socket or i-f transformer. A rectangular bar tip is used to heat two or more terminals in a row, as for resistor-capacitor units.

All special tips should be brought up to temperature and freshly tinned before holding against the terminals. Always have solder ready to apply immediately as a small dab between the tip and each terminal, to bridge the gap and improve heat conduction. This is necessary because terminals are rarely in a perfect straight line or circle.

Removing and Replacing Tips. Pencil-type soldering irons are made in two ways. On some, the tip and heating element form a single unit that is screwed into the handle of the iron. In others, the heating element screws into the handle and has removable tiplets that screw onto it in turn. You may also encounter a variation of this, in which the heating element is a permanent part of the iron and only the tiplet is removable.

Screw-in tips should have their threads thoroughly greased with a special high-temperature lubricant such as Led-Plate, to prevent freezing of the threads due to formation of oxide. If your parts jobber does not handle this lubricant, ask him where it can be obtained.

The tips of pencil-type soldering irons should be removed and replaced only while the iron is hot. Use two pairs of pliers to do this, as shown in Fig. 2. Hold the soldering-iron barrel with one pair while rotating the tip with the other pair. Never grip the ceramic portion of the heating element with pliers.

Heating elements can be unscrewed or inserted while the iron is cold. Do not lubricate the brass threads that make the electric connection to the heating element.

Solder Pots. Small solder pots are available for unsoldering the terminals of a defective tube socket, transformer, or other multiterminal component all at once.

To use a solder pot, immerse the terminals of the component in the molten solder. The board can rest on the rim of the solder pot. Pull gently on the component with pliers or your fingers, as shown in Fig. 3, as soon

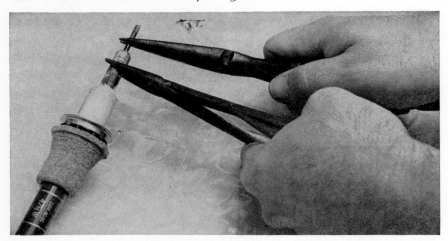

Fig. 2. Method of using two pairs of pliers to remove tip from pencil-type soldering iron. (Ungar photo)

Fig. 3. Using special solder pot to unsolder all terminals of multiterminal resistor-capacitor unit simultaneously. (Motorola photo)

Printed Circuits

as the solder has melted around the terminals. The component should then come out easily.

A board can be safely immersed for about 10 seconds. If the part cannot be pulled loose in that time, take the board out of the pot and look for clinched leads or bent tabs that hold the part even when the solder is molten.

Solder pots should be used only for unsoldering. Solder the new leads or terminals conventionally one at a time.

Be careful when using a solder pot. Molten solder may look pretty, but it can cause serious burns. Never let water fall into molten solder. Even a single drop can cause a minor explosion that will scatter molten lead over your bench and perhaps on your face.

Scum should be scraped off the surface of molten solder from time to time. Fill the pot with 60-40 bar solder, and add more solder as needed to keep the pot full to the brim. A full pot is needed so the maximum length of the terminals is in the solder.

Add a 1-inch length of rosin-core solder to a solder pot occasionally to get in some flux that will make the solder flow more freely.

Handy Tools for Printed Circuits. A thin-blade pocketknife can be used for prying up leads that have been bent over against printed wiring. Use the knife also for scraping printed wiring to clean it before soldering.

A stiff-bristled toothbrush can be used to brush away excess molten solder. The small wire brushes made for suede shoes are even better for this purpose.

An ordinary broom straw or a wooden toothpick can be used to push molten solder out of a hole after removing a lead, in preparation for inserting a new lead. A more permanent tool for poking out solder is an ice pick, an awl, or a small plated crochet hook from which the hook has been cut off. Some servicemen prefer to drill out the cold solder after removing the old lead.

Most jobbers sell inexpensive soldering aids for work on printed circuits. These have a wood handle, with an awl-shaped point at one end for poking solder out of holes, and a slot at the other end for prying up bent-over leads and terminals. Two methods of using this soldering aid are shown in Fig. 4. When using a soldering aid on a board that is removed from its receiver, it is a good idea to support the board on blocks or cigar boxes. This prevents damage to components, connections, and the delicate copper foil.

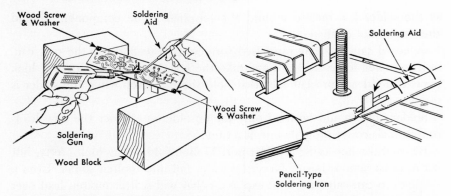

Fig. 4. Method of supporting printed-circuit board between blocks of wood while replacing a multiterminal television sound i-f transformer. Two-prong soldering aid is used to pry up a bent-over terminal while the solder is molten. Do not apply pressure to the board when it is supported by the corners in this way. If using a soldering gun, the lower-heat position of the switch is safest

Supplies for Printed-circuit Repairs. Always use 60-40 rosin-core solder on printed circuits, to get the lowest possible melting point. The thinnest size, which is about the thickness of the lead in a pencil, is preferred.

A small bottle of lacquer thinner or denatured alcohol is sometimes needed for removing the protective coating of special lacquer, varnish, or silicone resin on the portion of a printed circuit being repaired.

A jar of clear lacquer or silicone resin, having a small brush attached inside the cover, should be kept on hand for restoring the protective coating on repaired areas. A can of Krylon spray can be used instead, provided care is taken to keep the spray out of tube-socket holes and away from mounting holes having ground connections. Masking tape is handy for covering such areas temporarily. The protective coating serves to prevent dust and moisture from causing short-circuits between closely spaced wires on the board.

Check Your Work. After doing soldering work on a printed circuit, examine the board carefully for loose pieces of solder. These are often scattered over the entire board when molten solder is brushed from a joint being soldered. Small particles of solder embedded in the lacquer or silicone-resin coating can be easily wiped off with a clean cloth dipped in lacquer solvent or denatured alcohol.

Removing and Replacing Printed-circuit Boards. Most printed-circuit boards are fastened to the chassis or cabinet with self-tapping screws. If it is necessary to remove the board, remove these screws first and then study the connections that run to the board. If leads go to the loudspeaker, a control panel, a pilot-lamp socket, or some other separately mounted part, it is best to unmount these parts and remove them with the board. This minimizes cutting or unsoldering of leads. Use insulating or masking tape to prevent these loose parts from touching the printed wiring.

When leads run to another circuit board and are soldered directly to printed wiring on both boards, it is usually best to cut the leads at about their mid-points. This leaves colored insulation to aid in matching leads correctly when splicing them again after completing the repair work. Be sure to solder and tape each splice.

When interconnecting leads go to heavy standoff terminals, you can remove the lead from one terminal and resolder it later. Be sure to make a connection diagram when removing two or more similar leads.

If wire-wrap joints are used on terminals, simply unwind the wire from the terminal. Always solder the lead when replacing it, however; it is impossible to rewrap a lead properly by hand.

Be sure to replace all mounting screws when putting back a printed-circuit board. Some of these screws make contact with the printed wiring to provide ground connections to the chassis, so omission of a screw may cause trouble. Do not tighten mounting screws too much, because they can crack the printed-circuit board.

In table radios, the printed-circuit board often slides into molded grooves in the plastic cabinet and is held in position by the back cover. This simplifies removal and replacement of the board.

Locating Bad Printed-circuit Joints. Poor connections can occur just as often in printed circuits as in hand-soldered joints. The causes are the same —a dirty wire or terminal, a trapped bubble of air in the solder, or unmelted flux. The flux is the commonest cause, because a film of rosin flux provides an insulating gap under the solder. The result is a joint which may look perfect but is not.

A badly soldered joint usually is intermittent. The set works for a while, then goes bad. Sometimes jarring the set will make it work again. There is thus a good technical basis for the success of many housewives in making a set work by "giving it a kick."

When a customer gets tired of kicking his set, the serviceman is called. But finding this invisible bad joint can be a real problem. Here is how you start the hunt.

Remove the back cover of the set, apply power with a cheater cord, and tune in a station. When the trouble develops, take an insulating rod and apply pressure first to one end of the printed-circuit board, then the other. An old toothbrush handle is fine for this job. Push on terminals, sockets, and large mounted parts. Keep poking, easing up on the pressure as you get closer and closer to the bad joint, until you have isolated the trouble to a small region on the board. Sometimes the trouble can be located more quickly by making up a shorting lead having needle-point test prods. Use it to bridge each joint that might be bad, while listening to the set.

Resoldering Joints. When you think you have isolated the bad joint, turn off the set and resolder each joint in that area. Hold your pencil-type iron on each joint long enough to make the solder reflow.

If the first attempt did not bring success, probe some more to isolate another suspected area, then resolder the joints there.

Often a lot of time can be saved by resoldering all the joints on a printed-circuit board when a bad joint is suspected. Wipe the tip of your iron and retin frequently when doing this, to make sure the new joints are perfect. If the trouble still exists after resoldering joints, you are pretty sure it is in one of the components.

Locating Breaks in Printed Wiring. Since the copper strips of printed wiring are no thicker than metal foil, they can easily break if the board is bent too much when inserting a tube. The crack closes when pressure is released, leaving an open circuit that is almost invisible.

A magnifying glass may help in locating breaks in printed wiring. Pressing the board with an insulating rod at various points may help to isolate the region containing the break. If the wiring side of the board can be reached while the set is operating, try dragging a sharp-pointed tool lightly along the length of each suspected wire while power is on. Hold the tool only by its insulated handle. At the instant when the pointed tool bridges the tiny break, the trouble will clear up. A jumper wire with needle-point prods is even better for this purpose. Hold the prods at opposite ends of a suspected wire, to see if this clears up the trouble.

If the trouble can be isolated to a small number of printed wires, all of which are coated with solder, you can run a small soldering iron along the length of each wire to reflow the solder. This will usually bridge the break. Pull out the line-cord plug for the set first. If the wiring is covered with varnish, remove the coating first with a solvent and brush. After making the repair, recoat the entire board with brushed-on lacquer or a protective spray coat.

Repairing Breaks in Printed Wiring. Methods of repairing broken printed wiring are shown in Fig. 5. Small breaks can be bridged with molten solder. Remove the protective coating from the region of the break first with solvent. If the printed wiring is dirty at the break, scrape it lightly for about $\frac{1}{4}$ inch on each side of the break. Hold the soldering iron so it heats both ends of the break, then immediately apply solder. This must be done fast enough so the ends of the foil do not loosen from the board.

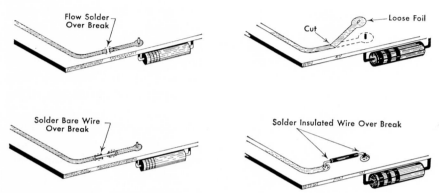

Fig. 5. Methods of repairing breaks in printed wiring

A stronger repair can be made by placing a $\frac{1}{2}$-inch length of tinned copper wire across the break. Heat the wire first until solder melts on it, drop the wire over the break, then apply the soldering iron and solder to the wire to anchor it to the printed wiring.

Repairing Loose Printed Wiring. A loose section of foil can be cemented back in place with ordinary service cement if not broken. If broken, it is best to peel off the loose section and bridge the missing foil with ordinary wire.

If a long run of printed wiring breaks off, solder a length of insulated hookup wire in its place. Lap joints can be made to the ends of the good wiring after scraping the wiring clean for about $\frac{1}{4}$ inch from each end. Use service cement to anchor the hookup wire in the exact position of the old wiring.

Repairing an Arc-over between Printed Wiring. Occasionally the insulating board of a printed circuit becomes carbonized between adjacent wires because of heat or a spark. If the burned area is small, the damage can be repaired by drilling a hole to remove the carbonized path. Apply lacquer to the hole with a toothpick to seal the edges of the hole and prevent absorption of moisture.

Locating Bad Parts in Printed Circuits. Use your eyes first. Examine each part carefully for signs of overheating or mechanical damage. Prod each part in turn with a plastic rod while power is on, to see if a broken lead inside a part is the cause of the trouble.

Whenever possible, do your work on printed circuits without removing the board from its chassis or cabinet. This is particularly important for troubleshooting, because the mounting bolts for the board generally serve also as ground connections for the printed circuit.

If you do have to remove the board, use jumper wires to restore these ground connections when making voltage or resistance measurements on the board. There will also be other connections that may need extending if you want to operate the set with the printed-circuit board out of the cabinet.

Making Measurements in Printed Circuits. Ohmmeter measurements of circuits containing two or more parts are generally made from the component side of the board. Voltage measurements can be made from either side of the board. Of course, the service manual for the set is needed to interpret all measurements properly. A 25-watt lamp behind a board makes wiring visible through the board, to speed identification of parts.

When working on the wiring side, needle-point prods are needed to pierce the protective lacquer or silicone coating. If your multimeter does not have these, get an extra pair of test leads from your jobber. Ask for prods that have chucks in which old-fashioned steel phonograph needles can be inserted. Apply the prods at soldered points when checking components, to avoid the risk of having a prod slip and tear foil wiring.

You will find that tube sockets are mounted on either side of the board in printed circuits. Remember that socket terminals are numbered in clockwise sequence when viewed from the bottom of the tube or socket. When working from the top of the socket, then, you will have to count tube pin numbers *counterclockwise* from the index space or key.

How to Use Adapters in Tube Sockets. Adapters are handy for measuring socket voltages from the tube side of a printed-circuit board while the tube is still in the socket. Adapters are available from jobbers for each type of socket found in radio and television sets. To use one, remove the tube from its socket, plug the adapter into the socket, then plug the tube into the adapter. The adapter has one exposed terminal for each socket contact, on which a test prod can easily be held or clipped.

Locating Breaks in Wiring with a Multimeter. When looking for an intermittent break in printed wiring, unplug the line cord or remove batteries, then place the needle-point prods of the ohmmeter at opposite ends of a suspected circuit. Hold the handles of the prod with one hand and apply gentle pressure to the board at various points with the other hand, while watching the meter. If the pointer flickers, look for a break in that piece of wiring. A magnifying glass will often help. If you cannot see the actual break, move the points closer to each other and repeat the test until you isolate the break in the wiring.

Testing Parts in Printed Circuits. It is usually more convenient to make resistance measurements on parts by working from the side on which the parts are located. Most parts have leads exposed enough to reach with test prods.

Use the same precautions in printed circuits as you would ordinarily. Always pull out the line-cord plug on a-c sets. Remove or disconnect batteries on portable sets when making ohmmeter tests. When working with printed circuits having transistors, follow the precautions given in the transistor and multimeter chapters, to avoid burning out a transistor when measuring the resistance of a nearby part.

When leads or terminals cannot be reached for testing on the component side of the wiring board, measure between the corresponding terminals on the wiring side of the board. To locate a terminal, bend a 10-inch length of solid copper wire for use as calipers to transfer distances. The distances between a terminal and two adjacent edges of the board locate its position.

Unsoldering a Resistor or Capacitor for Test. There are times when you will need to unsolder one lead of a resistor, capacitor, or other double-ended part in a printed circuit in order to check its condition positively with an ohmmeter. This presents no particular problem if you remember just two things: (1) you will get into trouble if you use force; (2) you will get into trouble if you use too much heat.

If the lead to be unsoldered is clinched, straighten it first. Now grasp the body of the part with one hand to pull it gently away from the wiring board, and hold a small soldering iron against the joint on the other side of the board. Remove the iron as soon as you feel the lead pulling free of the joint. You do not have to pull the lead entirely out of the hole in the board; just be sure it is free of the soldered joint.

If an ohmmeter test shows that the unsoldered part is good, push the part gently back against the board while holding the soldering iron against the joint to melt the solder. When you feel that the lead is going back into the blob of solder at that joint, remove the soldering iron and let the solder cool. If the joint now appears to need more solder, heat it again and add a small amount. You have thus made a test without weakening the leads by bending them excessively back and forth.

Removing a Bad Resistor or Capacitor. Resistors, capacitors, and other parts that have leads at opposite ends can often be replaced without re-

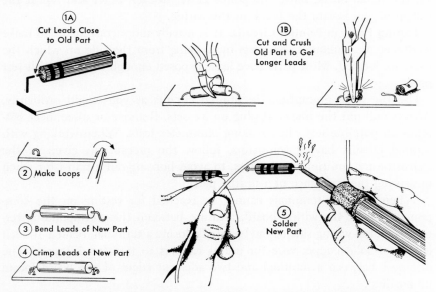

Fig. 6. Method of replacing axial-lead components on printed-circuit boards. Solder the new component to the projecting stubs of the old leads

moving the printed-circuit board from the cabinet. After verifying that the part is bad, clip its leads as close to the body of the part as possible. With resistors you can get extra lead length by crushing the resistor body with pliers as shown in Fig. 6. The leads of the new part can then be soldered to the projecting stubs of the old leads.

Installing New Resistors or Capacitors. Scrape each projecting wire of the old part until clean, then bend it into a hook or loop. Run the leads of the new part through the loop, cut off surplus wire, crimp to get a strong mechanical joint, and solder. Be careful not to overheat, because this may loosen the joint between the old lead and the printed wiring on the other side of the board.

After soldering a new part to projecting leads, always check the other side of the board for short-circuits. Overheating here can make the solder run out of the hole and spread over the wiring side of the board.

A new resistor or capacitor may be installed in the same holes as the old part, if you are careful not to overheat the printed wiring. After pulling the leads of the old part out of the holes in the board, apply the soldering iron momentarily to each joint and poke surplus solder out of the hole. Now bend the leads of the new part to match the holes, cut off surplus lead length, and insert the new part in the holes. Push the part down against the board, solder its leads on the other side of the board, then cut off any remaining surplus lead length.

There is no need to clinch the leads of a new part against the wiring. Clinching is done by manufacturers to prevent parts from falling out while the board is being handled on assembly lines prior to dip-soldering.

Some paper capacitors for printed circuits have both leads at one end. An ordinary capacitor of the correct size can be used as a replacement if you bend one lead back along the body and tape it to the body.

Unsoldering Multiterminal Parts for Test. Transformers, volume controls, some types of capacitors, and combination resistor-capacitor units have all their leads projecting from one end. Once these parts are soldered in place on a board, it is generally impossible to unsolder one lead at a time for test purposes. The entire part must therefore be removed. Before doing this, however, make a simple diagram showing how the part is inserted in its holes in the board, so you can replace the old part or put in a new part with the correct polarity and position. If the leads are bent over against the printed wiring, they must be straightened first.

With a special soldering-iron tip or a small solder pot, you can heat all the leads or terminals of the part simultaneously. Pull out the part as soon as the solder has melted at all the joints. Two types of special tips are shown in Fig. 7.

With an ordinary soldering iron, first remove as much solder as possible from each joint by lifting off the solder with a freshly tinned and wiped iron, or by brushing it off while molten. Now alternately heat each lead in turn while pulling or rocking the part from the other side of the board, until all leads are free of the board.

Removing Bad Multiterminal Parts. If a multiterminal part is bad and will be thrown away, cut off the leads or terminals flush with the board first to speed up the removal. Better yet, try to cut the leads on the component side of the board or break up and cut away the part itself. You can then remove the terminals one by one.

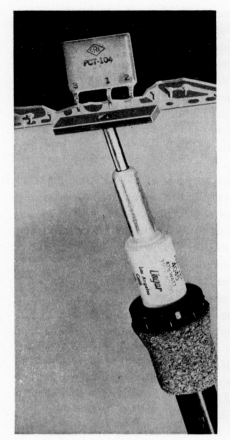

Fig. 7. Examples of rectangular and circular tips used on pencil-type soldering iron for removing multiterminal components from printed-circuit board without damage. Rectangular tip is used when terminals are in straight line. Circular tip is used for tube sockets and other circular arrangements of terminals. (Ungar photos)

Removing Modules. A few sets use individual preassembled stages called modules. These consist of stacks of small ceramic printed-circuit plates connected together by vertical wires, usually with the tube socket on the top plate. Some of the vertical wires are longer, for insertion in a printed-circuit board.

Completed modules are usually coated with a protective plastic insulating material which hides the parts in the module and makes repairs difficult. For this reason, the entire module must usually be replaced when one part in it goes bad. Replacement modules are expensive and hard to obtain, so eliminate all other possible causes of trouble first. Modules are removed the same way as other multiterminal parts.

Some of the parts in a module can be tested by measuring between the leads of the module with an ohmmeter. When the defective part is found, careful study of the circuit and the module will tell whether it is possible to isolate only the bad part in the module and connect an ordinary replacement part in its place. This part can be located outside the module and connected to the terminals or riser wires of the module.

Installing New Multiterminal Parts. After removing the terminals of the bad part and cleaning out the holes, try inserting the new part. Be sure you position it exactly the same as the old part. If the part goes into the holes easily, you are ready to solder it. Check terminals carefully when putting in a new i-f transformer, because they may be different even though the new transformer looks the same as the old one.

If the holes are too small for the new part, enlarge them first with a drill or other tool. Do not use force to push a new part into its holes, as there is too much risk of cracking the printed-circuit board.

With the new part in position, solder each of its terminals in turn to the printed wiring. Use a small soldering iron and only a small amount of solder. It is not necessary to twist or bend over any of the mounting lugs or terminals.

Replacing Volume Controls. Volume controls that are mounted directly on printed-wiring boards generally use spear-type terminal and mounting lugs like those shown in Fig. 8. These go through the board and are dip-

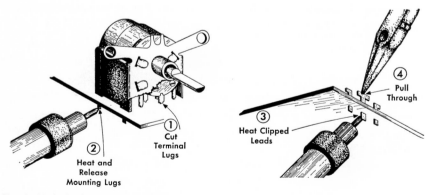

Fig. 8. Method of removing defective volume control from printed-circuit board

soldered to the wiring along with all the other leads and terminals. When either the control or the switch goes bad, the entire unit must be replaced. Cut each lead of a bad volume control above the board to remove the part, unsolder and remove the terminal lugs individually, clean out the holes, then install and solder the new part.

Unsoldering Transistors in Printed Circuits. In most sets, transistor leads are soldered directly to the printed wiring.

Modern transistors last so long that there is no longer a need to make them easily removable. A transistor should not be removed except as a last resort when definitely proved bad, because it is far more likely that some other part is causing the trouble.

Heat is one big enemy of a transistor. Never hold a soldering iron on or near a transistor lead any longer than necessary. When you do have to unsolder or resolder a transistor lead, always grip the lead with long-nose pliers somewhere between the soldering iron and the transistor body. The pliers will absorb any heat coming up the wire, and thus keep it out of the transistor.

Repairing Tube Sockets. Poor or intermittent connections between tube pins and tube-socket contacts can occur in printed circuits just as in ordinary sets. Bad sockets are hard to replace, so always try to fix the socket first.

A loose contact can generally be tightened by inserting an awl or other sharp-pointed tool between the contact and the socket hole as in Fig. 9, to bend the contacts closer together. Do this from the top of the socket after removing the tube.

If only one terminal of a socket is bad, unsolder and remove only that one terminal from the socket, as also shown in Fig. 9. You can then slip in a good terminal taken from a spare socket.

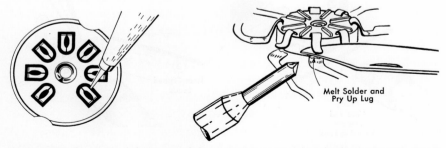

Fig. 9. Repairing loose socket contact with awl, and method of unsoldering defective socket contacts

Printed Circuits **281**

Fig. 10. Example of grounding foil that loosened during removal of a tube socket. Repair is made by cutting away the loose foil and replacing it with short lengths of hookup wire. (Admiral photos)

Replacing Tube Sockets. In those rare cases where a tube socket is bad and cannot be repaired, the bad socket must be removed from the printed-wiring board and a new one is installed. First of all, make a sketch showing the exact position of the index key or space on the old socket, so you can install the new socket correctly.

With wafer-type sockets, break up the socket with side-cutting pliers so you can unsolder and remove the contact terminals one by one. Brush away surplus solder. If any holes are closed with solder, melt and poke it out. Install the new socket. If some holes are too small, enlarge them with a drill until the terminals of the new socket go into the holes easily. Soldering the terminals then completes the job.

With solid molded plastic sockets, heat each socket lug in turn and brush away the solder, then reheat and pry the lug up away from the wiring foil with a knife. After all lugs are free on the wiring side of the board, apply the soldering iron to the grounding lug on the component side, then grasp the socket and pull it away from the board.

Carefully inspect the wiring in the socket location before installing a new socket. Cut out any films of solder that have bridged between the pieces of foil. If the grounding-lug foil is broken, restore the connections with hookup wire, as in Fig. 10. The new molded plastic socket is pushed into position with correct orientation while holding the back of the board. If too tight, enlarge the socket hole first. Each lug in turn is then soldered back to the printed wiring, after which solder is applied to the grounding lug on the other side of the board.

Fortunately, sockets do not fail any more often in printed circuits than in ordinary sets. A full-time serviceman may not get more than one socket-replacement job in a year.

Repairing a Cracked Board. When a crack occurs in a printed-circuit board as well as in the wiring, the repair is easy if the board is not actually broken in two. Install staples across the breaks in the printed wiring, then solder the staples to the wiring on one side and solder the ends of each staple together on the other side. Do not try to do this with a stapling machine, however. Drill a tiny hole through both the wiring and the board

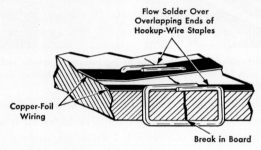

Fig. 11. Method of repairing crack in printed-circuit board

about ¼ inch away from each side of a break. Clean the wiring by scraping carefully or applying a solvent.

Now take a 1-inch piece of bare hookup wire and bend it into a U having ¼-inch legs. Insert this home-made staple in a pair of holes from the wiring side of the board. Bend the projecting ends together on the other side as in Fig. 11 and solder them, then flow solder between the staple and wiring. Repeat for each other broken wire on the board. Be sure that no solder flows over to adjacent wiring and causes a short-circuit. Apply a coat of protective lacquer to the repaired area to complete the job.

Repairing a Broken Board. If a corner breaks off a board, as in the vicinity of a mounting hole, only a few wires are affected. Here you may be able to cement the broken pieces together and bridge the broken wires with staples. Insert half the staples from each side of the board. Use additional staples in areas having no wiring, to give mechanical strength to the repair.

Drill the holes for the staples first. Apply cement to both edges of the broken board, insert the staples and clinch their ends with a screwdriver, then solder the staples and apply lacquer to complete the repair.

Speaker cement can be used if you hold the broken pieces together for a couple of minutes until the cement sets, but the staples will have to be put in afterward. Here there is great risk of breaking the cemented joint when clinching the staples. It is better to use an epoxy resin adhesive that comes in two tubes and is mixed just before use. This will build a strong bond if you allow it a full 24 hours to set. The staples can be inserted and clinched after applying the adhesive. Soldering can be done either before or after the bond has set.

Replacing a Broken Board. A full-time serviceman will get a badly broken printed-circuit board only about once in 2 years. This is handled by obtaining a new empty board from the distributor and transferring parts one by one from the broken board to the new board. Solder each part in position on the board, or at least bend the leads so the part cannot fall out. Work slowly to avoid mistakes.

13

Testing and Replacing Transistors and Crystal Diodes

What Transistors Do. A transistor is a tiny device that does the work of a vacuum tube yet requires no vacuum and no filament current. Signal amplification is obtained by making three connections in a tiny pellet of germanium or silicon called a semiconductor. With two connections to the same pellet we get a crystal diode that does the work of a vacuum-tube diode and likewise requires no filament current.

A transistor will operate from voltages as low as that of a single flashlight cell. It can equal the performance of tubes yet lasts much longer and requires only a fraction of the space and weight. This is why practically all portable radios today use only transistors and crystal diodes. Many auto radios, remote-control systems, home audio systems, portable television sets, and other types of equipment are likewise using these parts in some or all stages.

Although the theory of transistor operation by means of electron flow in one direction and holes (equivalent to positive charges) in the other direction is extremely complicated, it has nothing to do with servicing problems. Just as with tubes, you only need to know how to find the bad transistor, how to order the correct new type, and how to put it in the set.

Transistor Electrodes. The transistors that you encounter in servicing are technically known as junction transistors. They will have three terminals or leads, each going to one of the transistor electrodes. These electrodes are known as the base, the emitter, and the collector. When the terminals are identified on a transistor, the base will be marked B, the emitter E, and the collector C.

Transistors and Crystal Diodes 285

Transistor Symbols. Transistors have entirely different symbols from tubes on circuit diagrams. The symbols that you will see most often are shown in Fig. 1. The base is always a line at right angles to its connecting lead, exactly like the anode symbol for a vacuum tube. The emitter is represented by a line that meets the base at an angle and has an arrowhead on it. The collector is represented by a line that intersects the base at a similar but opposite angle and has no arrowhead.

When the emitter arrow points away from the base, it identifies an n-p-n

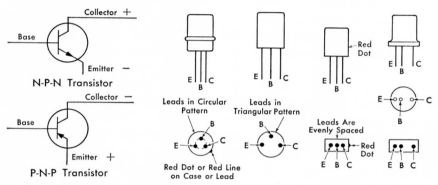

Fig. 1. Transistor symbols and base connections. When leads are equally spaced, a red dot or other mark will identify the collector

transistor. Here the collector requires a positive voltage with respect to the base.

When the emitter arrow points toward the base, it identifies a p-n-p transistor, in which the collector requires a negative voltage with respect to the base. The p-n-p transistor is by far the commonest in transistor radios, chiefly because it is easier to build and lower in cost.

The only time you need to know whether a transistor is n-p-n or p-n-p is when you use a transistor tester.

Transistor Base Arrangements. Typical arrangements of the three leads on the bases of common types of transistors are also shown in Fig. 1. These diagrams allow you to identify the leads of the transistors that you will encounter in servicing work. In most cases the unequal spacing of the leads identifies them. When leads are equally spaced, there will be a red dot or other marking near the collector lead. Correct identification of leads is important, because a new transistor can burn out in a flash if connected wrong.

Power Transistors. Ordinary transistors can handle the signal currents in r-f and i-f stages without overheating, and can even serve in the first a-f stage of a radio. The audio output stage of an auto radio requires a larger transistor, however, called the power transistor. Here the entire housing serves as the collector terminal, so there may be only two leads, as in Fig. 1. Some stages of transistor portable television sets also use power transistors.

Power transistors are always bolted to a metal chassis, in order to get rid of the heat developed in the collector electrode. There will usually be a thin layer of insulation between the transistor housing and the chassis, however, since the collector is rarely grounded in an output stage.

Transistor Type Numbers. Most transistors have a type number that starts with 2N, followed by a two-, three-, or four-digit numeral. There may be a modification letter at the end also, usually an A or B. Examples are 2N36, 2N186, 2N241A, and 2N1086. The number will usually be printed directly on the transistor.

Transistor Interchangeability. Hundreds of different types of transistors are made by different manufacturers for use in transistor radios and television sets. Sometimes two or more different types will be identical in performance or will differ so little that they can be used interchangeably. If your jobber does not have in stock the particular transistor that you need, ask him to let you see a transistor-interchangeability chart. This will tell whether some other transistor can be used instead.

Japanese Transistors. Many Japanese transistor radios have been sold in this country. These use Japanese-made transistors having their own type numbers. You may be able to obtain certain types of Japanese transistors from your jobber or from the local distributor for the Japanese set, but there will be times when you will have to use an American equivalent. Your jobber may have a list of equivalents for at least some of the Japanese types.

The circuits, parts, and construction features of Japanese transistor radios are similar to those used by American manufacturers. Be sure you can get a service manual for a Japanese set before promising a repair job, however. Look up the make and model number in the latest cumulative index to your collection of service manuals.

Transistor-radio Service Manuals. You will almost always need the service manual for a set that uses transistors, as a guide for troubleshooting to locate a defective connection on the printed circuits in which transistors are usually connected. These service manuals can be obtained in exactly the same manner as for other sets, by writing to manufacturers or by buying collections of service manuals.

As an added convenience for those who do a lot of transistor-radio servicing, collections of transistor-radio service data in book form are published from time to time by Howard W. Sams under the title, "Servicing Transistor Radios." These can be obtained from most radio jobbers. The manuals contain material extracted from Photofact Folders, along with useful circuit-troubleshooting information applying specifically to transistor radios.

Transistor Testers. Transistors go bad so seldom that they are just about the last thing to be checked when trouble occurs in a transistor receiver. This is fortunate because transistors must usually be unsoldered before they can be checked by a transistor tester. For troubleshooting, a serviceman needs to know whether the transistor is open or shorted, whether its leakage current is excessive, and whether its current gain is approximately satisfactory. There are a number of relatively inexpensive transistor testers on the market that give this information. These testers will not ordinarily detect transistors that have changed their characteristics. Trying a new transistor is the best procedure when the tester indications are doubtful, just as for tubes in television sets.

Examples of transistor testers are shown in Fig. 2. All will detect definitely bad transistors and give meter readings for current gain (often called beta) and leakage. They can also be used for checking crystal diodes. When you are ready to consider the purchase of a transistor tester, discuss the available instruments with your jobber and ask his advice.

New and improved instruments are continually being brought out by manufacturers, so there is no need to learn operating details until you have your own tester. The instruction manual for the tester will give instructions for making the tests and interpreting the results.

A transistor tester will usually have one tiny socket for plug-in transistors, along with three jacks marked C, B, and E in which individual collector, base, and emitter leads can be inserted.

Some transistor testers are battery-operated, while others operate from an a-c line. Battery-operated models will have a calibrating adjustment. When the adjustment can no longer be made, the batteries should be replaced.

Transistor testers generally have a three-position on-off switch, with the positions marked NPN, OFF, and PNP. Before testing a transistor, you will have to determine whether it is an n-p-n type or a p-n-p transistor.

You can get this information from the literature put out by transistor manufacturers, but it is usually more convenient to look in the service manual. Here it is usually given in the parts list. The circuit diagram will also identify the transistor type; remember that the arrow on the emitter symbol points toward the base symbol for a p-n-p transistor, and points away from the base for an n-p-n transistor.

Fig. 2. Examples of transistor testers designed for use by television and radio servicemen. Left—Triplett model 690-A, operating from internal batteries. Lower right—Simpson model 260 adapter, designed for plugging into the jacks across the bottom of the Simpson model 260 multimeter at upper right.

Some transistor testers are not designed for testing power transistors. If you feel that you need this feature, ask your jobber for a tester that has it.

Charts are not ordinarily furnished with transistor testers. One manufacturer states in the instruction book that the literature of a transistor manufacturer should be used to determine whether the leakage current is within limits for a particular transistor.

Testing Power Transistors with an Ohmmeter. If a transistor tester is not available, a power transistor can be safely checked with an ohmmeter after being unsoldered. Measurements are made as shown in Fig. 3, using the $R \times 1$ scale of the ohmmeter.

When checking between emitter and base, a good power transistor should read around 2 ohms for one position of the multimeter leads, and close to the high-resistance end of the scale for the other position of the multimeter leads.

When checking between the emitter and the collector, the readings should be well above 100 ohms for both positions of the multimeter leads.

Do not rely too much on ohmmeter readings of power transistors, because they vary with the type of transistor, the battery voltage in the ohmmeter, and the temperature of the transistor. The chief value of an ohmmeter test is in showing up a transistor that is definitely open or definitely shorted between two of its leads.

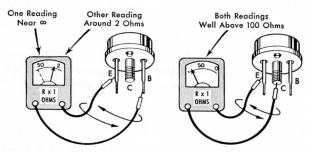

Fig. 3. Four-measurement method of checking a power transistor with an ohmmeter. For each pair of transistor terminals, reverse the ohmmeter leads after taking the first reading

Testing Low-power Transistors with an Ohmmeter. The battery voltage in an ohmmeter may actually damage some types of low-power transistors. The risk is minimized by using only the R × 100 range, however. Some transistor manufacturers say that an ohmmeter should never be used on low-power transistors. Others suggest using the test procedure shown in Fig. 4 if a transistor tester is not available.

After removing the transistor from its circuit, measure between the emitter and the base, then reverse the ohmmeter leads and repeat the measurement. One reading should be near infinity and the other less than about 1,500 ohms (near zero). The same combination of readings should be obtained between the base and the collector.

For a leakage test between the emitter and the collector, the readings should be above 5,000 ohms on the R × 100 scale for an r-f or i-f transistor, and above 500 ohms for an audio transistor. A lower reading indicates a leaky or partially shorted transistor.

Do not rely too heavily on any ohmmeter tests of transistors. Always try a new transistor when in doubt. If the receiver fails to work with a new transistor, the old transistor was probably good. You then have to look for trouble somewhere else or take the set to the manufacturer's authorized service agency for special tests. Be sure to replace all parts before turning a set over to anyone else for repair.

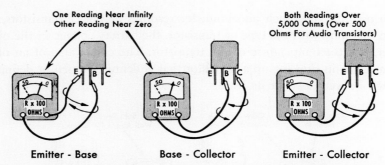

Fig. 4. Testing a low-power transistor by making six measurements with the R $\times$ 100 scale of an ohmmeter. For each pair of transistor terminals, reverse the ohmmeter leads after taking the first reading

Removing and Replacing Ordinary Transistors. Most transistors are today soldered directly to printed circuits, because they seldom go bad. When you do have to remove one for test replacement, be sure to make a sketch of lead positions before unsoldering a transistor. You can either

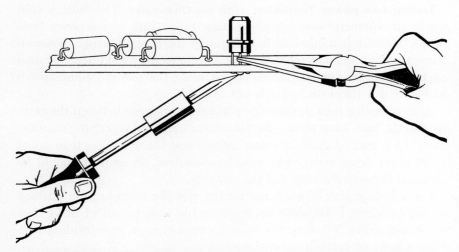

Fig. 5. Method of using long-nose pliers as heat sink to protect transistor while soldering or unsoldering a transistor lead. Pliers can also be used to pull the lead out of its hole while unsoldering

unsolder one lead at a time or unsolder them all at once with a solder pot or a special soldering-iron tip, as described in the chapter dealing with printed circuits. Always turn off the set before removing or replacing a transistor.

To prevent soldering heat from flowing up a lead and damaging the transistor, use long-nose pliers as a heat sink. Just grasp the transistor lead with the pliers between the transistor and the joint being unsoldered or soldered, as in Fig. 5. The pliers absorb any heat flowing up the lead, so essentially no heat gets past the pliers to the transistor.

When you need two hands for soldering or unsoldering, lock the pliers in position on the lead by placing a rubber band over the handles of the pliers. If there is not sufficient room to get the pliers on the leads, use an alligator clip on each lead.

After removing the transistor, remove surplus solder from each transistor hole in the printed wiring, then insert the leads of the new transistor. Be sure the new unit is oriented exactly the same as the old one, because the transistor may burn out immediately if inserted incorrectly. The panel-layout diagram in the service manual usually shows the correct transistor positioning, in case you overlooked making a sketch before taking out the old transistor.

Replacing Power Transistors. When replacing a power transistor, apply a coating of a heat-transferring silicone grease to both sides of the mica or anodized aluminum insulating washer. The grease fills the tiny depressions in the mating surfaces, thereby improving heat transfer to the chassis. Any excess grease that is squeezed out when the mounting bolts are tightened can be wiped off with a clean cloth, to prevent it from attracting conductive dust particles that would eventually short out the insulation. When soldering the connections for the new transistor, use only enough heat to get good joints. Typical mounting arrangements are shown in Fig. 6.

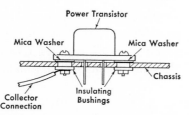 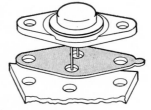

Fig. 6. Methods of mounting power transistors on a metal chassis that serves as a heat sink, with the transistor housing insulated from the chassis

292 Television and Radio Repairing

Use of Thermistor with Power Transistors. A temperature-sensing resistor called a thermistor is often mounted next to a power transistor and connected into the transistor circuit. It serves to prevent the power transistor from going up in smoke due to thermal runaway when the heat in the transistor is too great to be dissipated by the metal chassis that serves as a heat sink. You will usually find a thermistor in an auto radio having a push-pull output stage. Here the heat of the engine or the nearby

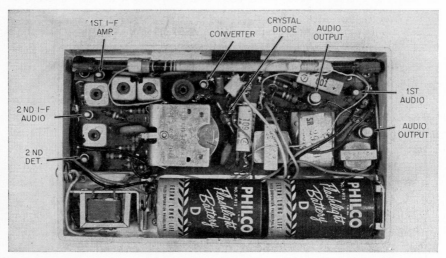

Fig. 7. Typical transistor portable radio with cover removed. Each transistor stage is identified. Audio output transformer is at right center, while audio coupling transformer is in lower left corner. (Philco photo)

auto heater adds to that produced by the power transistors themselves.

When temperature goes up, the resistance of a thermistor goes down. This lowers the forward bias voltage acting on the emitter of the transistor, thereby reducing collector current and preventing thermal runaway. The resistance of a thermistor can easily be checked with an ohmmeter, for comparison with the value indicated on the circuit diagram.

A circle is often drawn around a resistor symbol to indicate that it is a thermistor. On other diagrams the word cold is placed after the resistance value in ohms, to indicate that it is a thermistor and will have a much lower resistance when hot.

Special Features of Transistor Radios. A typical transistor radio, like that shown in Fig. 7, uses chiefly miniature versions of familiar parts. Each part has leads that go through holes in the printed-circuit board to the other side, where they are soldered to the printed wiring.

An audio coupling transformer is usually used ahead of the output stage in a transistor radio. This transformer serves to transfer the audio signal from the driver transistor to the output transistors while providing the correct impedance match. It thus performs much the same job as an output transformer, and looks very much like an output transformer. Continuity checks with an ohmmeter will reveal opens in the windings, just as for output transformers. The tests and replacement procedures are covered in the chapter dealing with coils and transformers.

When a push-pull audio output stage having two identical transistors is used in a transistor radio, the p-m dynamic loudspeaker may have a center-tapped voice coil. There will then be one extra lead coming from the loudspeaker, but no output transformer. A glance at the circuit diagram will always show you how to check continuity in this special loudspeaker circuit when you suspect trouble.

In some pocket-size transistor radios a single transistor is used to amplify signals at two different frequencies. This circuit technique is called reflexing. It reduces the number of transistors required in the set, thereby cutting manufacturing costs.

A reflexing transistor may be in an r-f or i-f stage. After the signal travels through this transistor the first time and is demodulated by the second-detector crystal, the audio signal is passed back to the transistor for further amplification. Capacitors and coils serve to keep the signals in their proper paths. A service manual is an absolute necessity when looking for a bad part in one of these sets.

All-wave Transistor Portables. Transistor radios designed for short-wave reception as well as the broadcast band may have nine or more transistors to provide the required higher selectivity and sensitivity. There will be a band-changing switch, with separate tuning circuits for each band. With the service manual at hand, however, there should be no unusual service problems.

Having additional bands often helps to isolate the trouble. If the set works only on some bands, look for trouble ahead of the i-f amplifier, in the tuning circuits for the bands that do not work. If the trouble occurs on all bands, look for defects between the first i-f stage and the speaker, after clearing the batteries of suspicion.

Transistors in Auto Radios. All-transistor auto radios use the same basic circuits as transistor portable radios. There will usually be an extra r-f amplifier stage ahead of the converter, however, to get the extra sensitivity needed for out-of-town reception while on highways. In addition, the output stage will have either one or two power transistors, to drive the larger loudspeakers needed to overcome wind noises at higher car speeds. All the transistors get their d-c operating voltages from the 12-volt auto battery.

An ordinary transistor radio having a built-in ferrite-rod antenna will not work well in an auto unless held up to a window, because the steel body of an auto blocks radio signals. This is why auto-radio antennas are always mounted outside the body.

In some cars, one section of the all-transistor auto radio can easily be taken out by the owner for use as an ordinary transistor radio. The removable section has its own batteries. The r-f stage and the power-transistor output stage that remain in the car get their power from the auto storage battery. The manual tuning dial on the removable unit may serve also for tuning in the car, or there may be a separate pushbutton tuner permanently mounted in the car.

Isolation of trouble is easy in these two-unit models. If the portable does not operate when out of the car, the trouble is in the portable. Always try new batteries first. If the radio is dead only when plugged into the car, the trouble is in the section mounted on the car.

Hybrid Auto Radios. An auto radio that has one or more power transistors in the output stage but tubes in all other stages is called a hybrid radio. The tubes are special low-voltage types designed to operate directly from a 12-volt automobile battery, eliminating the need for a power supply. The power transistors in the output stage likewise operate directly from the battery. A hybrid auto radio can deliver just as much undistorted output power as an all-tube set, but draws only about half as much current from the storage battery. There are no new servicing problems, once you get used to the idea that the highest d-c voltage in the tube stages is only 12 volts.

If the speaker makes a thumping sound the instant you turn on a hybrid auto radio, the transistors are very likely working satisfactorily. You then look for trouble in the tube stages. The explanation for this test is simple; tubes have an appreciable warmup time, whereas a transistor has none. Any sound heard immediately means the transistors are working.

Transistors in Remote Controls. The remote-control units used with some television receivers are low-power transistor-radio transmitters. The control unit will generally have a single transistor operating from a battery in an oscillator circuit. When the operating button is pressed, the control unit radiates a signal just strong enough to be picked up by the receiving unit in the television set.

Always check the battery first if a transistor remote-control unit does not work. The transistor is suspected last, just as in transistor radios.

Transistors in Home Audio Equipment. Transistors are used in the pre-amplifiers of some high-fidelity home audio equipment. They will have their own batteries, so always try a new battery first when trouble devel-

Fig. 8. Portable transistor television receiver using 19-inch picture tube. The rechargeable silver-cadmium battery block at bottom center provides 18 volts d-c for operating the set up to 5 hours. The battery is recharged by plugging the set into an a-c outlet overnight. (Motorola photo)

ops. If the batteries are good, proceed with troubleshooting just as for any other transistor audio amplifier.

Transistors are used in preamplifiers to eliminate a-c hum problems that sometimes occur with tubes in these stages. Transistors also are free from microphonics, which is important at the low signal levels existing in preamplifiers.

Transistors in Television Sets. Many portable television sets operate directly from batteries. Transistors and crystal diodes are used in most of the stages, to reduce battery drain. The batteries will usually be rechargeable. The only vacuum tubes required are the picture tube and the high-voltage rectifier tube. The example shown in Fig. 8 uses 23 transistors and 10 crystal diodes.

Servicing problems in transistor television sets are a combination of those for ordinary television sets and transistor radios. Television-tube troubleshooting charts can therefore be used to determine the stages likely to be causing the observed trouble. With the service manual for the set at hand, operating voltages on the transistors in these stages can then be checked just as in transistor radios, in search of the defective part. Transistors themselves are checked last, however, because they need replacement in less than 1 out of every 100 jobs.

Portable transistor television sets will usually have an a-c line cord, together with a power transformer and silicon-diode rectifiers. This a-c power supply serves two purposes—to recharge the battery, and to permit a-c operation of the set when desired. The battery-charging circuit will usually have a voltage-sensitive relay that automatically opens the charging circuit when the battery is fully charged and up to its correct voltage.

Transistor Receiver Troubleshooting. When working on transistor receivers, the following steps can serve as the framework for your troubleshooting procedure:

1. Remove the back cover, then turn on the set and check performance while using your eyes, ears, and nose to locate obvious defects.

2. Check the batteries, either by trying new batteries or by measuring battery voltage while the set is turned on.

3. Measure the current drain from the batteries and compare with the value specified in the service manual.

4. Isolate the defective stage or section by analyzing the trouble symptoms.

5. Make transistor electrode voltage measurements in the suspected stages and compare with the voltage values given in the service manual, to further narrow down the search.

6. Check the parts in the suspected stage or circuit with an ohmmeter.

7. Replace the defective part.

The new portions of this troubleshooting procedure will be taken up in turn. Study these instructions carefully, concentrating on the procedures and precautions that are different from those for ordinary tube-type receivers. You can then tackle your first transistor-receiver job with confidence, without worrying too much about damaging a transistor. If an accident does occur, find out why. The experience thus gained will be worth many times the cost of the ruined transistor, because you are not likely to make the same mistake twice.

Looking for Obvious Defects. When you first open up a transistor receiver, it is a good idea to turn the set on and let it warm up for a few minutes, then touch each transistor *lightly* with a finger while you look over the chassis carefully for obvious defects. If a transistor feels hot and is not a power transistor, it is probably defective.

An overheated transistor will always have to be replaced. Before installing the new transistor, however, check the resistors and capacitors in that transistor stage with an ohmmeter. One of them may have been the true cause of failure of the transistor.

Poor soldered joints and breaks in printed wiring are common troubles in the printed circuits used with transistors, so push each part with your finger as part of this preliminary inspection. If any such movement affects performance, you are on the track of the trouble.

Battery Problems in Transistor Sets. The general problems of battery testing and replacement are covered in the chapter dealing with power supplies. Some of the instructions deserve emphasis here because of their importance to transistor radios.

Always try new batteries first when a transistor radio begins distorting, develops a squeal, gets weak, or goes dead. Weak batteries can cause any of these troubles and a lot of others, so it is just a waste of time to start troubleshooting before putting in new batteries.

Check the polarity markings in the radio carefully before inserting each new battery. A transistor radio will not work if the batteries are reversed. Installation of a battery with the wrong polarity can actually ruin the transistors in the radio.

Be particularly careful when installing mercury batteries, because in these the metal case is the positive terminal and the central contact button is the negative terminal. This is illustrated in Fig. 9 for several sets that can take either type of battery.

Checking Transistor Battery Voltage. If the battery voltage is normal when a transistor set is off, but drops below three-fourths of normal value when the set is turned on, the battery probably needs replacement. Install a new battery and check voltages again. If the voltage of the new battery similarly drops when the set is turned on, look for a shorted capacitor or a short-circuit somewhere in the receiver.

Advantages of Mercury Batteries. Mercury batteries should be recommended for transistor radios wherever they can be used. They will last

298 Television and Radio Repairing

four to six times as long as ordinary flashlight cells of equivalent size, and have the added advantage of providing more uniform performance throughout their service life.

Although higher in first cost, mercury batteries usually cost less per hour of radio operation. They have the added advantage of reducing the number of times that new batteries must be installed each year.

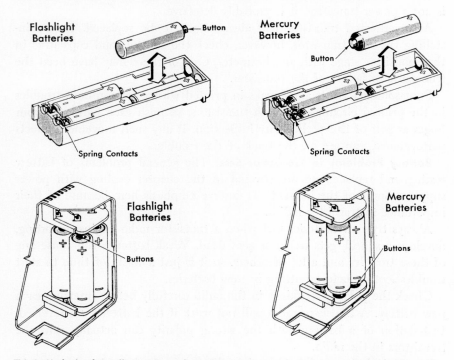

Fig. 9. Methods of installing ordinary batteries and mercury batteries in typical transistor radios

Rechargeable Batteries. Many portable television and radio sets use rechargeable nickel-cadmium or silver-cadmium batteries. Simple and relatively inexpensive battery chargers are available for charging the batteries from an a-c power line. Some battery chargers can be plugged into the cigarette-lighter socket of an automobile having a 12-volt storage battery.

In one 19-inch transistor portable television set, the battery will provide up to 6 hours of operation. Three hours of recharging are required for each hour of operation. When this television set is plugged into an a-c line, the battery is automatically transferred to the recharging circuit and the power transformer in the set is energized to permit operation from the a-c line.

A charger designed for a-c operation can be left plugged into the a-c line at all times, because it draws appreciable power only when actually charging. Radios designed for rechargeable batteries will usually have some means for making a quick connection to the battery charger. Overnight charging will usually restore batteries to full charge.

Checking Current Drain. When new batteries do not clear up the trouble in a transistor set, a logical first test is measurement of the total current being drawn from the battery. If this current is more than about 15 per cent higher than that specified in the service manual, look for shorted or leaky capacitors, shorts between wires, and other conditions that could increase battery current drain. If the current drain is over 15 per cent too low, look for open resistors, breaks in wiring, and other conditions that could reduce current drain.

To check current drain, the correct milliammeter range of a multimeter must be inserted in series with the batteries. Turn on the set and tune in a station at normal volume. Now clip the multimeter leads to the terminals of the on-off switch with correct polarity, then open the switch. The milliammeter in your multimeter will now complete the circuit across the switch and show the current drain.

Some service manuals give only the no-signal current drain, which is several milliamperes lower than for normal volume, so check the manual first. Tune between stations or turn down the volume control to get no-signal drain.

Analyzing Symptoms in Transistor Radios. Table 1 lists the common symptoms that you will encounter in transistor radios and gives in logical order of importance the probable causes for each. The recommended repair procedure for each cause is outlined only briefly in the table, because the detailed procedures are given elsewhere in this chapter and in chapters dealing with the specific parts mentioned.

The six-transistor circuit diagram accompanying this table shows most of the features that you will find in a transistor radio. It is rare that two manufacturers use exactly the same circuit, however. Use this circuit only to help you learn causes and cures for the symptoms. If the circuit of the transistor radio on which you are working does not have some of the parts mentioned, look for trouble in other parts in the same stage.

Evaluating Transistor Voltage Measurements. When installation of a new battery does not correct the trouble in a transistor radio, troubleshooting can begin with measurement of transistor electrode voltages. An ordinary 20,000-ohm-per-volt multimeter is entirely satisfactory for this purpose. Start with the stage suspected of being faulty, then check on each

Table 1. Common Trouble Symptoms in Transistor Radios

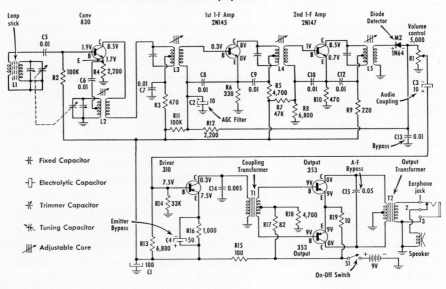

Symptom	Probable Cause	What to Do
No stations, no hum, no clicks	Dead battery	Replace battery
	Reversed battery	Check battery and receiver polarity
	Poor battery contacts in receiver	Clean battery contacts if corroded; increase contact spring pressure if weak
	Open earphone jack contacts	Short contacts 1 and 2 on jack; if set now plays, clean and adjust contacts
	Defective on-off switch	Short terminals of switch S1; if set now plays, replace switch
	Open in speaker voice-coil circuit	Check continuity with ohmmeter
	Open in power-supply wiring to output stage	Check continuity from switch S1 to electrodes of output transistors with lowest range of ohmmeter; examine printed wiring for hairline cracks
	Shorted a-f bypass C15	Unsolder one lead of capacitor C15 and try set; if it works, replace C15
No stations; hum when volume control is full on, and clicks when batteries are removed and replaced	Weak battery	Try new battery
	Open power lead to one or more transistors	Measure current drain. If too high or too low, check supply circuit continuity to each transistor visually or with ohmmeter

Table 1. (Continued)

Symptom	Probable Cause	What to Do
	Shorted capacitor in one of transistor stages	Measure electrode voltages with volume control at maximum. If one is more than 15 per cent off, check capacitors in that stage for shorts. Short in C1 will affect all stages
	Open capacitor in signal circuit	Check C5, C6, and C3 for opens by shunting each temporarily with good capacitor
	Open or short in L1, L2, L3, L4, L5, or T1	Check continuity of each winding in turn with ohmmeter
	Open in signal circuit	Check all wiring visually or with ohmmeter, to locate break or poor soldered joint
	Defective diode M2	Try new diode of same type
	Volume control open	Check volume control with ohmmeter
	Local oscillator not working	Turn on transistor radio and place near good broadcast-band radio tuned near high-frequency end of dial. Tune transistor radio about 455 kc lower and rock dial. If local oscillator in transistor radio is working, squeal will be heard in good set each time the signals beat together. If squeal is not heard, recheck all converter components and wiring, then try new converter transistor
	Shorted transistor in one stage	If an electrode voltage is off more than 15 per cent for a stage but no defective components or wiring can be found, try new transistor
No stations at one end of tuning range	Warped tuning capacitor plates	Determine which plates rub when set is tuned to dead end of range, and spread them out carefully with thin razor blade
All stations weak, with or without distortion	Weak battery	Try new battery
	Ferrite-rod antenna core is cracked	Examine core and bend it gently, looking for crack. If cracked, replace
	Open capacitor	Check C1, C2, C3, C4, C9, and C10
	Open or shorted i-f transformer	Check continuity of each winding of L3, L4, and L5 with ohmmeter
	Audio amplifier defect	Check components, wiring, and electrode voltages in audio stages

Table 1. (Continued)

Symptom	Probable Cause	What to Do
	Defective diode M2	Try new diode of same type
	Defective converter or i-f transistor	Try new transistor only as last resort after checking voltages and components in stages
Stations weak only at one end of dial	Defective transistor in converter stage	Try new transistor
Intermittent operation	Poor soldered joint	Resolder suspected connections
	Hairline break in printed wiring	Locate by pressing board, and solder break
	Poor battery contacts	Clean and tighten battery contacts in set
	Dirty or loose earphone jack contacts	Clean contacts and increase spring tension
	Dirty or loose contacts in on-off switch S1	If switch can conveniently be taken apart, try cleaning contacts and increasing contact spring tension; otherwise, replace switch
	Worn or dirty volume control	Try volume-control cleaner, or replace volume control
	Dirt on tuning capacitor	Clean out dirt on contacts or plates with strip of paper or brush
Distortion at all volume-control settings	Weak battery	Try new battery
	Shorted, leaky, or dried-out electrolytic C2, C3, or C4	Try new capacitor of about same size after disconnecting one lead of capacitor in set, or check capacitor in set with ohmmeter after unsoldering one lead
	High-resistance soldered joint	Resolder suspicious joints in audio section
	Carbon resistor has increased in value	Check values of carbon resistors in audio section
	Defective speaker	Try earphone; if sound is clear, replace speaker. If set has no jack, connect earphone temporarily in place of speaker
	Defective audio transistor	Try new transistor in driver and output stages in turn
Distortion only on strong signals	Shorted capacitor C2 or C13	Try new capacitor of about same size after disconnecting one lead of capacitor in set
	Defective diode M2	Try new diode of same type

Table 1. (Continued)

Symptom	Probable Cause	What to Do
Squealing or motor-boating	Weak battery	Try new battery
	Defective electrolytic C1	Try new capacitor of same size after disconnecting one lead of capacitor in set
	Defective bypass capacitor C13	Shunt C13 temporarily with new capacitor. If this clears up trouble, replace C13
	High-resistance joint in battery or tuning circuit	Clean battery terminals and contacts. Solder riveted leads. Tighten tuning capacitor mounting bolts
Squealing heard only when station is tuned in	Weak battery	Try new battery
	Antenna core is cracked	Examine core and bend gently, looking for crack. If cracked, replace
	Defective C2	Try new capacitor
One station heard over entire band	Local oscillator not working	Check components and wiring in converter stage first, then try new transistor
Noise heard along with all stations	Dirty tuning capacitor	Clean wiper contacts on tuning capacitor. Clean plates with strip of thin paper
	Tuning capacitor plates rubbing	Straighten warped plates with thin razor blade or knife
	Defective converter transistor	Try new transistor
Volume changes too much when set is tuned to different stations	Shorted capacitors C2 or C13	Unsolder one lead and check with ohmmeter, or try new capacitor
	Defective diode M2	Try new diode of same type
Volume goes down after set is on	Weak battery	Try new battery
Battery life is too short	User forgets to turn set off	Tactfully ask if some member of family is careless with radio
	On-off switch S1 is defective	Tune in station and operate switch several times. If it sticks in ON position, replace
	Shorted or leaky filter C1	Try new capacitor of same size, after disconnecting one lead of capacitor in set
	Defective audio transistor	Try new transistor in driver and output stages in turn

side of the stage. Use the d-c voltage range of the multimeter that is equal to or slightly higher than the highest battery voltage indicated in the service manual.

Electrode-voltage readings can be considered good if they are within 10 to 15 per cent of the value given in the service manual. For voltages under 1 volt, a reading within 0.2 volt of the indicated value can be considered good. If you get a backward reading, reverse the multimeter leads, because polarities are not always clearly shown on circuit diagrams.

When a transistor base voltage is too far off from the specified value, there will usually be trouble in that base circuit. When emitter and collector have the same voltage, look for an open collector circuit. When all three electrodes have the same voltage, look for an open ground lead, usually due to a break in printed wiring. Remember also that resistors can change in value and affect transistor electrode voltages. Resistors usually increase in value with age. Replace any resistor that is more than 20 per cent above its rated value.

When working on compact portables or in closely spaced areas of any set, conventional test prods are often the cause of accidental shorts between adjacent terminals. In transistor circuits a momentary short may ruin a transistor. To avoid this, place insulating tape on the metal shank of each test prod, so only the sharp tip is exposed.

Using an Ohmmeter in Transistor Circuits. A transistor operates on very low voltages. Even the voltage of the battery in an ohmmeter is enough to operate a transistor. This explains why the reversing of ohmmeter leads often gives a different reading when checking resistors and other parts in a transistor circuit without disconnecting leads. Reverse the leads and take another reading for each ohmmeter test in a transistor circuit. The higher of the two readings will be more nearly correct.

Changing the range of the ohmmeter will often also give a different reading for a part being tested with an ohmmeter, because this changes the value of the ohmmeter voltage that acts on nearby transistors.

The lowest range of an ohmmeter can under some conditions send enough current through an ordinary transistor to ruin it. Always use the $R \times 10$ or higher range when making resistance measurements in a circuit containing ordinary low-power transistors.

Be sure to remove the batteries from a transistor radio before making resistance measurements. The battery voltage may damage your ohmmeter. The ohmmeter may also provide a short-cut path from the batteries to a transistor that cannot withstand the full battery voltage.

Transistors and Crystal Diodes **305**

Polarity is important when using an ohmmeter to check the miniature high-value electrolytic capacitors used in transistor circuits. To avoid damaging these capacitors, follow carefully the test procedure given in the capacitor chapter.

Crystal Diodes. Many receivers use germanium or silicon crystal diodes as detectors or rectifiers in signal circuits. These diodes seldom go bad, so

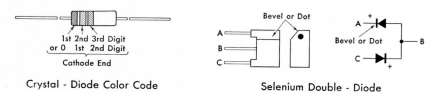

Crystal - Diode Color Code   Selenium Double - Diode

Fig. 10. Method of using resistor color code to indicate type number of crystal diode, and polarity identification on one type of selenium double-diode

you will usually find them soldered to their circuits. Examples of crystal diodes are shown in Fig. 10.

The video-detector crystal diode is sometimes mounted in the final i-f transformer shield. The top of the shield is then usually removable for easy access to the diode.

Diode Type Numbers. Most crystal diodes have a type number starting with 1N, followed by either two or three digits. Occasionally there will be a letter after the last digit. Examples of diode types are 1N34, 1N58A, and 1N295. If your jobber does not have the exact type number, ask to see one of his cross-reference lists of replacement diodes. There will often be another type which will work equally well.

When the type number is not printed on the diode, there will usually be a color code like that in Fig. 10. This gives the digits following 1N in the type number. Hold the diode so the end with the color bands is at your left, and read the colors from left to right according to the standard EIA color code given in the resistor chapter. If the first color is black, which represents zero, ignore it and read the next two colors to get a two-digit number.

If there is a fourth color band, it represents the letter at the end of the type number, as follows: brown—A; red—B; orange—C.

Testing Crystal Diodes. The estimated life of a crystal diode is 10,000 hours, as compared with around 3,000 hours for the average tube, but occasionally these diodes do go bad.

An ohmmeter is the best instrument for checking a crystal diode. Disconnect one lead of the crystal diode, being careful not to apply the pencil-type soldering iron any longer than necessary. Measure the diode resistance with an ohmmeter on the R × 100 range, then reverse the positions of the leads and measure again. For a good diode the reading should be at least 100 times greater in one direction than the other.

Replacing Crystal Diodes. Use pliers as a heat sink when soldering or unsoldering diode leads, just as for transistors. Be sure to note the polarity before removing a diode, because a new unit must be installed with the same polarity.

Diode Testers. Most transistor testers can also be used for checking crystal diodes. The instructions given in the operating manual for the tester should be followed here, since they differ with each make of tester. Two tests are usually made, one for leakage and one for forward current. An example of a transistor tester that can handle diodes is shown in Fig. 11.

Selenium Diodes. Some television sets use a plug-in selenium double-diode as a phase detector in the sync separator section. Exact-duplicate replacements for this are available at most jobbers, at about the same price as a single crystal diode. The center lead is common to both sections.

Fig. 11. Example of transistor tester that operates from an a-c power line (Triplett model 2590) and can be used also for testing crystal diodes

Each section of a selenium double-diode should measure 2,000 to 5,000 ohms minimum when using the R $\times$ 100 scale of an ohmmeter, and much higher when the ohmmeter leads are reversed. Ohmmeter measurements can be made without unsoldering the unit.

Correct polarity is essential when installing a new double-diode unit. A bevel or other marking near one lead should go in the same position as for the old part.

14

Power-supply Troubles in Home, Portable, and Auto Sets

Common Sources of Power. The power required by tubes and transistors in modern television and radio sets is obtained from the a-c power line in home sets, from dry cells in most portables, and from storage batteries in auto sets and a few large portables. Less common sources are solar cells and d-c power lines.

If power does not get to a set, the set does not work. It is usually easy to recognize power-supply trouble because the set is completely dead. There is no sound or hum, there is no light on the screen of a television receiver, all tubes are cold and dark, and there is no instantaneous thump in the speaker when a transistor set is turned on.

Locating and repairing power-supply troubles is just as easy as recognizing them, once you get familiar with the different types of power supplies. Each will be taken up in turn in this chapter, along with test and repair procedures for common troubles.

A-C Power-line Troubles. In the United States, alternating-current power-line systems operate at a voltage between 110 and 125 volts. The power company tries to hold the voltage to a center voltage plus or minus 5 volts. Thus, if the center voltage in a particular city is 115, the power company will attempt to keep the voltage in each home from dropping below 110 volts or rising above 120 volts. A common center-voltage value is 117 volts.

If the tubes do not light up when you first turn on an a-c or a-c/d-c set, check the connections to the power source. Be sure the line cord is plugged into the outlet and making good contact there. If there is a cube tap in the outlet as in Fig. 1, remove it and plug the set directly into the outlet. These cube taps frequently have poor or loose contacts and are troublemakers in other ways, so urge the customer not to use them on television and radio sets.

House-fuse Troubles. With an a-c power source, the source itself may be dead because of failure of a fuse. In modern home wiring, wall outlets are often connected to a separate circuit which has its own fuse. Therefore, these wall outlets may be dead while the ceiling lights in the house are operating properly. It is easy to check the wall outlet by plugging in a table or floor lamp that is known to be in good condition.

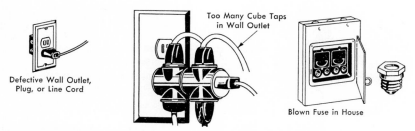

Fig. 1. Examples of power-line troubles for which a serviceman is often called. Sometimes the line-cord plug is simply knocked out of the wall outlet unknowingly while the housewife is using a vacuum cleaner

If the lamp does not light, the fuse box should be checked. This will be found near the electric meter. You can usually tell by inspection which fuse is burned out or whether a circuit breaker has opened. It is well for you to carry a few spare fuses (15-ampere rating) in your toolbox just in case the customer does not happen to have a replacement.

Never put a penny or a piece of tinfoil behind a burned-out fuse. Fuses are placed in the circuits for protection against defective wiring and to eliminate fire hazards that develop when lines are overloaded. Fuses usually burn out when the line has too many appliances connected to it, but occasionally fuses will fail because of old age.

If you replace a fuse and the new one burns out immediately, there may be a short-circuit in the receiver, in a lamp, or in some appliance. Unplug all devices in the house, then replace the fuse. If it blows again, the house wiring is defective. Recommend that it be checked for faults by a competent electrician. If the new fuse does not blow, plug the devices in again one by one and turn them on, waiting a minute for each to warm up. When the defective device is plugged in, the fuse will blow again.

If the fuse blows only when all devices are turned on, the line is overloaded. The customer will either have to use fewer appliances at a time or have the house wiring changed to provide more fused branches.

Defective Line Cords. If there is voltage at the wall outlet but no power at the set, look for a poor connection in the line-cord plug or a break in the line-cord wire itself. To check these with an ohmmeter, pull the line-cord plug out of the wall outlet, turn off the set, connect the prongs of the plug together with bare wire or a large clip, then measure the resistance between the other ends of the line cord, inside the set. If the resistance reading is infinity, there is a break somewhere in the line cord or in the

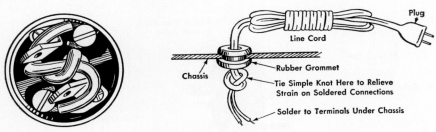

Fig. 2. Method of using simple knots to prevent strain on terminal connections at both ends of a line cord

plug. Do not try to repair it; replace the entire line cord and plug, because old line cords can be dangerous.

The proper way of connecting a line cord to a plug is shown in Fig. 2. The separated wires of the cord should go around the prongs inside the plug. The knot gives additional protection against loosening when someone jerks the cord to pull out the plug.

The line cord of a television or radio set runs through a hole at the back of the chassis to terminals under the chassis. A rubber grommet like that in Fig. 2 is generally used inside the chassis hole to prevent the sharp metal edges from cutting into the line cord. If this grommet is worn or missing, it should be replaced before installing the new line cord.

Rubber grommets for replacement purposes are available in various sizes. To install one, squeeze the grommet into an oval shape, insert one end in the chassis hole, then work the rest of the grommet into position with your fingers.

Line cords can be obtained with molded plugs, or plugs and wire can be bought separately. Always get the best quality, even if it costs a few pennies more.

If the spring contacts in the wall outlet are worn and do not grip the

Power-supply Troubles **311**

prongs of the plug tightly, try bending the plug prongs slightly outward with long-nose pliers. This is all that you can do, because replacement of a defective wall outlet is a job for a licensed electrician.

When installing a new cord in a set, knot the cord inside the chassis as in Fig. 2 before soldering the line-cord leads. Be sure the new wire is securely anchored with permanent hook joints or wrap-around joints before you do any soldering. Leave a little slack in the line, so no strain will be applied to the solder joints.

If the power connection to the set is satisfactory but the set still does not operate properly, check the voltage at the wall outlet with the a-c voltage range of your multimeter. If it is below 110 volts, report it to the service department of the local power company.

Receiver Fuses. A small cartridge-type fuse like that in Fig. 3 is often used in one line-cord lead of a home television or radio set. It serves to cut off all power to the set by melting open when a defect occurs that makes the set draw much more current than normal through the fuse. An S-shaped symbol represents a fuse on circuit diagrams.

A line fuse will usually have a rating of 1 ampere for sets drawing 40 to 60 watts; 1.5 amperes for 60 to 100 watts; 2 amperes for 100 to 150 watts; 3 amperes for 150 to 250 watts; 5 amperes for 250 to 350 watts.

Line fuses are generally mounted in spring clips to permit easy replacement. Always pull out the line-cord plug before replacing this fuse. When the fuse is open, full line voltage exists between the spring clips when the set is turned on. If you then touch both clips while removing or replacing the fuse, you will get a shock. Similarly, if you touch only one clip while the set is plugged in and some part of your body is grounded, you will get a shock even if the set is turned off.

In series-filament receivers a fuse may protect only the series-filament string, only the rectifier, or the entire set. Pieces of No. 24 to No. 28 bare copper wire from 1 to 3 inches long are used as fuses in some series-filament circuits. For replacement, use the same size and length of wire.

Circuit Breakers in Receivers. In place of a fuse, some sets have a circuit breaker that opens automatically when overheated because of excessive current flow. The circuit breaker is usually mounted at the rear of the chassis and has a manual reset button projecting outward, as shown in Fig. 4. When you encounter a set that is dead, look for a circuit breaker and push its reset button. If the button releases immediately or in a few minutes, look for the cause of the overload. A circuit breaker may protect the entire set or just the rectifier.

Thermal-delay Switches. Occasionally you will find a combination of a resistor and a bimetallic strip, as shown in Fig. 5. The 120-ohm resistor heats the bimetallic strip and makes its contacts close. This gives the tube filaments time to warm up before B+ voltage is applied to the tubes by

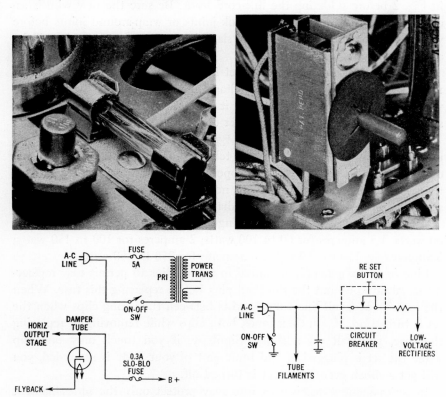

Fig. 3. Cartridge-type fuse in spring-clip holder. At its left is a special Slo-Blo fuse mounted vertically in a holder that prevents insertion of the wrong type of fuse. Circuits for both types are shown. (Figures 3, 4, 5, 6, 7, and 8 reprinted from April 1960 issue of *PF Reporter*, copyright 1960 by Howard W. Sams and Co., Inc.)

Fig. 4. Circuit breaker mounted at rear of television chassis in such a way that the manual reset button projects through a hole in the back cover. Here the circuit breaker protects the low-voltage rectifiers but not the tube filaments of the a-c input section of the receiver, as shown in the accompanying circuit diagram. (Howard W. Sams photo)

the contacts. If this unit is working, you should hear it click a few seconds after the set is turned on. If there is no click but the tube filaments are all heated, look for an adjusting screw. This rarely gets out of adjustment by itself but may need resetting if the switch has been accidentally bumped.

Power-supply Troubles **313**

Another type of thermal-delay switch is used to eliminate the initial surge of current when a set is turned on. This unit is sometimes called a Surgistor. It consists of an ordinary wirewound resistor mounted close to a bimetallic strip having open contacts when cold. The resistor is in series

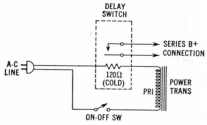

Fig. 5. Thermal-delay switch used in television receiver to allow tube filaments to warm up before plate voltage is applied to tubes. (Howard W. Sams photo)

Fig. 6. Thermal-delay switch known as Surgistor, used in television set to suppress initial current surge through cold tube filaments. (Howard W. Sams photo)

with tube filaments and cuts down the voltage applied to them when the set is first turned on. Current flow through the resistor heats it up. This makes the bimetallic strip bend and short out the resistor. Filament current flows through the bimetallic strip and keeps it warmed up sufficiently to keep the contacts closed. A typical unit and its circuit arrangement are shown in Fig. 6.

Fusible Resistors. In portable television receivers you will often see plug-in resistors of the type shown in Fig. 7. These also serve to limit surge current when power is first applied. At the same time they act as a fuse to protect the rectifier circuit against overloads. When fuse action occurs and the resistor burns out, always replace with an exact-duplicate unit.

Thermistors. Special temperature-compensating resistors called thermistors are sometimes used in series-string filaments to reduce the filament

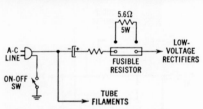

Fig. 7. Example of fusible resistor as used in a portable television set to limit surge current through low-voltage rectifiers when set is first turned on. (Howard W. Sams photo)

Fig. 8. Disk-shaped thermistor having leads soldered to opposite metallized faces, connected in series-string filament of television set to control surge current. (Howard W. Sams photo)

current when the set is first turned on. A typical thermistor may have a resistance of 200 ohms when cold but only 6 ohms when hot. These thermistors usually have a disk-shaped construction, as shown in Fig. 8.

You will also find a thermistor in series with the vertical deflection coils in some television sets. Here it compensates for the increase in the resistance of the vertical deflection coils as they warm up, to prevent the height of the picture from decreasing gradually.

Slow-acting Fuses. Television receivers have occasional current surges that do no harm but are large enough to blow ordinary medium-lag fuses. Use of a slow-acting fuse eliminates the nuisance of replacing fuses frequently. These fuses are available under such trade names as Slo-Blo and Fusetron. Some are used as line fuses, while others protect just one special circuit. They often have special mounting arrangements, to prevent accidental replacement with some other type or size of fuse.

Examples of slow-acting fuses are shown in Fig. 9. A spring is used inside each to apply tension to the fuse wire, so as to give a clean break with a minimum of arcing when the fuse does blow. Since these fuses seldom need replacement, some have pigtail leads for soldering into circuits.

A chemical-type slow-acting fuse, sometimes a fusible surge limiter, is used in high-voltage sections of television receivers. This fuse can with-

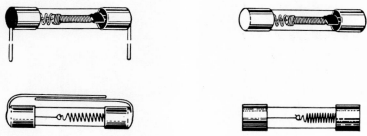

Fig. 9. Examples of slow-acting fuses used in television and radio sets

stand high current surges having a short duration. When a permanent short develops in the circuit protected by the surge limiter, the resistance wire in the fuse gets hot enough to ignite the chemical. The chemical then melts the wire to give a clean break.

Metallic Rectifiers. Selenium, silicon, and germanium rectifiers are widely used in place of vacuum-tube rectifiers in television and radio receivers. All three of these types of metallic rectifiers are usually soldered directly into the receiver circuits because their average life is much longer than that of a vacuum tube. When overloaded by the failure of some other part in the set, however, they can burn out or drop in efficiency. This is why you need to know how to test and replace each type.

Selenium Rectifiers. A selenium rectifier is easily recognized because it has a number of square or round plates that are coated on one side. The plates are spaced apart and bolted together through the center as in Fig. 10.

A selenium rectifier may be bolted to the bottom of the chassis as in Fig. 11, but will more often be bolted to the top of the chassis so air can circulate more readily between its plates. There will be two terminal lugs on a half-wave rectifier, and three lugs on a full-wave rectifier. Connecting leads can be soldered to these lugs with an ordinary soldering gun without fear of overheating the rectifier.

Do not let the soldering gun touch the metal plates of a selenium rectifier, because heat can damage the coating of selenium and reduce the operating efficiency of the rectifier.

Fig. 10. Example of full-wave selenium rectifier having three terminal lugs. (General Electric photo)

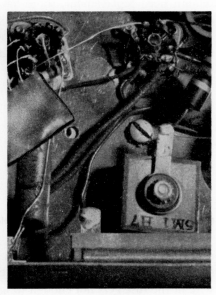

Fig. 11. Selenium rectifier bolted to bottom of radio receiver chassis having conventional wiring. (Centralab photo)

Selenium-rectifier Ratings. Only three factors need be considered when ordering the replacement for a selenium rectifier: (1) the voltage rating should be the same as for the old unit; (2) the current rating should be equal to or higher than that of the old unit; (3) the size and mounting should be similar to that of the old unit.

Selenium rectifiers for television and radio receivers used in the United States are rated at 130 volts. Output-current ratings are generally 50, 65, or 75 ma for a-c/d-c radios, 100 or 150 ma for large radios and small television sets, and 200, 250, 300, 350, 400, or 500 ma for larger television sets and high-fidelity home audio systems.

For a starting stock to go in your tube library, you can get along with one 150-milliampere selenium rectifier for radios and one 400-ma or 500-ma unit for television sets. Each will replace any of the smaller sizes adequately.

How Selenium Rectifiers Go Bad. When a selenium rectifier is overloaded, it becomes overheated. The coating of selenium on the plates then gives off an unpleasant odor like that of rotten eggs. This characteristic smell is a definite clue to the bad part. Do not breathe the fumes given off by an overheated selenium rectifier, because they are poisonous. Even a few good whiffs will cause a severe headache.

If the set is off and no smell is noticeable when you arrive on a service call, ask the customer if she noticed any odor when the set went bad. Look also for signs of burning on the coated surfaces of the plates, or other indications that the coating has melted and then rehardened. Continuous sparking on the plates while the set is on means that the rectifier is overloaded and probably damaged. It should be replaced after clearing up the cause of the overload.

The most common cause of selenium-rectifier failure is accidental overload by a bad filter capacitor or a shorted bypass capacitor. Always check these possibilities before installing a new rectifier, to prevent early failure of the new unit.

Discolored or burned spots on the coating are due to sparking. This may be the result of too high an applied voltage, but is more often produced when the set is first turned on after a long period of idleness. The spots are self-healing. They do not appreciably affect the operation of the rectifier unless about 20 per cent of the total coated area on all plates is burned, or unless the sparking continues. The rectifier should then be replaced.

When a selenium rectifier goes bad, its output voltage will drop and the electrode voltages of all tubes or transistors will therefore be low. A bad rectifier will also pass more hum voltage than can be handled by the electrolytic filter capacitors.

Testing Selenium Rectifiers. When the d-c output voltage of a selenium rectifier is appreciably lower than that indicated in the service manual, the rectifier can be tested with the ohmmeter section of your multimeter. Use the highest resistance range. Turn off the set, unplug the line cord, unsolder one of the rectifier leads, and hold the test prods of the multimeter on the rectifier terminals as in Fig. 12. Note the reading, reverse the multimeter prods, and read the meter again. If one resistance reading is at least ten times greater than the other, the rectifier is good.

318   Television and Radio Repairing

Fig. 12. Using × 100K resistance range of multimeter to test selenium rectifier in television set. (International Rectifier Corp. photo)

A typical good rectifier will read above 1,000 ohms for one position of the leads and below 100 ohms for the other position. If both readings are high and nearly equal, the rectifier is open or otherwise defective. If both readings are low and nearly equal, the rectifier is shorted.

The highest resistance range should be used so enough voltage is applied to get a true operating test. A multimeter generally has only a 1.5-volt battery acting for its lowest resistance ranges, and this is not enough to test a selenium rectifier properly.

An ohmmeter can give only a rough check of the quality of a selenium rectifier. If the B+ voltage is below normal and there are no other obvious defects, it is logical to try a new selenium rectifier.

Power-supply Troubles   319

Replacing Selenium Rectifiers. Install the new rectifier in the same mounting hole used before. Be sure to use a lock washer under the nut for the mounting bolt. Tighten the nut securely so the new rectifier cannot turn. Use the mounting bolt from the old rectifier if none came with the new unit.

A new selenium rectifier may act up for a while when the set is first turned on. If sparking occurs on the plates, turn off the set for a few minutes, then try it again. Occasionally as many as half a dozen on-and-off cycles are needed to revive a new rectifier that may not have been properly processed at the factory.

Selenium rectifiers have a definite polarity. The positive terminal will be indicated by a plus sign or a red dot of paint. The negative terminal will have a minus sign or a yellow dot. The terminals of the new unit must be connected with the same polarity as those of the old unit.

A low-value resistor is almost always used in series with a selenium rectifier, to serve as a fuse. This resistor is designed so it will burn out whenever excessive current flows through the rectifier. Always check the resistor when replacing the rectifier.

Silicon Rectifiers. The tiny silicon rectifier shown in Fig. 13 is often used in place of a selenium rectifier in receivers. This rectifier is a two-terminal version of a silicon junction transistor. Although a silicon rectifier

Fig. 13. Holding lower lead of silicon rectifier with pliers while soldering. (International Rectifier Corp. photo)

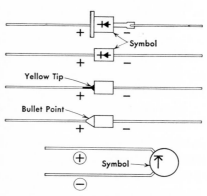

Fig. 14. Polarity indications on typical silicon diodes. Arrow or similar indicator always points to positive terminal

may be more expensive than a corresponding selenium-rectifier unit, its improved performance is considered worth the extra cost. A silicon rectifier rarely requires replacement unless overloaded because of failure of some other part. Up until the time that the silicon rectifier is burned out, it will maintain the operating voltages of the set at the same values as when new.

Silicon rectifiers are often used in pairs as half-wave voltage doublers, to give a higher output voltage than can be obtained with one rectifier alone. There will usually be a low-value resistor in series with each silicon unit.

Replacing Silicon Rectifiers. Be sure to observe correct polarity when replacing a silicon rectifier. To prevent soldering heat from damaging the unit, make a circular loop in each lead and use long-nose pliers as a heat sink on the lead being soldered, as shown in Fig. 13. Polarity markings for several types of silicon rectifiers are shown in Fig. 14.

Some manufacturers make silicon rectifiers available with special mountings for use in place of selenium rectifiers. Order new silicon units with the correct voltage and current ratings, just as when ordering selenium rectifiers.

Germanium Rectifiers. A few television power supplies use germanium rectifiers in place of selenium rectifiers. These two types are usually interchangeable, so the instructions for selenium rectifiers apply also to germanium rectifiers. Remember that a surge-limiting resistance of at least 4 ohms should be used in series with each of these rectifiers.

Safety Precautions. Many sets have a paper capacitor connected between one side of the a-c power line and the chassis. If this capacitor becomes shorted, there is no noticeable effect on performance but the chassis can be extremely dangerous for one position of the line-cord plug. It is a good idea to test this capacitor for shorts with an ohmmeter after completing a repair job.

When a television or radio set having a metal cabinet is used in a room having a concrete floor, a tingling shock may be felt when the cabinet is touched. Exposed metal parts of plastic and wood cabinets can give this same slight shock. It is due to the very small amount of alternating current that flows through the line capacitor even when this unit is good. Reversing the line-cord plug will sometimes eliminate the shock, but not always. A rubber mat on the floor will eliminate the shock. Sometimes it is easier to spray exposed metal surfaces with an insulating plastic or cover them with insulating tape.

Power-supply Troubles 321

Always check a customer's complaint of receiving a shock. In most cases it will be due to this normal harmless condition, but there is always the possibility that the capacitor has gone bad and can give a potentially dangerous shock.

If a customer calls after you have returned a repaired radio, saying that she now gets a shock when touching the set, suggest over the phone that she try reversing the line-cord plug. If this clears up the trouble, you have saved yourself a no-charge callback.

Polarized Line-cord Plugs. A few receivers have polarized line-cord plugs, which can be inserted only one way in a wall outlet because one

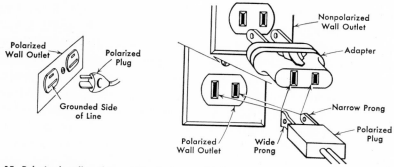

Fig. 15. Polarized wall outlet, polarized plug, and adapter

prong of the plug is wider than the other. These plugs ensure that exposed metal parts of the receiver are grounded, eliminating the shock hazard. Most homes have polarized wall outlets, in which one slot is slightly wider than the other.

The wide slot of a polarized a-c outlet is supposed to be connected to the ground side of the a-c line. To verify this, measure between the wider slot and the grounded screw on the wall outlet with an a-c voltmeter; the voltage should be zero. The wider prong on the line-cord plug goes to the chassis of the set, to achieve automatic grounding of the chassis and exposed metal parts.

The adapter shown in Fig. 15 permits use of a polarized plug in a nonpolarized wall outlet. Of course, such use eliminates the safety feature of the polarized plug.

D-C Power Lines. In the older sections of some large cities 115-volt direct-current power lines are still in use. Here you will encounter universal a-c/d-c sets which can have the same line-cord, plug, wall-outlet, and fuse troubles as do a-c sets.

Sets designed especially for use with d-c power lines may have a polarized line-cord plug in which one prong is set at right angles to the other. This type of plug can be inserted only one way in the corresponding d-c wall outlet.

More often in d-c power districts you will find conventional plugs on radio line cords. Here the plug must be inserted the right way to make the set operate. No harm is done if the plug is inserted wrong, but the set will not work. If the tubes light up but no sound is heard, try reversing the plug in the outlet.

Auto-radio Power Supplies. At one time practically all auto radios used vibrators. These interrupted the d-c voltage of the auto storage battery to give an a-c voltage that could be stepped up by a power transformer. The high a-c voltage was rectified by a vacuum tube, gas tube, selenium rectifier, or additional contacts on a synchronous vibrator. Filtering then gave the required d-c plate supply voltage needed by the tubes in the auto radio.

Many auto radios with vibrators are still in use, and some are still being manufactured, even though the majority of new cars have hybrid auto radios that operate directly from the 12-volt auto storage battery.

Replacing Vibrators. Vibrators in auto radios look like metal tubes but are larger and usually have a bright aluminum housing. They are easy to recognize, as they are the only plug-in parts that hum loudly when working.

Vibrators go bad just about as often as tubes in auto radios, and are replaced just as easily. When replacing a vibrator, it is a good idea to replace the buffer capacitor also. This is connected across the secondary of the power transformer, and is usually damaged by vibrator failure. Replacing the capacitor prevents it from ruining the new vibrator.

Fig. 16. Assembled and cutaway views of fuse holder used in battery lead of auto radio

Auto-radio Fuses. When a radio is installed in a car, there is usually only one power lead. It runs from some point in the battery system to the radio set itself. This lead is usually a shielded cable which has somewhere in its length a small fuse holder. The two telescoping sections of the auto-radio fuse holder are locked together with a bayonet connection, as in Fig. 16. Inside the cylinders is a glass cartridge-type fuse.

In many auto-radio failures, the fuse has burned out. This may be because the vibrator in the radio sticks and draws a very heavy current, or it may be due to a shorted capacitor or to normal aging of the fuse. Ordinarily the fuse is easily replaced. Use the same current rating as for the old fuse. Usually this is 7½ amperes for 12-volt sets and 14 amperes for 6-volt sets. Since this fuse is in a concealed place, your customer may not be aware of its existence.

If the replacement fuse burns out almost immediately when the auto radio is turned on, there is probably trouble inside the set. Remove the vibrator, put in a new fuse, and try the set. If the second fuse does not burn out, a new vibrator is needed or there is other plate-supply trouble.

Auto-radio Power Connections. A single-wire power connection is all that is needed for an automobile radio, since the metal radio cabinet is bolted to the bulkhead of the car and is therefore connected to the frame. The storage battery in the car has one terminal connected to the frame; hence the auto frame serves as the return path for battery current. If the mounting bolts get loose, the path may be broken.

If the power connections are secure and the radio set does not operate properly, measure the d-c voltage between the car frame and the terminal from which the radio gets its power. Do this first with the radio turned on and the motor running, then with the radio on but the engine stopped. The generator furnishes charging power to the battery when the engine is running fast, increasing the voltage of the entire system to as high as 8 or 16 volts in modern cars, depending on whether a 6- or 12-volt battery is used. This high voltage explains why a car radio often sounds louder when the motor is running. If the voltage is much below 6 or 12 volts with the engine stopped, the battery is defective or needs charging. It should be checked at a service station or a garage. If the voltage is above 7.2 or 14.4 volts when the engine is running at normal driving speed, suggest that the customer have the voltage regulator in his car checked by a garage.

Auto-radio Spark Plates. The battery lead of an auto radio generally goes to a spark plate located as close as possible to the hole through which the lead enters the housing of the set. A spark plate is simply a small metal plate that is insulated from the chassis by heavy paper or sheet mica, to give a small capacitance that acts with an r-f choke to prevent ignition interference noise from getting into the receiver through the battery lead.

A small selenium rectifier is sometimes used between the battery lead and the chassis in place of a spark plate. Trouble in these parts is so rare that it is enough for you to understand how they work. Typical connections are shown in Fig. 17.

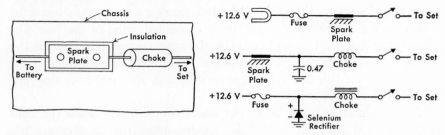

Fig. 17. Spark-plate construction and location in auto-radio circuits, and method of using selenium rectifier in place of a spark plate

Hybrid Auto Radios. The changeover from 6- to 12-volt storage batteries for automobiles, combined with the development of tubes and power transistors requiring no voltages higher than 12 volts, made it possible to eliminate the vibrator-type power supply from auto radios. The resulting auto radio, using both tubes and transistors, is commonly called a hybrid auto radio.

Hybrid sets offer no particularly new servicing problems. The same battery polarity precautions must be observed as for transistor radios. Terminals must not be grounded while power is on, because of possible damage to the transistors and melting of the test prod by the heavy current that can be delivered by a 12-volt storage battery.

There will be no vibrator buzz when a hybrid radio is turned on, but there should be an instantaneous thump in the speaker. This is produced by a surge of current through the output transistor, which requires no warmup time.

Hybrid and all-transistor auto radios should never be operated when the speaker is disconnected. Without the speaker as a load on the output transistors, the voltages on the collectors of these transistors can rise high enough to damage the transistors.

Testing Auto-radio Tubes. Tubes in hybrid radios are best tested by substitution, just as in television sets. Older tube testers apply voltages that can ruin tubes designed only for storage-battery voltages. If you prefer to use a tube tester, be sure it is a new model having a low-voltage shorts-test circuit and other features required for testing these tubes without ruining them.

A type 0Z4 metal tube or 0Z4G glass tube is often used as the rectifier in a vibrator-type power supply. This is a gas tube and requires no filament voltage. It can go bad just like any other tube, and generally goes bad suddenly. The tube can be tested in practically all tube testers; if it does not strike or light up in the tester, to show that an arc has formed in the gas, it should be replaced.

The type 0Z4G glass tube will generally be located in a metal shield called a hash box, which also contains the power transformer and vibrator. Be sure that the cover of this box is replaced, because the tube can cause interference noise if not shielded.

Other tubes in vibrator-type auto radios can be tested conventionally in tube testers and replaced the same as in ordinary radios. In a high percentage of auto radios, you can get at the tubes and the vibrator by removing the bottom cover of the set, without taking the set out of the car.

Polarity in Auto Radios. In some automobiles the positive terminal of the storage battery is grounded to the frame, while in others the negative terminal is grounded. Hybrid auto radios and all-transistor radios require a definite polarity. Reversing the polarity of the voltage applied to these sets can immediately damage transistors and low-voltage electrolytic capacitors, blow the fuse, and possibly ruin other parts.

All-tube auto radios having ordinary vibrators will work with either polarity. Only older sets having synchronous vibrators require a particular polarity. The synchronous vibrator had extra contacts that served in place of a rectifier but became unpopular because the contacts did not last long.

Some auto radios have a switch or other provisions for reversing the polarity. You will generally need the service manual as a guide when changing the polarity switch.

Tube-type Portable Radios. Older portable radios generally have 1.4-volt tubes that get filament current from a 1.5-volt dry cell. In addition, these sets usually require either a 67½- or 90-volt B battery to supply the plate voltage for the tubes. The two batteries may be separate or may be combined in a single AB battery pack, depending on the type of set and its design. Older three-way portables have in addition a rectifier and switching arrangement that permits operation from a 115-volt a-c or d-c line as well as from batteries.

The AB battery pack shown in Fig. 18 is used in large tube-type portables and three-way portables. The pack generally has a single socket that fits a corresponding plug at the end of the battery cable in the set, so there is no tedious connecting of individual wires when the battery pack is changed.

Small tube-type portable radios usually have separate A and B batteries, as in Fig. 19. Usually from two to five sets of A batteries will have to be replaced before it is necessary to put in a new B battery. With these sets, then, the first thing to try is new A batteries.

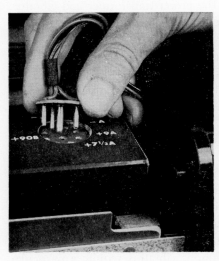

Fig. 18. Inserting plug in typical AB battery pack for older tube-type three-way portable radio. Unequal spacing of pins makes it impossible to get the plug in wrong. (Ray-O-Vac photo)

Fig. 19. Connections are made automatically when flashlight cells are pushed into this tube-type portable radio and cover is replaced. B battery has snap connections. (Ray-O-Vac photo)

Three-way Portables. A few models of older three-way portables have the 1.4-volt tube filaments wired in series much as in universal a-c/d-c sets. The series connection simplifies changeover from battery operation to power-line operation. The required A-battery voltage is therefore 6, 7½, 9, or 10½ volts, depending on the number of tubes in the set.

In most three-way portables, the line-cord plug must be inserted in a special outlet on the chassis when battery operation is desired. Inserting the plug generally serves to operate a switch that completes the connections to the batteries. Removing the plug disconnects the batteries to permit power-line operation.

In some three-way receivers the batteries are left in the circuit at all times. This tends to lengthen battery life, because on power-line operation the voltages are usually higher than battery voltages and send a reverse current through the batteries that has some rejuvenating effect.

Troubles in Three-way Portables. Power-supply defects in a three-way portable receiver can be localized by determining whether defective operation exists at all times or only when the set is used on one of the power sources. If the set will not work on either type of power supply, a signal circuit, a bad tube, or a defective power-supply circuit should be suspected. Trouble that occurs only on power-line operation is confined to the a-c/d-c power-supply system, its switching, or circuits that are in use only on power-line operation.

Tube-type receivers designed for operation from either batteries or a power line are often sensitive to line-voltage changes. When the a-c line voltage goes below a critical point, the local oscillator stops and the set goes dead. As the set ages, or as it warms up during use, this critical point goes higher and higher.

The failure of the oscillator is due to a reduction in the oscillator-tube filament voltage, combined with the normal aging of the selenium rectifiers used in these sets. The three parts to suspect when you encounter a three-way radio that works intermittently are the oscillator tube, the selenium rectifier, and the voltage-dropping resistor in series with the tube filaments.

To find out which part is guilty, try a new oscillator tube first. If this does not clear up the trouble, remove the chassis, connect your d-c voltmeter to measure the d-c output of the selenium rectifier, turn the set on, and note the voltage. Let the set operate for about half an hour. If the voltage is the same at the end of this warmup period, the rectifier is okay. If the voltage drops, put in a new rectifier.

If the rectifier is good, increase the filament voltages on all tubes by reducing the value of the voltage-dropping resistor in series with the filaments. This can be done by connecting a 15,000-ohm 2-watt resistor in parallel with the resistor in the set. After doing this, check filament voltages on a few tubes to make sure they are not higher than rated values. If too high, try a larger resistor value than 15,000 ohms.

Three-way portables may have high-capacitance electrolytic capacitors across tube filaments. These often cause squeals or gurgles when they go bad.

Effect of Weak Batteries. As a battery is used, its voltage gradually falls. The radio first loses its ability to receive distant stations, and the sound becomes distorted.

The drop in performance in a battery-radio set starts from the time a new battery is connected and gets progressively worse, until finally the set fails to operate at all. This reduction in performance due to dropping battery voltage is so gradual that it is not usually noticed by the customer as he uses the set. He probably would not consider replacing the battery as long as local-station reception was satisfactory.

There comes a time, however, when the set will play for only a short period when it is turned on. After the batteries have had a short rest the set may play again for an hour or so and then stop. At this point the customer will probably decide that he should replace the battery. This is usually done without bothering to check with the serviceman or measure the voltage of the battery.

Tubes in a radio should be checked whenever a battery replacement is necessary. It may be found that the tube failure has caused the set to stop operating and that there is some battery life left in the old battery. *Caution:* Always turn off a battery set or three-way portable before removing or replacing tubes, to protect the other tubes from burning out.

In battery receivers you cannot see when the tubes are lighted unless the set is in a dark room and you carefully examine each tube. By looking at the bottom of the tube elements in a dark room you may be able to see a deep cherry-red color indicating that the filament is operating. When in doubt, check the tube in the tube tester.

When battery installation is somewhat complicated, replacement and connection instructions will usually be found on a label pasted inside the cabinet. Look for it and read it carefully before putting in new batteries. Better yet, study the positions and connections of the old batteries before you take them out.

Replacing Batteries in Transistor Radios. Most transistor radios today use ordinary flashlight batteries. Since these are available at hardware and drugstores and are replaced almost as easily as in flashlights, most people buy and install the batteries themselves.

An example of a transistor portable radio that uses four flashlight batteries is shown in Fig. 20. All four batteries should be replaced at the same time. Just pull out the old batteries, clean the contacts in the radio if they look corroded, then insert the new batteries one by one.

You are most likely to get transistor radios for service when the customer accidentally puts in the batteries with wrong polarity. In some cases this can damage transistors and other parts, so the set may not work when you do put in new batteries correctly. Here is one time where you can

Power-supply Troubles 329

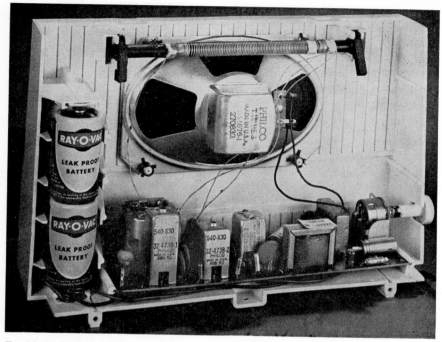

Fig. 20. Transistor portable radio in which four flashlight batteries (two are behind cells at left) provide the required 6 volts. Ferrite-rod antenna is mounted above oval speaker. Transistors are permanently wired into printed circuit. The two leads from the speaker go directly to the output transformer. Slot in plastic cabinet at right permits removal of printed-circuit board without taking knob off volume control. (Ray-O-Vac photo)

logically suspect damaged tranistors first if the set does not work when new batteries are put in.

Many of the larger transistor radios are designed for 6-, 7½-, or 9-volt operation. Some of these sets use individual flashlight batteries as in Fig. 21, while others use a single special battery having the correct total voltage. Single batteries will usually have button-type terminals as in Fig. 22.

Battery terminal strips in a receiver are pushed over the battery buttons with correct polarity. The terminals are so designed that a wrong connection cannot be made permanently. Study the terminals carefully before putting in a new battery, because you can ruin transistors if you accidentally touch the terminals together with wrong polarity while the set is turned on. Here is another good reason for turning off a set before installing new batteries.

330 Television and Radio Repairing

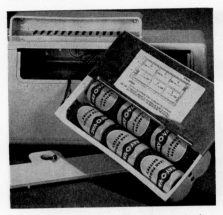 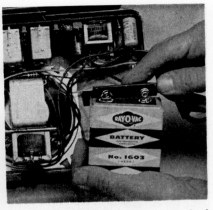

Fig. 21. Example of 9-volt transistor radio having six individual flashlight batteries in removable battery compartment. Diagram on cover gives correct polarity. (Ray-O-Vac photo)

Fig. 22. Placing terminal strip on new 9-volt battery for transistor portable radio. Battery terminals have different shapes. (Ray-O-Vac photo)

Rechargeable Batteries. Transistorized television receivers and a few large transistor radio sets use miniature storage batteries that can be recharged many times.

For portable television receivers, recharging is recommended after 4 hours of use or oftener. If the battery is used longer than this, a chemical reaction occurs that permanently reduces the battery's rechargeability. Recharging instructions are usually displayed prominently on the receiver. If the user ignores these instructions, the voltage may drop.

Rechargeable batteries must be charged much more slowly than they are discharged. In one receiver, it takes 20 hours of charging to restore the full charge after 4 hours of use. A moderate amount of overcharging will not damage the battery. Charging units that operate from an a-c power line are built into the receiver or sold as an accessory.

Checking Batteries. The best way of checking the condition of a battery is to measure the d-c voltage across its terminals after the set has been operating for about 5 minutes. If the voltage is appreciably lower than that specified in the service manual, the battery needs replacement if it is not rechargeable.

If you do a lot of work on battery sets and batteries, it may be desirable to have a battery-testing meter like that in Fig. 23. This is just a voltmeter with built-in load resistors. A switch allows you to select proper loads for various battery voltages.

If a battery is rechargeable, ask if it has recently been charged for the required period. If not, put it on charge overnight and recheck the voltage the next day. If still low, the rechargeable battery needs replacing or the receiver has a defect that is placing an overload on the battery.

To determine whether overloading is the cause of trouble, measure the current being drawn from the battery. If it is appreciably higher than the

Fig. 23. Example of battery tester in use. An offer to check batteries free often brings new customers to a shop. (RCA Electron Tube Division photo)

value specified in the service manual, look for trouble in the receiver rather than the battery.

Mercury Batteries. Some transistor radios use a mercury battery that has much longer life than the conventional zinc-carbon flashlight cell. A mercury cell has a no-load voltage of about 1.345 volts when new. This voltage holds up well throughout the useful life of the battery, so only a slight decrease in the measured voltage is a clue that the battery is near the end of its useful life.

Solar Cells. It is possible to operate a transistor radio directly from the sun's energy, just as is being done in space satellites. In radios, the solar cells are mounted on top of the cabinet in such a way that they can be tilted and rotated to face the sun. The cells then generate a d-c voltage. This voltage operates the set on sunny days, and usually also charges a small storage battery that is used as the power source at night. Solar-cell radios are as yet an expensive novelty, but customers often ask about them.

Shelf Life of Batteries. Ordinary zinc-carbon flashlight cells have a shelf life of 1 to 2 years when stored at ordinary room temperature and humidity, but much less at the higher temperatures and humidities of southern seaboard locations. This is why many batteries have printed dates after which they can no longer be considered new. Keep your battery stock as small as possible and always sell the oldest units first, or order batteries only as needed. Do not store batteries near a steam radiator, near a hot-air duct, or in a damp basement.

Can Batteries Rejuvenate Themselves? Often a customer brings in a portable radio and says that the sound got weaker or faded out entirely after the radio had been playing for a few hours. To check performance, you turn on the radio and find that it is perfectly normal. The explanation here is that the battery was exhausted and became polarized. When allowed to rest for several hours, depolarization took place and the battery returned to normal. Check battery current drain. If it is higher than the specified value, look for trouble in the set. If the current drain is normal or lower than normal, the battery is nearing the end of its useful life and should be replaced.

Battery Rejuvenators. It is possible to prolong the life of dry cells as much as four times by using a battery rejuvenator. This is a simple battery-charging circuit that operates from the a-c power line and is built into some portable radios that use costly B batteries. For each hour that the batteries are used, the set should be plugged into the a-c line 1 hour for rejuvenation. The changeover to transistor circuits for portable radios makes these rejuvenator circuits unnecessary, because it does not pay to use the technique on inexpensive flashlight cells.

15

Testing and Replacing Carbon Resistors

Why Resistors Are Important. In all receivers, from the simplest radio set to the biggest and finest television set, only a few different types of parts are used. These building blocks—coils, capacitors, and resistors—work together with tubes or transistors to make possible the modern miracle of seeing and hearing actions that are occurring hundreds or even thousands of miles away.

If just one of these parts fails, the receiver may go completely dead, so not even a whisper comes from the loudspeaker and not a flicker of light shows on the screen; it may look and sound like a thunder-and-lightning storm; it may hum; it may oscillate or squeal like a stuck pig.

No matter how the misbehaving set acts, only one thing is required to make it operate again. You have to find and replace the part that has failed.

In many of the sets coming to you for repair, the part that has failed will be a resistor. It is pretty important, therefore, to learn how to test resistors, find the bad one, order the correct size replacement, remove the bad one, and install the new one as in Fig. 1. It is important, yes, and it is also quite easy because resistors are the simplest electric part used in receivers.

What Carbon Resistors Look Like. Most of the resistors used in television and radio sets are small carbon resistors. These are usually much thinner than a pencil and about half an inch long. They have only two wire leads, one coming from each end of the resistor. The leads usually come straight out of each end axially (in line with the body of the resistor). Older sets will have some carbon resistors in which the leads come

334 *Television and Radio Repairing*

off at right angles to the body of the resistor. The axial-lead resistors usually have a tan-colored plastic insulating covering around the body, whereas side-lead resistors have only paint on the body.

Most carbon resistors are circled with bands of different colors. Some capacitors also have color bands or dots and look like carbon resistors, but

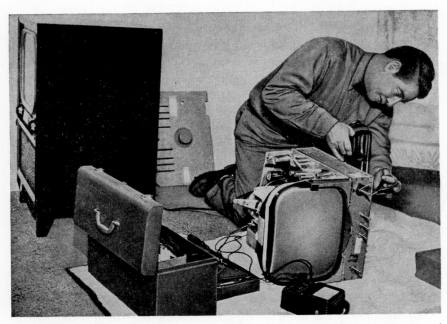

Fig. 1. Installing new carbon resistor in television receiver in customer's home. Note use of blanket-type pad under receiver and toolbox

when you study capacitors, you will learn how to recognize them by their color markings.

Review of Resistor Action. As you already know, resistors provide an electrical path that has a definite amount of opposition to current flow. This opposition is called *resistance*. Increasing the resistance in a circuit increases the opposition to current flow, thereby reducing the current. On the other hand, decreasing the resistance in a circuit allows more current to flow.

In some circuits only a small amount of resistance is required for proper operation. In other circuits, however, a lot of resistance is needed. The unit used to specify the exact amount of resistance is called the *ohm* (pronounced like ome in the word home).

Carbon Resistors 335

The resistance of any resistor is stated as being so many ohms, just as the voltage of any battery is stated as being so many volts. Large values are expressed in millions of ohms, called megohms. Thus, 5,000,000 ohms is 5 megohms, and 2,500,000 ohms is 2.5 megohms. A quarter-megohm, written 0.25 megohm, is therefore 250,000 ohms.

Abbreviating Resistance Values. The symbol used on diagrams to represent a resistor is a zigzag line. On most circuit diagrams there is not much space alongside a resistor symbol for printing the resistance value. Therefore, instead of writing out OHMS or MEGOHMS after the value, certain easily recognized abbreviations are used.

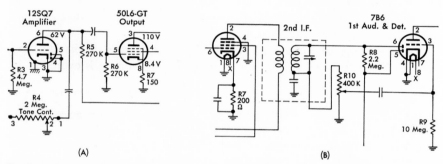

Fig. 2. Methods of specifying resistor values on circuit diagrams. Here the Greek letter omega represents ohms, K represents thousands of ohms, and Meg. represents millions of ohms.

The simplest abbreviation is nothing at all, and that is exactly what is used to represent ohms on many circuit diagrams. Therefore, if you see a numeral alongside a resistor symbol with nothing after it, assume that the numeral is the resistance value in ohms. An example of this is indicated in Fig. 2A for 150-ohm resistor R7.

For megohms the abbreviation is logically Meg or MEG, written with or without periods. This abbreviation is so logical that it has even become a part of the conversational language of a serviceman. Instead of saying two megohms, he will simply say two megs. Similarly, he will call out to his assistant, "Hey, Mac, hand me a two-meg resistor."

Sometimes it is even difficult to find space for lettering three zeros on a circuit diagram. For this reason, and also because it saves time, the letter K is used to represent a thousand ohms. Thus, 5K alongside a resistor symbol means 5,000 ohms; 250K means 250,000 ohms; 1.5K means 1,500 ohms. Note that K is used for this purpose in Fig. 2A.

Although the abbreviation system just described is widely used and entirely clear, some manufacturers use the capital Greek letter omega on diagrams to represent the word ohms, as shown in Fig. 2B. Read Ω as ohms whenever you see it, just as you would if Ω were not there. Sometimes Ω is used after MEG and after K also.

Other Abbreviations for Ohms. Occasionally you will find the letter M after a resistance value on diagrams. It may represent either a thousand ohms, like K, or a million ohms, like meg. You will have no trouble figuring out their meaning if you first look at all the resistor values on the diagram. You will always find that thousands of ohms are represented by K, M, or the actual three zeros of the complete value. The remaining abbreviation will therefore be megohms. Another clue is the fact that values in megohms are never larger than 22 in receiver circuits, because 22 megohms is the largest resistor used.

Power Ratings of Resistors. The ability of a resistor to handle and get rid of heat is called its power rating. This power rating is expressed in

Fig. 3. Actual sizes of ½-watt, 1-watt, and 2-watt resistors of one make. Units made by other manufacturers may vary slightly from these sizes

watts, the same as the power ratings of electric toasters, electric-light bulbs, and all electrical appliances. The commonest wattage ratings for carbon resistors are ½, 1, and 2 watts.

As you will recall, the heat-producing power drawn by a resistor increases when the applied voltage is increased or when the current increases. If heat is generated by a resistor faster than it can be passed off to the surrounding air, the resistor gets hotter and hotter. The resistor may even smoke and eventually burn up.

The larger (longer and fatter) a resistor is, the higher is its power rating in watts and the more heat it can handle without getting too hot. Replacement resistors with various power ratings are shown actual size in Fig. 3. You can use this illustration as a rough guide for finding the power rating of any carbon resistor.

Occasionally you will come across larger carbon resistors, having power ratings up to 5 watts. You will also find much smaller carbon resistors in receivers, rated as low as $\frac{1}{10}$ watt because they cost the manufacturer less and are in circuits where there is practically no current to produce heat. Use ½-watt resistors for replacing these.

Carbon Resistors 337

Finding Sizes of Burned-out Carbon Resistors. How do you find the correct size for the new resistor if the old one is burned out and there is no value printed on it or the value is charred beyond recognition? This problem has a simple answer—look up the circuit diagram of the receiver, and locate the resistor on the diagram.

If the service manual that you are using contains a photograph or a pictorial wiring diagram showing the exact position of every resistor under the chassis, your job is easy. Just locate the bad resistor on the illustration with respect to easily recognized large parts, as in Fig. 4. Note its identifying number, look up this number on the accompanying parts list (Fig. 5) or on the circuit diagram (Fig. 6), and read its resistance value in ohms. Usually the correct wattage rating is also specified.

As an example, assume the resistor at the right of the shield can in Fig. 4 is charred black. The arrow line and number identify it as R77, and Fig. 5 tells you that to replace it you need a 100,000-ohm 1-watt resistor. This table even gives the corresponding IRC (International Resistance Co.) part number for this resistor as BTA-100K, to simplify ordering a replacement. The circuit shows the parts to which R77 is connected, but need not even be looked at for this job unless you forget how the old resistor was connected after you have removed it.

Use of Circuit Diagram. When no parts-locating photo or diagram is available, you can work from the circuit diagram itself. This has many landmarks that are easily identified both on the diagram and on the set. The best landmark is a tube-socket terminal.

If the bad resistor in your set is connected to a certain terminal on a particular tube, locate the socket for that tube on the diagram and look for a resistor connected to the same terminal. This is illustrated in Figs. 7 and 8.

If more than one resistor goes to the terminal in question, you will have to find out where the other lead of the bad resistor goes. If the other lead goes, for instance, to a 100,000-ohm resistor in your set, you would then choose the resistor on the diagram that also went to a 100,000-ohm resistor.

Whenever tracing resistor connections in a set or on diagrams, consider wires as direct connections no matter how long they are. In a set, a resistor may be at one end of the chassis and connected by a long wire to a tube-socket terminal at the opposite end of the chassis. On a diagram the same resistor may be drawn right next to that tube terminal or placed in some other convenient location.

338   Television and Radio Repairing

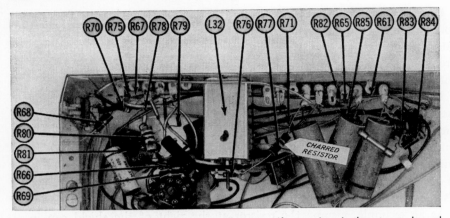

Fig. 4. Portion of parts-locating Photofact Folder illustration. If one resistor in the set were burned out, as indicated by arrow, this illustration would quickly identify it as **R77**. (Howard W. Sams photo)

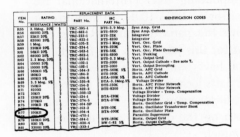

Fig. 5. Example of resistor replacement data in Photofact Folder for set shown in Fig. 4. Value for resistor R77 (encircled) is 100,000 ohms. (Howard W. Sams illustration)

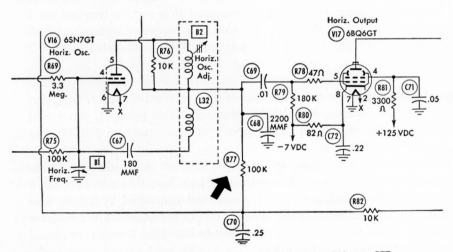

Fig. 6. Portion of circuit diagram for receiver of Fig. 4, with arrow pointing to **R77**

Carbon Resistors 339

Fig. 7. Bottom of chassis of a-c/d-c superheterodyne radio. Practically every resistor here goes to a tube terminal or other easily recognized landmark. Assume the problem is to find the value of the bad resistor at top center, the left lead of which goes to pin 7 of the 35Z5 socket. (Howard W. Sams photo)

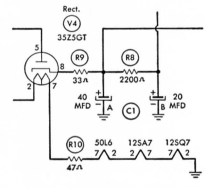

Fig. 8. This portion of the circuit diagram for the receiver of Fig. 7 shows that the only resistor connected to pin 7 of the 35Z5 tube is 47 ohms

How Carbon Resistors Go Bad. Although carbon resistors by themselves ordinarily last a lifetime, occasionally one will fail because of mishandling, overheating, or failure of another part in the radio. No matter what the cause, however, they can fail in only three ways. They either break, burn out, or change their resistance value because of aging or overheating.

Broken Carbon Resistors. When a carbon resistor breaks, it no longer offers a path for current flow. A broken resistor is just like no resistor at all.

Any number of things can cause carbon resistors to break. The leads of the resistor may have been pulled too tight when the resistor was being connected into the circuit originally. The leads may have been bent too close to the body of the resistor. Too much force may have been put on the resistor when the set was being checked and repaired in a shop. Vibration or ordinary jarring may have caused the break at a weak point in the body of the resistor or in one of its wire leads.

Ordinarily you will not be able to find a broken carbon resistor just by looking at it. As long as the insulation does not break, there is no way to tell that the resistor is broken internally except by checking with an ohmmeter.

Sometimes a carbon resistor does not stay broken. Any little jarring of the resistor causes the two ends of the break to come together again. The next little jar may cause the two ends of the break to fall apart again, though. It may take days or weeks for the trouble to occur again, or the resistor may break and fix itself every few seconds. A resistor that will not stay broken is known as an intermittent defect.

Burned-out Carbon Resistors. If an excessively high current flows through a carbon resistor for any length of time or if the resistor develops a heat-producing internal flaw, the resistor begins to smoke. If this overheating is continued long enough, the resistor may actually burn in two and become open. The resistor no longer provides a path for current flow, so the excessively high current stops.

Rarely will a carbon resistor overheat and burn up because of failure of the resistor itself. Ordinarily the overheating will occur either because one of the resistor leads touches another part in a crowded chassis or because of failure of some other part associated with the resistor in the circuit. If another part is at fault and drawing too much current through the resistor, replacing the resistor will not fix the set. If the defect that caused the resistor to overheat is not also fixed, the new resistor will overheat and burn out too.

A shorted capacitor is the commonest cause of resistor failure. That is why servicemen say that troubles most often come in pairs when one is a resistor.

Changes in Resistance Values. When carbon resistors overheat, their resistance value changes. Sometimes the resistance may go back to its original value when the resistor cools off, but more often the resistance stays at its new and wrong value.

If the resistance of a resistor changes sufficiently, the receiver goes bad and a serviceman is called. The set may sound weak, distorted, or even dead, just as for failure of any other part.

Finding Bad Carbon Resistors. Often you do not even have to use a test instrument to find a bad carbon resistor. Your ears, eyes, and nose are enough to tell you which one it is, particularly if there has been overheating.

Broken resistors and those with poor connections are found by the noise test, which involves tapping each resistor in the set with a wooden stick and noting which one makes the most noise when jarred while the set is turned on.

If these quick and easy methods do not work, you still have the always-reliable ohmmeter test to fall back on, for checking the resistance values of the suspected resistors. The three methods of finding a bad resistor will now be taken up in turn.

Ear-eye-nose Test. Most carbon-resistor failures are due to overheating. The overheating may cause the resistor to burn itself up or change its value of resistance. Most likely the resistor will be blackened, blistered, or entirely burned in two.

The first thing to do in checking for a bad resistor, therefore, is to inspect all the resistors under the chassis of the set. If you find one that is blackened, suspect it and check its value with an ohmmeter as described later. If you find one that is entirely burned in two, however, find what caused it to burn apart before replacing it.

If you do not find a bad resistor just by looking at the set, you may be able to find one when the set is turned on. If an excessively high current is passing through one of the resistors when the set is on, the resistor in trouble will give itself away. It may smoke like a house afire; it may crackle like butter frying in a hot frying pan; it may produce the strong pungent smell of an overheated resistor.

Checking for a defective part first with the ears, eyes, and nose is one of the most important procedures in servicing. Often the defect can be located by this means alone. If it is, a lot of time is saved, so always remember to look first.

Noise Test for Carbon Resistors. Sometimes a receiver becomes noisy when the resistance material in a carbon resistor cracks or one of the leads pulls loose from the resistance material. The noise occurs when the two ends of the break in the resistor temporarily come together and then fall apart again. Each time this make and break occurs, you will hear a crackling noise or a loud crash from the loudspeaker. If the action occurs often enough, it will sound like frying instead.

If you suspect that a resistor is causing the set to be noisy, check for the bad resistor with the set on and at normal sound volume. While listening to the sound coming from the loudspeaker, wiggle and tap each resistor and its leads with an insulating stick or piece of wood. If wiggling a particular resistor increases the noise, you have probably found the bad one.

To make sure that the resistor you suspect is actually the one causing the noise, keep all the other parts from moving while you wiggle the one resistor back and forth.

Measuring Resistance of Carbon Resistors. When you cannot find the bad resistor with your ears, eyes, and nose or with the noise test, measure the resistance of each resistor you suspect. Compare each measured value with the specified correct value.

Before connecting the ohmmeter, always pull the receiver line-cord plug out of the wall outlet. If the receiver is a battery-operated set, always disconnect all the batteries. If the power source is not disconnected, you may burn out your meter or bend its pointer way out of line. This can happen in a fraction of a second. Meter repairs are expensive, so never use an ohmmeter where there is a possibility of getting it connected across a voltage.

If there are possible shunt paths across the resistor being measured, disconnect one end of a resistor that gives a suspicious low reading and measure the resistor again. If you do not disconnect one end of the resistor, the resistance you measure may be the combined resistance of the resistor and some part connected directly across it. The measured value will then be too low and may be meaningless.

When checking, connect the ohmmeter directly across the resistor. Choose a resistance range that puts the pointer near the center of the scale, where the ohms scale is easiest to read and has greatest accuracy.

Finding Open Resistors with an Ohmmeter. If there is no deflection of the pointer when you connect the ohmmeter across a resistor, even on the highest resistance range, you have found a bad resistor. It is broken or burned out internally, and is said to be open. The highest range of your multimeter is at least 20 megohms, and resistors larger than this are rarely if ever found in receivers.

Carbon Resistors 343

Finding Changed Values of Resistance with an Ohmmeter. When the resistance of the resistor being checked is more than 20 per cent larger or smaller than its rated value, the resistor may be the cause of the trouble and hence should be replaced. Thus, a 100-ohm resistor that measures less than 80 ohms or more than 120 ohms should be replaced. Replacement of an off-value resistor is the only way to find out if it is the troublemaker.

Finding Noisy Resistors with an Ohmmeter. If wiggling the leads of the resistor whose resistance is being measured causes the pointer on the ohmmeter to move back and forth, the resistor is bad. When the two broken parts in the resistor are touching, the meter will show the resistance of the resistor. When the two broken parts come apart, the meter will show an open circuit.

Of course, the ohmmeter leads must be making good contact with the resistor leads during this wiggling; if they make poor contact, the meter pointer will flicker and a good resistor may be falsely accused. Always use alligator clips on the ohmmeter leads to get good tight connections.

Getting the Correct Replacement Carbon Resistor. Having found the bad resistor, the next step is getting a new resistor with the same or a higher power rating and the same value of resistance. The way these two ratings are found for the bad resistor will now be taken up.

Power Rating of Replacement. To find the power rating of a carbon resistor, compare its size with those shown actual size in Fig. 3. Resistors having the same physical size will have essentially the same power rating. When in doubt, choose a higher power rating for safety. Sometimes the power rating in watts is given after or under the resistance value on circuit diagrams, using the letter W to represent watts. Thus, ½W means a power rating of one-half watt.

In many receivers the smallest-size resistor that will not burn up is used by the manufacturer to save a few cents. Whenever a resistor burns out, it is a good policy to replace it with one having a larger power rating. The extra power-handling capacity may prevent a future failure.

Replacement with resistors having a higher power rating gives you the added selling point of being able to say that the parts you put in are better than the original ones. Half-watt resistors are generally the smallest handled by jobbers for replacement use, anyway.

Resistance Rating of Replacement. The resistance value of a resistor is given on the circuit diagram or in the parts list of the receiver.

If the resistance value is printed on the bad resistor and can still be

344 *Television and Radio Repairing*

read, this is of course the easiest and fastest way to get the information. Printed values can be read directly, but more often the only markings are color bands and dots that represent the numerals of the resistance value. These color codes are easily read and are used often. It is well worth while for you to memorize the list of ten colors and their corresponding values after you have mastered the following explanation of the codes.

Color Coding of Carbon Resistors. The color code on a resistor specifies the resistance of the resistor in ohms just as well as if the value were printed. It is a standard EIA (Electronic Industries Association) code.

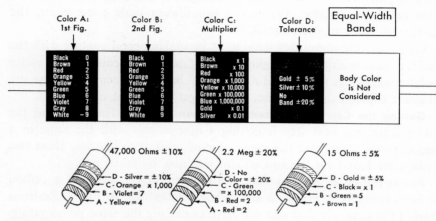

Fig. 9. Color code for carbon resistors, with three examples of how equal-width bands are read. A double-width band for color A indicates that the resistor is wirewound

Color-coded resistors are painted with different colors in a definite manner to show resistance value. One method uses three equal-width bands of color around the body of the resistor. Another common method uses the body color of the resistor, an end color, and a central dot or band of color. In both methods, the three colors are read in a definite order, and numbers are substituted for colors according to the code in Fig. 9 to get the resistance of the resistor.

Occasionally a fourth color is used to show the tolerance or accuracy of the resistance marking. This color is easily identified because it is either gold or silver. The tolerance color is ignored when reading the value of the resistor. The meaning of resistor tolerance is taken up later in this chapter.

Reading Resistors Having Equal-width Bands of Color. When the resistor has three or more equal-width bands of color, the colors are read

Carbon Resistors **345**

in order, starting with the band at the end. The first color is the first numeral in the resistor value. The second color is the second numeral in the resistor value. The third color gives the multiplier by which you multiply the first two numerals to obtain the resistor value in ohms. The fourth color band, if there is one, will be either gold or silver, and gives the tolerance of the resistor.

Reading Resistors Having Body-end-dot Color Code. When resistors have leads coming off at right angles to the body, a different method of painting the colors on the resistor is used. In this body-end-dot method, the body color of the resistor is the first figure in the resistance value. The end color is the second figure. The center dot is the multiplier. The gold or silver tolerance color, if used, will be on the other end of the resistor.

The body-end-dot method of color coding is occasionally found on older resistors having in-line leads, as at the lower right in Fig. 9. The method of reading these is the same as for side-lead resistors.

Just remember that, when color bands are of different widths, you read body color A first, then end color B, then center dot or band C. When color bands are all the same width, you read the colors in order from the end and ignore the body color.

Resistor Color-code Example 1. Suppose you have a resistor with equal-width bands of red, green, and yellow, reading from the end. Since the bands are equal in width, red represents the first numeral, which is 2. Next is green, which is 5. The third color is yellow, which stands for $\times$ 10,000, so you simply multiply 25 by 10,000 to get the complete resistance value in ohms—250000 ohms. Adding a comma is the final touch to make it easier to read, giving 250,000 ohms.

Note that yellow represents the number 4 for colors A and B, and the four zeros of the multiplier for color C. This holds true for all colors, so that brown is 1 or one zero, red is 2 or two zeros, and so on. Thus you need only memorize the following list of ten items to memorize the color code:

Black is zero
Brown is one
Red is two
Orange is three
Yellow is four

Green is five
Blue is six
Violet is seven
Gray is eight
White is nine

Resistor Color-code Example 2. Suppose the background color of a resistor is brown, the end color is red, and the dot in the center is orange.

This is read according to the body-end-dot method, so brown is 1, red is 2, and orange is 1,000. The resistance value thus is 12,000 ohms.

Resistor Color-code Example 3. Suppose color A is brown, color B is black, and color C is black. Here brown is 1, black is 0, and black for C means $\times$ 1, so the value is 10 ohms.

Missing Colors. If one of the three color markings on a resistor appears to be missing, the missing color may be the same as the body color of the resistor. Therefore, try reading the body color when there is no other color in a location where one should be. The perfect example of this is an all-red resistor having leads coming out at right angles as in Fig. 9. Here body color A is red for 2, end color B is red for 2, and dot color C is red for 100, so the value is 2,200 ohms.

Tolerances of Carbon Resistors. The fourth color marking, either gold or silver, is used on some carbon resistors to specify the tolerance of the resistor. Tolerance means how much the actual value of resistance may deviate from the numerical value indicated by the color code.

In resistors, as in so many other things, accuracy costs money because it means more careful adjustment during manufacture. A 100-ohm resistor that is accurate to within 20 per cent (actual value somewhere between 80 and 120 ohms) may cost 10 cents, whereas a 100-ohm resistor that is accurate to within 1 per cent (actual value between 99 and 101 ohms) may cost seven times as much.

Most of the resistors used in radio receivers have a tolerance of 20 per cent. This means that their actual resistance may vary as much as 20 per cent above or below the value specified on the resistor. In circuits where a more accurate resistor is required, a resistor with 10 or 5 per cent tolerance is used.

Reading Tolerances. When no tolerance color is painted on a resistor, the tolerance of the resistor is 20 per cent. This means that, when you measure this resistor with an ohmmeter, you will get any value from 20 per cent below to 20 per cent above its rated value. Yes, strangely enough, a brand-new 100-ohm resistor may measure anywhere from 80 to 120 ohms and still be considered entirely good if it has 20 per cent tolerance. Here 20 per cent of 100 is 20, so the resistor can be as much as 20 ohms below or above 100 ohms.

If the tolerance band or color is silver, the resistor has a 10 per cent tolerance. This means that the resistance value can be anywhere between 10 per cent below and 10 per cent above its rated value. A 100-ohm resistor with 10 per cent tolerance could be anywhere between 90 and 110 ohms.

If the tolerance band or color is gold, then the tolerance of the resistor is 5 per cent. The resistance value now is somewhere between 5 per cent below and 5 per cent above its rated value, which for carbon resistors is pretty good accuracy. For a 100-ohm resistor, the tolerance range would be 95 to 105 ohms.

Locations of gold and silver tolerance markings on carbon resistors are shown in Fig. 9, but you will have no trouble spotting these because they are always gold or silver for resistors.

Figuring Tolerance Limits. It takes only a little simple multiplication to find what the actual limits of resistance are for a resistor having a particular tolerance. Knowing these limits, you can easily tell whether or not a resistor you are checking has changed its resistance too much. In addition, knowing the permissible limits of the original resistor allows you to use a replacement resistor anywhere within the limits, if you do not have a new resistor on hand with the same resistance as the old one.

To find the tolerance range of a resistor, multiply the resistance value in ohms by the per cent tolerance expressed as a decimal. Thus, for 20 per cent, multiply by 0.2 to get the tolerance value; for 10 per cent, multiply by 0.1; for 5 per cent, multiply by 0.05. Subtracting this tolerance value from the rated resistance value gives the lower limit, and adding gives the upper limit.

As an example, for a 50,000-ohm resistor with 20 per cent tolerance, the tolerance value is 0.2 times 50,000 ohms, or 10,000. Its lower limit of resistance would then be 50,000 minus 10,000, which is 40,000 ohms; its upper limit would be 50,000 plus 10,000, which is 60,000 ohms. Thus a 50,000-ohm resistor with 20 per cent tolerance may actually have any value between 40,000 and 60,000 ohms.

If the 50,000-ohm resistor had 10 per cent tolerance, it could vary only 5,000 ohms above and below its specified value—between 45,000 and 55,000 ohms.

In a similar manner, a 50,000-ohm resistor with 5 per cent tolerance could be between 47,500 and 52,500 ohms.

For replacing bad carbon resistors in television and radio sets, a tolerance of 10 or even 20 per cent is satisfactory practically all the time.

Tolerances of Replacement Resistors. In replacing a carbon resistor, the goal is to get a new one with the same resistance value in ohms, the same tolerance or a *lower* per cent tolerance, and the same or a *higher* power rating in watts.

When you do not have the right resistance value on hand, figure out the tolerance range and use any resistor that is within this range when measured with an ohmmeter. This check with an ohmmeter is highly desirable, because the new resistor will have its own tolerance range and its actual resistance value may be outside the acceptable limits. Of course, you can always connect a questionable resistor temporarily; if it cures the trouble, fine—leave it in.

Removing a Defective Carbon Resistor. Once you locate a defective carbon resistor and have obtained a suitable new resistor for replacement, the next step is to remove the bad resistor. Apply the hot soldering-gun tip to the end of one of the resistor leads. As soon as the solder on the joint has melted, grasp the resistor lead with long-nose pliers near the joint, and gently pull and wiggle the lead until it comes out of the joint. Sometimes you may have to pry open the hook in the resistor lead first with pliers or a screwdriver so you can pull it out. If the lead is hopelessly jammed into a terminal lug with a lot of other wires, do not try to get it out; just cut it off as close as possible to the joint.

Repeat the unsoldering procedure for the other resistor lead, and your bad resistor is out. If you do a good neat job of unsoldering, you will not even be able to tell where the resistor was connected, and this brings a warning: Before removing a resistor, make a little diagram showing accurately the terminals to which it was connected, so that you cannot possibly make a mistake when putting the new part in. Make a mistake just once, and you will see why this warning is so important. Whereas it takes only a few minutes to locate a defective part, it can take hours to locate a good part that is connected wrong.

Making a diagram of connections is a good idea *before removing any part*. Suppose you were relying on your memory and someone came into your shop just as you had pulled out the bad part. You might not get back to work on the set for several hours, and you would really have to be good to remember where that resistor came from after such a long interruption.

After removing the defective resistor, examine the terminals in the set carefully to make sure that you did not also loosen some other leads. Unsoldering often results in lumps of solder running down between terminals, so look for these also and remove all surplus solder. If the leads of the new resistor are to go through the holes in the soldering lugs, you

Carbon Resistors 349

may have to spend a little time cleaning the solder out of the holes with your freshly wiped hot soldering-gun tip.

Installing a New Carbon Resistor. The new resistor should always be installed just as soon as you have removed the defective one. This means that you should have the new resistor on hand before you remove the old one, to minimize chances for mistakes in installing the new one.

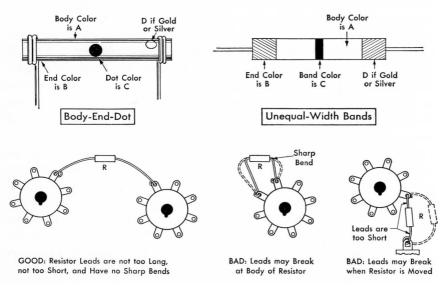

Fig. 10. Good and bad installations of replacement carbon resistors. Dotted lines show how bad installations should be rearranged

The leads of a new resistor will usually be longer than necessary. You can cut these leads down to the correct length either before or after installing the resistor. In both cases, leave the leads long enough so you can move the resistor around a bit after connecting it, but not so long that there will be danger of the bare leads touching other terminals. Examples of good and bad lengths of leads for carbon resistors are shown in Fig. 10.

Many servicemen prefer to insert the leads of the new resistor through the holes in the soldering lugs, pull the leads far enough to get the resistor in the desired position, then bend back the leads to form hooks and cut off surplus wire.

Having installed the new resistor and adjusted the leads to the correct length, form permanent hook joints by crimping with pliers, then solder the joints. Remember—hold the hot soldering-gun tip on one side of the joint, and apply rosin-core solder to the new wire until the hook in the joint is filled with solder that flows smoothly over the soldering lug.

When there are a number of wires at the terminal to which a new resistor must be connected, it is sometimes difficult to get the resistor lead through the hole in the terminal. In this case it is perfectly all right to bend the resistor lead around one of the leads already connected to the terminal. Of course, make a permanent hook joint here also, and be just as careful to do a good soldering job.

Installing High-voltage Resistors. The carbon resistors used inside the shielded high-voltage compartment of a television receiver are often special types designed to withstand high voltages. When ordering a replacement from your jobber for one of these, be sure to specify that it is for the high-voltage power pack of the set. Better yet, get the correct replacement resistor from the manufacturer's distributor for that make of set.

Before removing a burned-out high-voltage carbon resistor, study its connections carefully. Notice how long each lead is. Notice how smoothly rounded are its soldered connections. Notice the exact position of the resistor. These things are important because a high-voltage discharge called corona can occur if the new resistor is put in carelessly. Corona occurs at sharp points in high-voltage circuits. The resulting colored glow in the surrounding air makes a buzzing sound and interferes with the television picture.

Check Your Work. After installing the new resistor, make one last careful check to be sure you have connected it to the correct terminals and to be sure you have put in the right value of resistor. Poke each soldered joint of the resistor with a screwdriver to be sure you have good tight joints. Adjust the position of the resistor so it does not touch other parts, and adjust its leads so they are well away from adjacent terminals and other leads.

Though practically all end-lead carbon resistors made today for replacement use have insulated bodies, it is still good practice to arrange resistors so they do not touch other parts. With noninsulated side-lead resistors, this spacing between parts is of course essential. Finally, inspect adjacent parts and joints carefully to make sure no bare leads are touching.

Using Spaghetti on Resistor Leads. Very frequently, the underside of a chassis is so crowded that bare resistor leads are likely to touch adjacent parts and cause trouble. Here it is best to slip a short length of spaghetti

Carbon Resistors 351

insulation over each lead before installing the new resistor. You can buy spaghetti insulating tubing in various sizes from your parts jobber. Allow only enough wire to project from the spaghetti so you can form the hook joint. Avoid touching the spaghetti with your soldering-gun tip, because heat will make the spaghetti smoke and smell like any other burned insulating material.

Trying Out the Set. Having replaced the defective resistor, you are ready to test your work by trying out the set. You can often do this without replacing the chassis in the cabinet. First turn on the set, with or without putting the knobs back on the shafts of the controls. Now plug the line cord into the power outlet on your bench.

If you are working on a universal a-c/d-c set, remember that the chassis can be electrically hot for one position of the line-cord plug. Use an isolation transformer between the set and the power line to protect yourself from a dangerous shock.

Sometimes Resistors Are Not Guilty. If the receiver works after the new resistor is installed, fine. Put the chassis back in the cabinet, replace all knobs, screws, and gadgets, wipe off the cabinet so it looks like new, then collect your money for the job.

Do not be discouraged if the set does not work after the new resistor is put in. Sometimes a resistor goes bad because of failure of one or more other parts, such as a capacitor. If there are indications that your new resistor is running too hot and getting ready to smoke, look for a shorted tubular paper capacitor somewhere near the resistor. You will learn how to test capacitors like this in a following chapter. Look also for two leads that are touching together and causing a short-circuit that overloads the resistor. Remember—troubles do not always come singly.

A lot of other parts can go bad along with resistors, making it necessary to use troubleshooting procedures. Just put the set aside until you are ready to tackle it again, or have another serviceman finish the job if the set has to be returned to its owner promptly. There is a lot more to fixing sets than can be learned all at once, so do not expect to be able to fix every set right from the start.

Recommended Stock of Resistors. Since it is both impractical and unprofitable to make a trip to a parts jobber every time you need a new carbon resistor, you will want to start with a good assortment of these resistors.

Some resistor manufacturers sell collections of the most-used values in a handy cabinet, as shown in Fig. 11. Prices depend on the number of resistors and their power ratings. Your minimum starting stock can be all 1-watt resistors, and should include 27, 47, 150, 220, 390, 470, 680, 1,000, 2,200, 3,300, 3,900, 5,600, 10,000, 12,000, 15,000, 22,000, 27,000, 33,000, 39,000, 47,000, 56,000, and 68,000 ohms, and 0.1, 0.27, 0.47, 1.0, and 2.5 megohms.

Of course, larger assortments are still more desirable, but the above will give you a good start at a big saving over the cost of individual resistors. Start with 1-watt assortments because these can be used to replace lower-

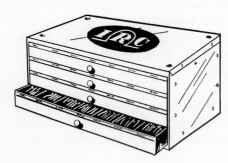

Fig. 11. Example of carbon-resistor stock cabinet having sizes most needed for repairing television and radio sets. The larger the kit, the more likely it is to have the value needed

wattage resistors just as well. Later you may also want to have an assortment of ½-watt resistors, as these are smaller and sometimes easier to install in a crowded television chassis.

As your business expands, you will find it profitable to buy the most-used values of resistors in lots of ten of a kind, to eliminate the nuisance of frequent reordering and save money.

If you have a bit of spare room in your toolbox, put in the following ten resistors: 1,000, 5,600, 10,000, 22,000, and 47,000 ohms, and 0.1, 0.27, 0.47, 1.0, and 2.5 megohms. Hold them all together with a rubber band. Put in a few more sizes if you have room, because having the right resistor on hand will save you a trip to the shop when you encounter a burned-out resistor on a house call.

Printed Resistors. Carbon or metallized resistance material can be printed, painted, or sprayed directly on a plastic chassis or on a block of insulating material to give what is known as a printed resistor. This technique has been used for some time in the manufacture of hearing aids and vest-pocket-size portable radio receivers, because resistors made in this way take practically no space.

Printed resistors are widely used in resistor-capacitor units, some of

Carbon Resistors 353

which are also known as rescaps, couplates, modules, and printed-circuit units. An example is shown in Fig. 12. The resistors are printed directly over terminal rivets or over previously applied strips of silver ink that serve in place of wiring. The entire unit is then coated with an insulating material.

Testing Printed Resistors. Printed resistors can burn out or open just like ordinary carbon resistors, though failures are rare because they are used in circuits that carry little current. A printed resistor can be tested with an ohmmeter if its leads are accessible, just as you would test any

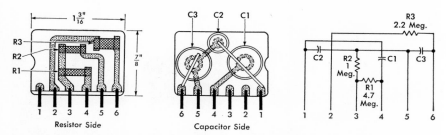

Fig. 12. Front and back views and circuit diagram of printed-circuit unit in which resistors can be tested individually with an ohmmeter because they go directly to external leads. Dots at line intersections on circuit diagram indicate connections; no dots then mean no connections

other resistor. Their values are usually about 1 megohm in television and radio sets, so use the highest ohmmeter range. Just as with other carbon resistors, the correct value can generally be found on the circuit diagram or parts list for the set.

The printed resistors in Fig. 12 can easily be checked with an ohmmeter because every one is connected directly to external leads. Thus, measuring between leads 3 and 4 checks R1, which should be 4.7 megohms within 20 per cent; measuring between 3 and 5 checks R2, which should be 1 megohm within 20 per cent; measuring between 2 and 6 checks R3, which should be 2.2 megohms within 20 per cent. Tolerance of printed resistors is generally 20 per cent. For a 4.7-megohm resistor, the tolerance range then is 3.8 to 5.6 megohms. For a 2.2-megohm resistor, the tolerance range is 1.8 to 2.6 megohms.

When a capacitor is in series with a printed resistor in a printed-circuit unit, the resistor cannot be tested with an ohmmeter. A good capacitor is the same as an open circuit to an ohmmeter. Trying an entire new unit is the only test you can make if an open printed resistor is suspected here.

354 Television and Radio Repairing

Replacing Printed Resistors. When a printed resistor fails, it is usually best to replace the entire printed unit with a new printed unit or with new individual parts.

If you can get at the terminals or if they go directly to external leads, it is sometimes possible to connect the correct resistance value of a ½-watt

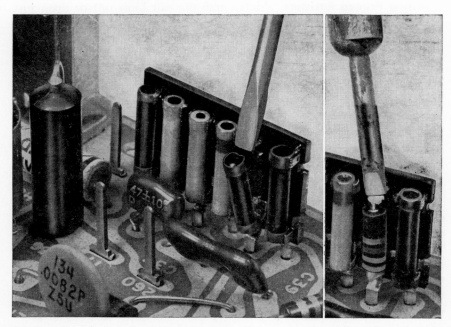

Fig. 13. Prying defective resistor out of its mounting clips on a component pack with a screwdriver (left), and soldering conventional new resistor in its place without removing resistor-capacitor unit from printed-circuit board of television receiver. Note vertical mounting of molded paper capacitor at left, use of bare jumper wire at bottom center, and use of vertical terminals on printed-circuit board as test points to aid in servicing. (Philco photos)

or smaller carbon resistor between the terminals of the bad resistor. First scrape away a path across the bad printed resistor to make sure it stays open. Hold the tip of the soldering iron on the joints only long enough to make the solder flow smoothly, because excessive heat may damage nearby parts or unsolder adjacent joints in printed circuits.

In the component pack shown in Fig. 13, individual resistors or capacitors can be removed and replaced without removing the pack from the printed-circuit board. The components can be tested individually because their ends are accessible. When installing a replacement, choose a unit having the same electrical value, then shorten and bend its leads so they can be soldered to the exposed terminals on the component pack board. Crimp the leads before soldering, if possible.

Thermistors. As you know, heat is a troublemaker in television sets. Many parts become warmer and change slightly in their characteristics after a set has been operating for a while. The deflection yoke that surrounds the neck of the picture tube is an example. As this yoke heats up, its resistance increases and its current goes down. The most noticeable effect in some sets is a reduction in picture height after the set has been operating 5 or 10 minutes.

A special type of resistor called a thermistor is used in series with the deflection yoke to compensate for heat. Examples are shown in Fig. 14.

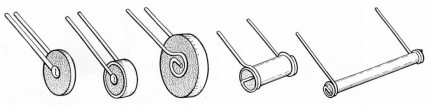

Fig. 14. Examples of thermistors used in television receivers. All are soldered directly into receiver circuits, since they seldom go bad

The thermistor drops in resistance when it warms up, thereby compensating for the increase in yoke resistance.

A thermistor rarely fails, but when it does, be sure to get an exact-duplicate replacement from the manufacturer's distributor. An ordinary resistor will not do.

Reading Circuit Diagrams. Now that you are beginning to see the need for using circuit diagrams to get servicing information for fixing radios, a bit of extra information on reading these diagrams should prove useful at this time.

Several different ways are used on diagrams to show whether or not wires join when they cross each other. In that shown in Fig. 12, a dot is used to indicate that wires are connected together where they cross, and the dot is omitted when there is no connection.

In another widely used diagram-drawing method, a curved jumper or half-circle is used at a crossover to indicate that wires do not touch, and a dot is used to indicate a connection at a crossover.

In rare cases the dot is omitted when jumpers are used. If you see curved jumpers on the diagram, then you can assume that there is a connection wherever lines cross but do not have this curved jumper.

Combining Resistors in Series. Even though you try to keep a good stock of resistors on hand, there will be times when a needed value is missing. Instead of holding up the repair job, you can combine two smaller-value resistors in series as in Fig. 15 to give the required value. Just remember that resistor values add when in series. Wrap a bit of insulating tape over the splice between the two resistors if there is a possibility that the splice will touch a bare terminal.

It is a good idea to check the combined resistance value with an ohmmeter, to make sure it is somewhere near the specified value. If it is too far off, try other combinations. Resistors can have up to 20 per cent tolerance. If the tolerances of individual resistors happen to be in oppo-

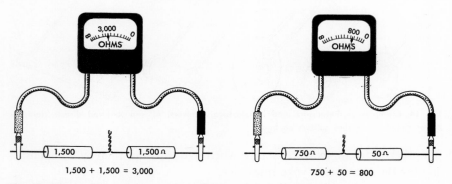

Fig. 15. When resistors are in series, add their values to get the total resistance

site directions, the combined resistance will also be as much as 20 per cent off from what you expect.

Combining Resistors in Parallel. Resistors can also be combined in parallel to obtain a value that is not in your stock. The easiest way is to combine resistors that have the same value, as in Fig. 16. Just divide the resistor value by the number of resistors you put in parallel. Thus, if you need a 1,000-ohm resistor, connect two 2,000-ohm resistors in parallel. Similarly, three 1,500-ohm resistors in parallel give 500 ohms.

When resistors of different values are connected in parallel, the combined resistance will always be lower than that of the smallest resistor in the group. Try connecting different combinations in parallel as in Fig. 17, and you will see that this always holds true. When a particular value is needed, start with the next higher resistor value that you do have, and try different resistor values in parallel with it until the ohmmeter shows that you have the right combination. With a little practice, you will be

Carbon Resistors 357

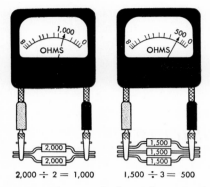

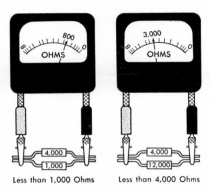

Fig. 16. When equal-value resistors are in parallel, divide the value of one of them by the number of resistors in parallel to get the combined resistance

Fig. 17. When two resistors of unequal value are in parallel, the combined resistance is less than that of the smaller resistor, as shown by the ohmmeter readings here

able to get close enough to the required value in two or three tries, without any complicated mathematical figuring at all. Remember that coming within 20 per cent is usually close enough.

When equal-value resistors are used in parallel, their wattage ratings add; thus, if 2,000-ohm 1-watt resistors are paralleled as in Fig. 16, they act the same as a 1,000-ohm 2-watt resistor.

16

Testing and Replacing Wirewound Resistors

Wirewound Resistor Facts. Though no two look alike, all the resistors in Fig. 1 are made the same way, by winding a length of Nichrome resistance wire around a ceramic, glass, fiber, or asbestos form. For this reason, they are called wirewound resistors. The wire used is just like that in the heating elements of electric stoves, but usually much thinner.

Some of the smaller wirewound resistors, which have low wattage ratings, look exactly like end-lead carbon resistors. Other low-wattage types are wound on flat forms and encased in molded plastic. Larger fixed wirewound units, with wattage ratings of 5 watts and up, have a variety of shapes and terminals, with the resistance wire wound on hollow tubular cores so heat can escape from the inside as well as the outside. The re-

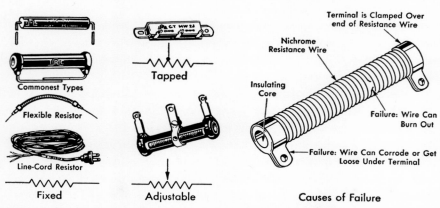

Fig. 1. Examples of three types of wirewound resistors, along with their symbols and common causes of trouble in any wirewound resistor

sistance wire is usually covered with a heat-resisting insulating coating. Flexible resistors and line-cord resistors use the same kind of wire, wound around asbestos string and covered with braided insulation.

Some special types of wirewound resistors have fixed taps for making extra connections, hence are called tapped resistors. Others have adjustable taps that can be slid to any desired position along the length of the resistor and tightened there with a screwdriver.

Ordinarily you will find wirewound resistors only in power-supply circuits. In the power supply the current is usually considerably larger than anywhere else in the receiver, so the resistors often must have higher wattage ratings than the 5-watt limit of carbon resistors. Wirewound resistors can be made large enough to have any required wattage rating.

How Wirewound Resistors Go Bad. Only two things can ordinarily make wirewound resistors go bad. They either burn out or one of their terminals gets loose.

If an excessively high current passes through a wirewound resistor for any length of time, the wire may melt. When this happens, the resistor is said to be burned out or open, and it no longer offers a path for current flow.

Sometimes a bad contact develops between the resistance wire and a terminal of the resistor, as shown in Fig. 1. This may actually open up the resistor, just as does a burned-out wire. More often, though, a loose or corroded terminal connection makes the resistance value go way up, with the action often occurring intermittently. The resistor may be good when cool and go bad as soon as it warms up after the set is turned on, or may alternately be good and bad.

How Bad Wirewound Resistors Are Found. Overheating and consequent melting of the resistance wire is the main trouble encountered in wirewound resistors. You can usually spot an overheated resistor by its charred appearance or discoloration.

If you suspect a particular wirewound resistor as being bad, measure its resistance with an ohmmeter. Remember that to measure the resistance the line-cord plug must be pulled out of the wall outlet. One end of the resistor must be disconnected from the circuit if there are other paths for direct current between the resistor terminals.

If the resistance measured is infinite, the resistor is open and should be replaced. If the measured resistance is more than one and a half times as high as it should be, or if it changes when one of the resistor terminals is wiggled, the resistor is likewise bad and should be replaced.

Resistance Values of Wirewound Resistors. Once you have located a bad wirewound resistor, the next job is to get the correct size of replacement. The resistance value of the new unit should be the same as for the old one if at all possible. Wirewound resistors are generally used in parts of sets where the resistance value is somewhat critical. Standard tolerance for wirewound resistors is 5 per cent for this reason.

Values of wirewound resistors used in television and radio sets are generally well under 100,000 ohms. For higher values, carbon resistors are used. Always use the same type of resistor for a replacement—carbon for carbon and wirewound for wirewound.

When a wirewound resistor is burned out, you will usually have to refer to the circuit diagram or parts list to find the resistance value, because printed markings and color codes generally disappear after a few years of use. Where printed markings are still clear, by all means use them. Likewise, if the resistor is color-coded and the colors are not faded, read the code to get the resistance value.

Color Codes for Wirewound Resistors. When a wirewound resistor is marked according to one of the EIA systems shown in Fig. 2, you can read its resistance value directly just as for carbon resistors.

Small end-lead wirewound resistors look exactly like carbon resistors and are color-coded the same way except that the first color band is twice as wide as the others to indicate wirewound construction. With wirewound resistors below 10 ohms, the regular color code does not work because the lowest it can specify is brown-black-black for 10 ohms. To take care of this, gold and silver colors are used for color C to indicate decimal multipliers.

How to Use Decimal Multipliers. For values from 1 to 10 ohms, gold is used for color C to represent the decimal multiplier 0.1. This means that the value represented by the first two colors must be multiplied by 0.1 to get the final value in ohms. Thus a 5-ohm resistor would be green-black-gold; here green is 5, black is 0, and gold means that 50 is multiplied by 0.1 to get 5 ohms. Similarly, red-green-gold is a 2.5-ohm resistor; here red is 2, green is 5, and gold means that 25 is multiplied by 0.1 to get 2.5 ohms.

For values from 0.1 to 1 ohm, silver is used for color C to represent the decimal multiplier 0.01. Thus, a 0.47-ohm resistor would be yellow-violet-silver (a pretty combination); yellow is 4, violet is 7, and silver means that 47 is multiplied by 0.01 to get 0.47 ohm.

Wirewound Resistors

Color Code for Flexible Resistors. Instead of using bands of color, flexible resistors are color-coded by means of different colored threads woven into the insulating covering.

The body color of the woven fabric covering is color A and gives the first number of the resistance value. The triple-thread color woven into

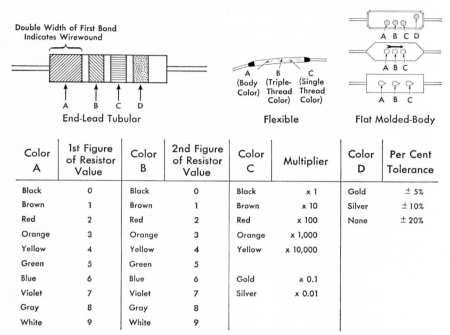

Color A	1st Figure of Resistor Value	Color B	2nd Figure of Resistor Value	Color C	Multiplier	Color D	Per Cent Tolerance
Black	0	Black	0	Black	x 1	Gold	± 5%
Brown	1	Brown	1	Brown	x 10	Silver	± 10%
Red	2	Red	2	Red	x 100	None	± 20%
Orange	3	Orange	3	Orange	x 1,000		
Yellow	4	Yellow	4	Yellow	x 10,000		
Green	5	Green	5				
Blue	6	Blue	6	Gold	x 0.1		
Violet	7	Violet	7	Silver	x 0.01		
Gray	8	Gray	8				
White	9	White	9				

Fig. 2. Standard EIA color code for wirewound resistors, using same colors as for carbon resistors but different arrangements of the colors

the fabric is color B, for the second number in the resistance value. The color of the single thread woven into the covering is color C, for the multiplier.

As an example, suppose a flexible resistor has a yellow woven fabric covering, a green triple thread, and a single black thread woven into the covering. Color A is yellow for 4, color B is green for 5, and color C is black for × 1, so the resistance is 45 ohms.

Dot Color Codes. Some wirewound resistors have a molded plastic housing. These resistors are often marked with colored dots that give the resistance value. There will sometimes be an arrow or other marking to show the direction in which you read the colors of the dots. If there are four dots, the fourth will be gold, silver, or unpainted to indicate tolerance; you then know that you should start reading from the other end of the resistor even when there are no arrows to indicate this.

Power Ratings for Wirewound Resistors. Ordinarily, you will have to look on the circuit diagram or in the parts list of the set to find the power rating of a bad wirewound resistor. The values printed on the resistor usually fade out quickly because of heat.

Sometimes the power rating is placed under the resistance value on a circuit diagram. Here W represents watts, so 5W would mean 5 watts, and 10W would mean 10 watts.

The power rating of the new unit should be the same as for the old unit, or higher. This means that the new resistor should be the same physical size as the old, or even larger. Remember, however, that if the new resistor is very much larger, you may have trouble fitting it into the available space in the chassis.

The power P in watts that a resistor must handle is equal to the current I in amperes multiplied by the voltage E across the resistor in volts. If the resistance value R in ohms is known, the power can also be found by multiplying $I \times I \times R$, or by multiplying E by E and dividing the result by R. The corresponding formulas are $P = EI$, $P = I^2R$, and $P = E^2/R$. The wattage rating of a new resistor should be at least twice the actual power value.

When in doubt about the power rating, you can safely order a 10-watt unit for the average television or radio set. Rarely are higher power ratings needed for single fixed resistors, even in power-supply circuits.

For convenience in installing, try to get a new resistor having essentially the same size and the same kind of leads or terminals as were on the old unit. Five-watt and ten-watt wirewound resistors for replacement sometimes have both leads and terminals, but it is usually best to use the leads. Just ignore the terminals or cut them off.

Installing New Wirewound Resistors. Follow the same procedure as for carbon resistors. Make a diagram of the connections, then unsolder the leads of the bad resistor and connect the new resistor in its place. Keep the leads short but not too short, just as for carbon resistors. With resistors, it does not matter which end goes to a particular terminal.

After installation of the new resistor, check for other defects in the circuit that might cause the new resistor to fail too. Look for wires that are touching each other or the chassis. Look for another bad part that could cause excessive current to flow through the new resistor. As with carbon resistors, shorted capacitors are frequently the real cause of the trouble. Remember that wirewound resistors rarely fail by themselves—they are usually overloaded by failure of another part in the circuit.

Finally, adjust the position of the new wirewound resistor so it is well away from paper capacitors or other parts that might be damaged by heat.

Fuse Resistors. More and more, resistors are being used as fuses in television circuits. These are special wirewound resistors, designed to burn out when the current through them exceeds the rated value.

Some fuse resistors have plug-in pins as terminals, to permit easy replacement. Others have tinned bare leads much like ordinary wirewound resistors, and must be replaced by soldering and unsoldering.

Fuse resistors generally blow only when a short-circuit develops in some other part. It is therefore essential to check electrolytic filter capacitors and bypass capacitors before replacing a fuse resistor.

A fuse resistor is widely used in series-string filament circuits. Here it protects the tubes by acting as a slow-blow fuse that opens the filament circuit when current becomes excessive for any reason.

Some television sets have a special fuse resistor that breaks open when an excessive surge of current flows through it. A springlike terminal pulls the resistor apart when it breaks, as shown in Fig. 3, to give fuse action. An exact-duplicate resistor is of course required here. Its leads should be cut to the same length as those on the old resistor. Servicemen often call these resistors grasshoppers because of their spring action and the appearance of the mounting.

Surge Resistors. Many television sets have in series with the power line a surge resistor like those shown in Fig. 3. This serves to reduce the heavy initial current that flows through tube filaments when a set is turned on. The identifying clue is a bimetallic contact strip mounted on top of the resistor.

The action of a surge resistor is entirely automatic. The resistor is connected in series with the tube filaments, and the thermostat contacts are connected to its terminals. The contacts are open when cold. The resistor warms up as the set is turned on, making the contacts close and short out the surge resistor in about 30 seconds. By that time all the tubes have warmed up and can take the full line voltage.

Surge resistors rarely go bad. When they do fail, an exact-duplicate replacement is required.

Multivalue Resistors. Wirewound resistors come in many different sizes, types, and wattage ratings, and cost much more than carbon resistors.

Fortunately they go bad so seldom that you can get along without any stock of them, at least at the start.

One answer to the stock problem here is a series of five different multi-value wirewound resistors. Each unit contains four 10-watt resistors mounted in one housing and having eight leads, as shown in Fig. 3. These four resistor sections can be connected together in various series and

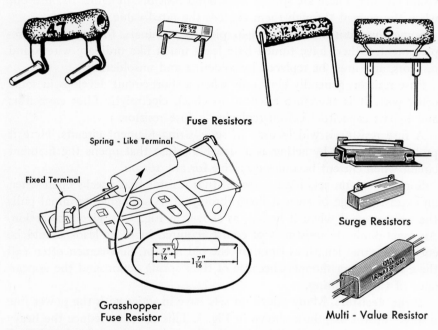

Fig. 3. Examples of wirewound resistors made for special purposes

parallel combinations to give many different resistance values, by following the simple instructions furnished with the units. The five different units thus cover the entire range from 0.5 ohm to 50,000 ohms. A good starting point is a kit containing two of each, or ten in all, costing around $7.

Tapped Resistors. Only two types of tapped wirewound resistors are used to any extent in television and radio receivers. One is round like the ordinary wirewound resistor but is usually thicker and longer, with extra terminals along its length. This type is held rigidly in position by brackets riveted or bolted to the chassis. The extra terminals or taps connect to the resistance wire along its length. The positions of the taps determine the amounts of resistance between the taps and the end terminals of the resistors.

The other type of tapped resistor is flat rather than round, and is generally enclosed in a metal housing or can, as shown in Fig. 4. The metal-encased types are sometimes called Candohms, this being the trade name of one manufacturer of these units. The metal can protects the resistance wire and its insulation against damage and helps to get rid of the heat

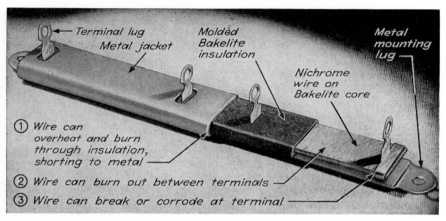

Fig. 4. Ways in which a metal-encased wirewound resistor can go bad. Be sure excess solder does not flow down between a terminal and the metal jacket when soldering or unsoldering

generated by the resistor. Metal-encased tapped resistors are generally riveted directly to the side of the chassis, so that heat is transferred directly from the metal can to the heavy metal chassis.

Tapped resistors are used chiefly between B+ and B− of a power supply. The taps allow various fractions of the total voltage across the resistor to be taken off and used for tubes that require lower voltages.

Tapped resistors use the standard resistor symbol on circuit diagrams, with lines going to various points along the zigzag symbol to represent the taps, as in Fig. 1. Sometimes dots are used at the connecting points on the resistor symbol to indicate a connection, but more often no dots are shown.

How Tapped Resistors Go Bad. Since tapped resistors are exactly the same as ordinary wirewound resistors except for the extra terminals, they have the same defects. They can burn between any pair of terminals, or one of the terminals can get loose. In addition, the metal-encased types can become grounded to their metal can and the chassis when the insulation burns up, as shown in Fig. 4.

Determining If a Tapped Resistor Is Bad. Whenever a tapped resistor is blackened or discolored, has a tarry mess around it, or smokes when the set is on, suspect it as being bad. It may be burned out or may be shorted internally to the metal shield.

To check a tapped resistor, first make a simple but accurate diagram of all the leads going to its terminals, somewhat as in Fig. 5. Now you can unsolder all the leads connected to the terminals of the resistor, to per-

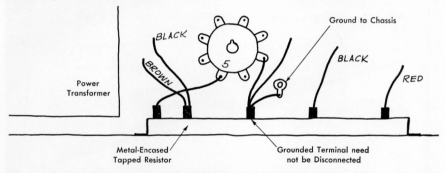

Fig. 5. Before unsoldering the leads of a defective tapped resistor, always make a rough sketch of the connections as a guide for connecting the new unit

mit ohmmeter measurements. Leads connected to the grounded resistor terminal (the terminal connected to the chassis of the set) need not be removed for these measurements.

After removing the wires from the terminals, measure the resistance of each section (between two adjacent terminals) of the tapped resistor. Compare each measured value of resistance with that specified for the corresponding section of the resistor on the circuit diagram of the set.

If the measured resistance is essentially the same as that specified on the diagram for a particular section, then that section of the resistor is good. If an infinite resistance (no needle deflection) is measured for one section of the resistor, that section is burned out. If the resistance measured is quite a bit less than the resistance specified for that section and the resistor is of the Candohm type, the resistor is shorted internally to the metal case.

If the resistance of every section of the resistor measures right, you have cleared the tapped resistor of suspicion, unless it is metal-encased and has no terminal connected directly to ground (the metal can or chassis). Here you will also have to measure the resistance between each terminal of the resistor and the metal can. A low or zero-resistance reading between any terminal and the can means there is an internal short or ground between the resistance wire and the can.

Examples of ohmmeter readings pointing to an internal ground are given in Fig. 6. The ground fault in each case provides a zero-resistance parallel path across part of the resistance, causing the observed decrease in one resistance reading. When the fault is several sections away from

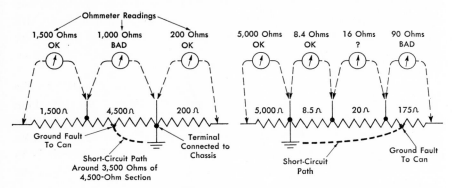

Fig. 6. Two examples of ohmmeter readings obtained when checking metal-encased tapped wirewound resistors that have ground faults

the grounded terminal, the good intervening sections may show slightly lower resistance values because of the parallel path around them through other resistors and the short-circuit path, but the reduction is usually small. Thus, in the second example the 8.5-ohm section reads 8.4 ohms and the 20-ohm section reads 16 ohms, but by themselves these are not definite indications of trouble. The real clue here to a ground fault is the 90-ohm reading for the 175-ohm section.

When a metal-encased resistor is grounded through a fault, replace the entire unit. When there is a fault to ground, it is not practical to replace just the grounded section because you would first have to cut it out with a hacksaw to eliminate the unwanted ground.

Getting Replacement Tapped Resistors. Whenever you find a defective tapped resistor, always try to get the exact replacement for it. Exact replacements can usually be obtained from the set manufacturer's local distributor if the set is not more than five years old. Rarely if ever will you be able to get a correct replacement tapped resistor from a parts jobber or mail-order parts-supply firm.

To order the replacement, you need to know the manufacturer's stock number for the part, as given in the parts list for the set. When ordering, always give the model number of the set too.

Removing the Bad Tapped Resistor. With an exact-duplicate replacement tapped resistor on hand, the next step is removal of the defective unit.

If the tapped resistor is tubular, it will ordinarily be mounted on the chassis with mounting brackets that push against each end of the resistor to keep it in place. To remove the resistor, merely bend each spring bracket back slightly and allow the resistor to fall out. If any leads are still soldered to the unit, remove them first. These tubular tapped resistors are rarely used by set manufacturers but are occasionally used by servicemen for replacing other types.

Tapped resistors are usually of the metal-encased type, riveted to the chassis of the set. The easiest way to remove the rivets is to drill them out, using a drill slightly larger than the body of the rivet. Prop up the chassis rigidly for this. Make a starting hole for the drill with a center punch. Sometimes it is easier to drill out the rivet head, because it is on the outside of the chassis where there is more room to work. When the rivets have been removed, the unit will fall out.

Installing the New Tapped Resistor. If an exact-duplicate replacement is obtained, mounting will be simple. It is the same as for the old unit, except that machine screws and nuts are used instead of rivets. A popular size of machine screw for such replacements is a 6-32 screw $\frac{3}{8}$ inch long, with corresponding nut. This will fit most of the holes originally made for rivets, making it unnecessary to enlarge the mounting holes.

Before reconnecting the wires to the terminals of the new resistor, check for another bad part that may have caused the original resistor to overheat and fail. The bad part may be just a lead that was touching the chassis or another lead, or it may be another defective part. A shorted paper capacitor is often the guilty part. Unless this other defect is found, the new resistor may overheat and fail within a short time.

After reconnecting all the leads to the new resistor according to your diagram, check the connections carefully. One little mistake may ruin the new resistor.

What to Do When New Tapped Resistor Is Unobtainable. In many cases the exact-replacement tapped resistor may not be available. The manufacturer may have gone out of business, or he may not have a replacement on hand, especially if the set is more than five years old. In such cases it will be necessary to replace the defective resistor in some other manner or to replace only the defective portion of it.

Bridging the Defective Section of a Tapped Resistor. When the tapped resistor is merely burned out and is not internally grounded to the chassis, it is possible to bridge the burned-out section with an ordinary 10-watt wirewound resistor, as in Fig. 7. The new resistor should have the same resistance value as the burned-out section. The leads of the new resistor are soldered to the terminals on each side of the burned-out section. The good portions of the tapped resistor are still used.

Fig. 7. Bridging a burned-out section of a tapped resistor with an ordinary 10-watt wirewound resistor, after making sure there is no ground to the metal case

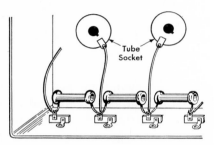

Fig. 8. Installation of three 10-watt wirewound resistors on individual insulated terminals that have been bolted to the chassis, to replace a three-section tapped resistor

The wattage rating of any one section of a tapped resistor is usually only 4 or 5 watts. A 10-watt replacement resistor thus gives an ample margin of safety.

Since replacing only the burned-out section is simpler and much cheaper than replacing the whole tapped resistor, many servicemen do this even when an exact replacement can be obtained. There is one drawback to this type of repair, however. The two ends of the burned-out wire in the tapped resistor may touch together again and cause trouble, or the burned insulation around these ends may give way and cause a ground to the metal case. This is why it is best to put in an entire new tapped resistor whenever one is available.

After putting in the new resistor, check as usual for the part that caused the original resistor to burn out. Finally, adjust the positions of leads to the new resistor. Check also for accidental shorts to other terminals and to the chassis before turning on the set to see if the repair is successful.

Replacing a Tapped Resistor with Ordinary Wirewound Resistors. When a tapped resistor is internally grounded to the metal can and no exact replacement is available, a different method of repair must be used. One method is to use individual 10-watt wirewound resistors to replace each section of the tapped resistor, after removing the tapped resistor. These resistors are mounted on insulated terminals that you will have to install first. Connect the resistors in series in the proper order to give the equivalent of the defective tapped resistor. An example of a finished job of this type is shown in Fig. 8. The resistance of each separate resistor should be as close as possible to the resistance of the section it replaces. The correct values are obtained from the circuit diagram of the set.

To support the separate resistors off the chassis, fasten individual insulated terminals to the chassis with machine screws and bolts. Sometimes the original rivet holes can be used, but generally you will have to drill a new hole for mounting each terminal.

With the terminals in position, connect the new resistors to the lugs with permanent hook joints, in the same order as the sections of the tapped resistor, but do not solder them yet. First connect the circuit wires to the correct terminals of the new resistors, working carefully from the connection diagram you made previously.

It may be necessary to extend some leads by splicing on short lengths of insulated wire. When doing this, slip spaghetti insulation over the splice or tape the splice to prevent the bare wires from shorting against some other parts.

Since more room is required for mounting resistors in this manner, be sure that there is ample room on the chassis before starting. Staggered positioning of the resistors as in Fig. 9 often allows the replacement resistors to fit in about the same length of chassis as the old tapped resistor.

Soldering comes next. Be sure there is no solder dripping down from the insulated terminal lug to the metal mounting bracket of the terminal.

Flexible Resistors. In a flexible resistor, the Nichrome resistance wire is wound around braided glass or other heat-resisting material and covered with similar insulation.

Flexible resistors can be distinguished from ordinary wires because they are fatter and because they have metal terminal caps at each end. These caps fasten the resistance wire to the wire leads of the resistor.

Finding a Bad Flexible Resistor. To check the resistance of a suspected flexible resistor, unsolder one end of it from the set and measure with an ohmmeter. If a definite resistance reading is obtained and it does not change as the resistor is bent back and forth, the resistor is probably good. Values of flexible resistors are generally under 100 ohms.

Sometimes you will find the insulation on the flexible resistor broken so that the resistance wire is exposed. Although there is no actual break in the resistor, it should be replaced to prevent later troubles.

The resistance value of a flexible resistor can be found on the circuit diagram of the set. The colors of the resistor insulation may also tell you

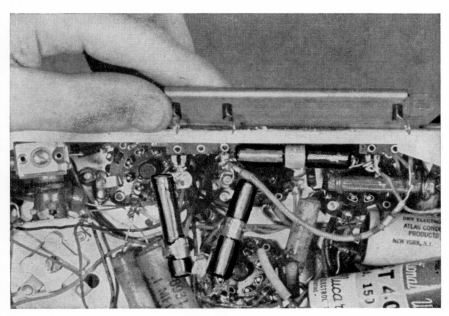

Fig. 9. Staggered arrangement of three 10-watt wirewound resistors to replace the defective two-section metal-encased unit being held above the chassis. A resistor of the right value for one of the sections could not be obtained, so two smaller-value resistors were used in series to add up to the required value

the value if they follow the standard color code and are not faded beyond recognition.

Replacing Flexible Resistors. Replacement flexible resistors cannot ordinarily be obtained from jobbers. Just put in an ordinary 5- or 10-watt wirewound resistor having the required resistance value.

Ballast Resistors. Iron, nickel, or Nichrome resistance wire is sometimes mounted in a glass or metal radio tube housing to give what is known as a *ballast resistor* (also called a ballast tube or simply a ballast). The tube housing plugs into a tube socket the same as any other radio tube.

Mounting the resistance wire in a tube housing provides a simple means of getting the resistor in the open air above the chassis of the set, where heat can be dissipated much faster than for an equivalent resistor under the chassis. In addition, since the resistor is mounted above the chassis, the heat produced by the resistor cannot overheat parts beneath the chassis.

Where Ballast Resistors Are Used. A ballast resistor is connected in series with the filaments (or heaters) of the tubes in a television or radio set, to reduce the power-line voltage to the lower voltage required by the tube filaments. An example is shown in Fig. 10. Part of the line voltage

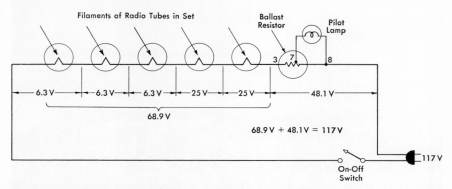

Fig. 10. How a ballast resistor is used to reduce the line voltage to the lower value required by the filaments (heaters) in an a-c/d-c radio set

is dropped or wasted across the ballast resistor, so that only the correct amount is left to be applied to the filaments of the tubes.

In many ballast resistors a tap is provided on the resistance wire to furnish the correct voltage for a 6- to 8-volt (6–8v) pilot lamp. This tap is connected to one of the pins on the base, so that connection to it is made automatically when the ballast resistor is plugged into its socket. Thus, in Fig. 10 the pilot lamp in the set would be connected between terminals 7 and 8 of the ballast resistor.

Current-regulating Action of Ballast Resistors. Some types of ballast resistors have an additional regulating feature. The resistance of the wire changes as the current through the resistor varies, in such a way as to counteract the change in current; this is true ballast action. When the current decreases because of a lower line voltage, the resistance wire cools off, and its resistance decreases just enough to bring current back up to the correct value. When the current increases because of a higher applied voltage, the resistance wire gets hotter and goes up in resistance, thereby reducing the current automatically to its correct value. Thus the ballast resistor maintains a fairly constant current through the filaments of the tubes, even though the line voltage applied to the set may vary over quite wide limits.

Checking Ballast Resistors. Since a ballast resistor is connected in series with the filaments of the tubes in the set, it breaks the entire filament circuit when it opens or burns out. When this happens, none of the tubes light up. It is better to check the tubes first, however, since they fail much more often than do ballast resistors.

Ballast resistors can be checked easily with an ohmmeter. Most ballast resistors have somewhere between 135 and 400 ohms of resistance. If the resistance is infinite as measured between the tube-base pins that connect to the ends of the resistance wire, the ballast resistor is burned out and must be replaced. Pin connections for ballast resistors can be determined by noting which terminals of the resistor socket are connected into the filament circuit.

Replacing Ballast Resistors. Exact replacements for ballast resistors are not usually available. Instead, you merely order the correct type of universal adjustable ballast and adjust it according to the instructions furnished. Your jobber will tell you which one to order.

Whenever replacing a ballast resistor, note which pins are missing on the old unit and cut off the corresponding pins on the new unit. This is important because the unused socket terminals of ballast resistors are often used as terminals for other circuits.

Line-cord Resistors. In many of the older universal a-c/d-c radios a line-cord resistor was used in series with the filaments. The line-cord resistor did exactly the same thing as a ballast resistor—it dropped part of the total line voltage so filaments of the tubes get the correct voltage value.

In a line-cord resistor the resistance wire is wound spirally on an asbestos cord located alongside the two insulated wires of the line cord. One end of the resistance wire connects to one of the prongs of the line-cord plug. The other end of the resistance wire comes out at the other end of the line cord inside the radio. This end is connected so the resistance wire is in series with the filaments of the radio tubes, as shown in Fig. 11. Thus, there is a complete path from one prong on the line plug through the line-cord resistor, through the filament of each tube in turn, and then through the on-off switch and back to the other prong of the line-cord plug.

Some line-cord resistors have a tap to provide the required 6 to 8 volts for a pilot lamp in the radio. In this case there will be four leads coming out of the line cord inside the radio: the two regular ones, one for the filament circuit, and one for the pilot lamp.

Recognizing Line-cord Resistors. Line-cord resistors can be distinguished from ordinary line cords by their appearance. First of all, they are usually fatter than an ordinary line cord. Second, they will be made with a woven fabric covering rather than a rubber covering. Third, they get quite warm when the set is turned on. This heating of the line cord is normal—it is

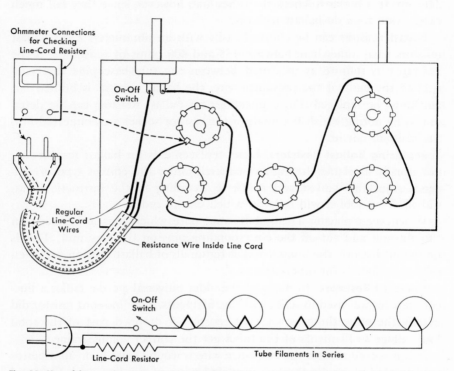

Fig. 11. Use of line-cord resistor to drop the line voltage to the correct value for five series-connected tube filaments

the same amount of heat that would be given off by an equivalent ballast resistor in the set itself.

Finding a Bad Line-cord Resistor. Since a burned-out or broken line-cord resistor will also break the filament circuit of the radio, the line-cord resistor is one possible source of trouble when the tubes in the set fail to light up or become warm.

To check the line-cord resistor, first pull the line-cord plug out of the wall outlet and turn off the radio. Now measure the resistance between the extra line-cord wire inside the radio and each of the prongs of the

Wirewound Resistors

line-cord plug with an ohmmeter, as indicated in Fig. 11. The on-off switch of the radio should be in the off position for this. For one plug prong the reading should be infinite, and for the other prong it should be less than a thousand ohms, just as for ballast resistors. If the resistance measured is infinite for both prongs, the line-cord resistor is bad.

Repairing Line-cord Resistors. When you find a line-cord resistor that is bad, inspect the connections to each of the prongs on the line-cord plug. One of the prongs will have one end of the resistance wire connected also to its terminals. This connection may be broken.

If you find that the resistance wire is broken right next to the terminal of the line-cord plug, carefully pull out a small additional length of the resistance wire and refasten it to the plug terminal. If the plug does not have screw terminals, it will be best to take off the old plug and put on a new plug having screws. Be sure the resistance wire is connected to the same line-cord wire as before. Screw terminals on the plug allow you to make a good tight mechanical connection to the thin Nichrome resistance wire, as this wire is exceedingly difficult to solder.

If the line-cord resistor is broken or burned out internally, it will be necessary to get an entire new replacement cord of the right type.

Replacing a Line-cord Resistor. Replacement line-cord resistors, having the two regular line-cord wires and the plug as well as the resistance wire, are usually easily obtainable. For ordering, all you need to know is the resistance of the old line-cord resistor. The simplest way of finding this is to look on the circuit diagram of the set. The resistor value in ohms will be found right next to the line-cord plug.

EIA Color Code for Line-cord Resistors. In most line-cord resistors, the insulation on the two regular line-cord wires is either red and blue or red and black. The insulation on the resistance wire is colored according to the resistance value in the line cord, as follows:

Yellow	135 ohms	Orange	260 ohms
Blue	160 ohms	Gray	290 ohms
White	180 ohms	Maroon	315–320 ohms
Green	200 ohms	Dark Brown	350–360 ohms
Light Brown	220 ohms		

This color code can be used for ordering purposes only if the color on the resistance wire is not faded beyond recognition.

Installing the New Line-cord Resistor. The new resistor is connected exactly like the old, using soldered connections in the set. Since the resistance-wire lead is the most fragile, allow plenty of slack for it inside the set so the strain goes on the other two leads if someone trips over the line cord.

Some line-cord resistors have the braided outer covering brought out as a tie cord at the receiver end. Fasten this cord to the chassis at some convenient point, pulling it tight enough so all the strain is on this tie cord when you pull on the entire line cord.

In no case should a line-cord resistor be shortened. To do so would also shorten the resistance wire and reduce the resistance, causing burned-out tubes or another burned-out line-cord resistor.

17

Testing and Replacing Controls, Switches, and Clock-radio Timers

Why Variable-resistance Controls Are Needed. All the resistors considered up to now have had essentially fixed resistance values. True, the position of the tap in an adjustable tapped resistor could be changed, but this required a screwdriver and could be done only with power off.

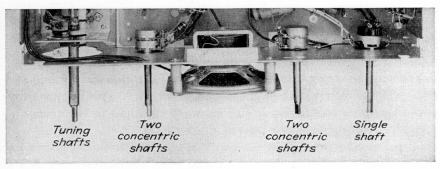

Fig. 1. Bottom of television-receiver chassis, showing front-panel controls. Left to right: dual station selector and fine tuning controls; dual contrast and brightness controls; dual vertical and horizontal hold controls; combination volume control and on-off switch on single shaft. (Howard W. Sams photo)

All television and radio receivers require variable resistors in which the position of the tap or the total resistance value can be conveniently controlled by turning a knob, such as for changing volume, tone, and picture size. These control units are called *rheostats* and *potentiometers*. They can change either current or voltage in a circuit.

Figure 1 shows the front-panel controls on a modern television receiver. It is common practice in television to use dual controls in place of two individual knobs and controls. Dual controls make the set appear less complicated to operate.

There are usually additional controls at the rear of a television set. Instead of knobs, these usually have slotted shafts that can easily be turned with a screwdriver or knurled shafts that can be turned with fingers, as shown in Fig. 2.

Tuning and band-changing controls will be covered in a later chapter since they do not involve resistance.

Potentiometer and Rheostat Symbols. The standard symbol used on circuit diagrams to represent a potentiometer is shown in Fig. 3. The arrow represents the movable contact.

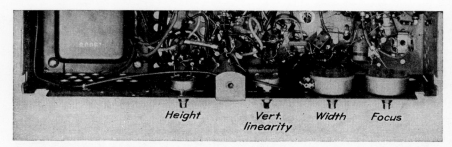

Fig. 2. Bottom rear of television chassis, showing screwdriver controls. Bottom of power transformer can be seen at left. (Howard W. Sams photo)

You will also see on circuits the symbols for a rheostat. This is the same as a potentiometer with one end terminal not used. The resistance between the other two terminals changes when the shaft of the control is rotated. A rheostat is thus a variable resistor.

One rheostat symbol is the same as that of a potentiometer except for an end terminal being unconnected. The other rheostat symbol uses a slant arrow through the resistance symbol to indicate that it is variable.

Rheostats are used chiefly for hold controls, sync controls, and focus controls in television receivers, and occasionally for tone controls.

How a Potentiometer Controls Voltage. In a potentiometer, moving the contact arm varies the fraction of voltage passed on to the next circuit, as indicated in Fig. 3. When the contact arm is at the high-voltage end of the resistance element, all the input voltage is passed on to the next stage. As the contact arm is moved toward the low-voltage or grounded end of the resistance element, the output voltage is lowered.

Construction of Potentiometers. Figure 4 gives a quick look inside some typical potentiometers. Each has a resistance element, a contact arm, and

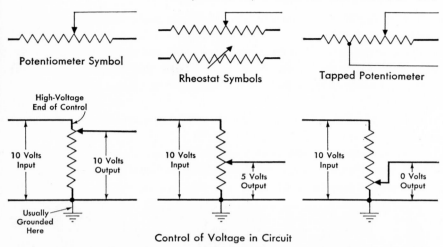

Fig. 3. Potentiometer and rheostat symbols, and manner in which a potentiometer controls the amount of voltage passed on to the next stage or circuit

a shaft. Two types of carbon resistance elements are used. One type has a horseshoe-shaped flat piece of insulating material coated on one side with carbon resistance material. The other type has a strip of insulating material bent to form an almost complete circle and coated on the inside with a carbon resistance material. Carbon units are used chiefly in signal circuits, where only small currents are handled.

Wirewound potentiometers have a similar bent strip of insulating material with the resistance wire wound around it. They are used chiefly in power circuits, where high currents are handled.

All types of potentiometers have a movable contact that is mounted on, but insulated from, the shaft that goes through the center of the potentiometer. This movable contact provides the desired movable connection with the resistance material. The position of the contact varies as the shaft is rotated. Wirewound potentiometers feel rougher than carbon units when rotated because the contact moves over the turns of wire.

Dual Potentiometers for Television Sets. A single potentiometer has a solid brass or aluminum shaft going through the entire unit. Attached to the inner end of the shaft is an insulating piece that supports the movable contact arm. To illustrate how dual controls work, suppose a hole is drilled lengthwise through the center of the shaft. This could be done without affecting operation of the control. Next, take another potentiometer with a long, thin shaft and push its shaft through the hole in the first, from the

380 *Television and Radio Repairing*

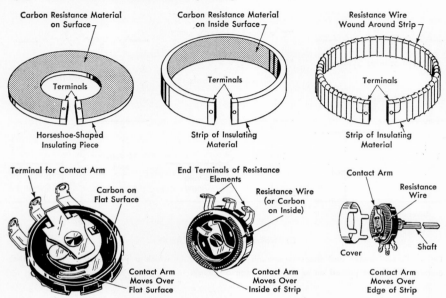

Fig. 4. Construction details of resistance controls that can occasionally be repaired when they go bad. The center terminal always goes to the movable contact arm, and the outer terminals go to the ends of the resistance element

rear. Mount both controls so their housings cannot turn, place a wheel-shaped knob on the larger shaft, and place an ordinary knob on the small shaft. The front knob turns the rear unit, and the rear knob turns the front unit.

Dual potentiometers are treated just like two individual units when testing or troubleshooting. Only when ordering replacements or assembling replacements from special kits do you need to consider the dual arrangement. Only main operating controls are combined into dual units. Back-of-set controls are always single units.

How Potentiometers Go Bad. In any moving mechanical part there is friction and resultant wearing away of material. In potentiometers this action can produce poor contact between the resistance element and the moving arm. Friction and motion can also impair the connection between the moving arm and its fixed terminal lug (the center lug).

With volume-control potentiometers, both types of wear cause noise to be heard from the loudspeaker when the control is rotated. In television sets, these poor contacts can cause flickering of part or of all of the picture when the control is rotated, but the control is still said to be noisy.

In general, if noise is heard or its effects are seen on the picture when a control knob is touched, turned, pulled, or pushed, the control is certainly defective and requires either repair or replacement.

A potentiometer opens when one of the terminals breaks away from the resistance material or when the resistance material burns out or breaks. This type of trouble is rather rare in radio sets, but burnout occurs quite often in television receivers when failure of some other part results in excessive current through the control.

If a volume control fails to change the volume when it is rotated, there is most likely a break in its resistance element. With tone controls and other television controls, however, the trouble is more likely to be in some associated part when a control has no effect.

Repair of Potentiometers. An open or burned-out potentiometer must always be replaced, as repairs are practically impossible. Noisy potentiometers should really be replaced also, but customers are often unwilling to pay the cost of this unless the control is so bad that the set can no longer be used. This is particularly true in television sets, where concentric controls double the replacement cost. (When one unit of a pair goes bad, both must usually be replaced.)

If a set has been standing unused for several months or more, the potentiometers in it may become noisy merely because of dirty contacts. Here, turning the control rapidly from one extreme of rotation to the other a number of times will often clean the contacts and eliminate the noise.

Cleaning Fluids for Noisy Potentiometers. A noisy potentiometer can often be repaired by applying one of the special contact and volume-control cleaners that are available at jobbers. Most of these cleaners come in convenient aerosol spray cans or in bottles having applicator brushes or medicine droppers.

First, try applying the fluid alongside each terminal of the control. The terminals should be pointing upward, so the fluid drips down them to the resistance element. Rotate the control back and forth through its entire range a few dozen times while the fluid is running down.

Plunger-type applicators are also available. These screw over the threaded portion of the control. A push on the plunger then forces the cleaning fluid around the shaft of the control under pressure.

Sometimes it is necessary to take the back cover off the potentiometer and apply the fluid directly to the resistance element, the contact arm, and the shaft. The cover can usually be removed by prying up two or three tabs.

Do not try bending the contact arm unless it is obviously loose at all positions. Bending to get more contact pressure may cure the trouble for a while, but eventually the excessive pressure may wear away the entire resistance element.

If the volume changes suddenly when a volume control is rotated, the resistance material is badly worn out somewhere along its length. The control will also be noisy, but here replacement is required because no liquid cleaner can bridge a gap over missing resistance material.

Testing Potentiometers with an Ohmmeter. You can usually tell when a volume control needs replacement by listening for noise and for sudden changes in volume while turning the control. Similarly, you can in some cases detect a noisy picture-control potentiometer by rotating it back and forth while watching the picture. The picture will jump unsteadily and have streaks or flashes of light.

At times you may want to measure the resistance of a potentiometer with an ohmmeter. Make a diagram of the leads going to the terminals of the control, then unsolder and remove the leads from the terminals. If the control has an on-off switch, leave the switch terminals connected since they are in a separate circuit. Place the ohmmeter test leads on the two outside terminals of the potentiometer. This measures the whole resistance element, because each end of it is attached to one of the outside terminals.

If the ohmmeter shows a definite value of resistance, the potentiometer is good. If there is no deflection of the ohmmeter (infinity reading) even on the highest range, chances are that the potentiometer is bad.

Where to Get Replacement Potentiometers. When a potentiometer goes bad, the best source for it is your parts jobber. There are now so many different sizes and types of potentiometers that few shops try to keep a large stock on hand. Later, when you are well established, you will want to keep on hand some of the commonest sizes of volume controls. Recommendations for this stock are given later.

If a special control is required, your jobber may not have it. You will then have to get an exact-duplicate replacement from the distributor handling parts for the make of set involved. If there is no local distributor for that set, you will have to mail in the order and get the part back by mail. This takes around a week at the very least, so do not promise a fast repair job if there is any possibility that you will need a special control.

Potentiometer-ordering Data. Rarely will you find complete data stamped on a potentiometer. The resistance value alone is not usually enough for ordering. You need to know whether it is carbon or wirewound, whether it has an unusually high wattage rating, whether it has a fixed tap, whether it has a taper in the resistance element, and whether it has any unusual mechanical features, such as a combined on-off switch, a combined concentric control, or a special shaft shape, size, and length.

Giving all these data is the hard way of ordering a control. It is much easier to give the set manufacturer's number for the original part or the number of the correct replacement made by a potentiometer manufacturer.

One of the most complete sources of potentiometer data is the Photofact Folder for the receiver. This folder will have a parts list giving the manufacturer's original part number for the potentiometer and the correct replacement part number for several different manufacturers of potentiometers. For some controls a shaft type number and a switch type number must be given along with the potentiometer number.

You may be able to obtain from your jobber a reference manual that lists sets by make and model number, and gives the correct replacement parts number for each control in each set. Such a manual is well worth its price when you do not have a full collection of service manuals yet only need to replace an obviously bad control.

At some parts jobbers, it is only necessary to give the make and model number of the set and the function of the control (whether volume, brightness, contrast, etc.). The salesman will look up the set in one of his own manuals to get the number of the correct replacement part. This takes time, however, so do not expect such service all the time or from every jobber.

Ordering Shafts. Most controls come with a universal shaft that fits the majority of control knobs. When a set has special knobs, however, it is necessary to specify the correct special shaft also when ordering. This is why shaft numbers are given for some replacement controls.

Common types of shafts are shown in Fig. 5. If the new potentiometer does not have the correct shaft, the knob may not fit. The shaft and potentiometer are often sold disassembled, and must be put together according to instructions furnished with the parts.

If an on-off switch is mounted on the defective potentiometer, order a new switch too. The switch on the defective unit will not ordinarily fit on the new control. The stock number of the correct switch is also listed in most manuals after the number of the replacement potentiometer.

When replacing a volume control in an older set, give consideration to replacements having a push-pull on-off switch. The customer will appreciate this extra feature that modernizes his set. Before ordering, however, be sure the knob from the old control will fit the new control tightly enough for push-pull operation.

Determining Type of Potentiometer. If the resistance of a potentiometer is above 10,000 ohms, you can be reasonably sure that it is carbon, even though a few wirewound controls are made in resistances up to 50,000 ohms. If the resistance is more than 50,000 ohms, you are quite safe in assuming that it is a carbon control. Controls under 10,000 ohms may also be carbon, so here you cannot judge from the value alone.

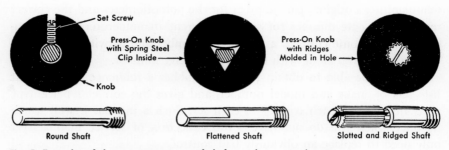

Fig. 5. Examples of the commonest types of shafts used on potentiometers

A sure way to determine the construction of a potentiometer is to remove its back cover. This can usually be done by prying out three or four metal tabs that are bent around the front of the unit. With the resistance element exposed, you can tell at a glance whether it is carbon or wirewound. If it is wirewound, the coils of wire on which the movable contact rides will be plainly visible. Carbon controls will have a smooth black resistance element.

Wirewound potentiometers ordinarily have a power rating of 3 to 5 watts. They are used, therefore, in circuits where appreciable current and power must be handled. Carbon controls have much lower power ratings, and for this reason are used chiefly in circuits where little or no current flows through the potentiometer.

If a carbon control is replaced with a wirewound control, the wirewound control may cause excessive noise as the shaft is rotated. The noise results from the fact that the resistance in wirewound controls does not vary smoothly; it varies in steps as the contact moves from one turn of wire to the next.

Always replace wirewound controls with wirewound, and replace carbon controls with carbon.

Dimensions of Potentiometers. In many small radios, especially personal portables, a midget-size volume control is used. Ordinarily these controls are about a half inch less in diameter than standard controls, and fit into a much smaller space. In ordering a replacement, therefore, give some attention to the space available. If the diameter of the old unit is less than about $1\frac{1}{4}$ inches, a midget or universal replacement control should be ordered. This is the size you usually get for replacements, anyway.

Taper of Potentiometers. In some potentiometers, the resistance is not evenly distributed along the resistance element. Instead, it has a low value per unit of length at one end and much higher values per unit of length as you move toward the other end. This variation in resistance per unit length inside the control is called *taper*. With linear controls the resistance per unit length is the same along the entire resistance element.

With left-hand taper, rotating the shaft in a clockwise direction causes the resistance between the center terminal and the left-hand terminal to increase slowly at first and then more and more rapidly as rotation continues. Volume controls in practically all television and radio receivers have left-hand taper.

Why Taper Is Needed. The reason for having a tapered resistance element in some carbon potentiometers is most easily understood by considering volume controls.

When listening to a program, you expect the set to be twice as loud at the full-on position of the volume control as at its middle position. Similarly, you expect to get twice as much volume at the middle position as at the quarter-on position. In other words, you expect the loudness to vary linearly with the position of the volume-control knob.

In order to double the apparent loudness, the strength of the audio signal must be more than doubled. It must, in fact, be increased almost ten times. This is based on a known characteristic of the human ear. The resistance of the volume control must therefore vary by factors of ten between the positions where the volume is to be doubled or halved. This is done by tapering the resistance of the volume control.

Several other kinds of taper are in use, because different tapers are needed when a control is used in different circuits. All, however, accomplish the same result—they make the action of the control change the way you expect it to.

There is no need to worry about taper when you can find the parts number for the correct replacement control.

Extra Taps on Volume Controls. You may find an extra terminal on some volume-control potentiometers. This terminal will usually be placed away from the regular three terminals of the control. The extra terminal connects to the resistance element somewhere along its length. This tap is usually used as part of an automatic-tone-control circuit that increases the loudness of the low notes at low volume. This compensates for a characteristic of the human ear that makes weak low notes sound weaker than they actually are.

Replacement volume controls are available with taps in various positions. The position or resistance value of the tap need be matched only roughly because it has no effect at ordinary volume and can be changed considerably in value before a difference in tone at low volume is even noticed. When ordering by parts number, you do not need to worry about taps.

If the correct tapped replacement cannot be obtained, use a standard volume control having the same total resistance value. The lead that went to the tap on the old control can then be connected to the center terminal. This gives bass boost at all volume-control settings. The change will rarely be noticed by the customer, because the original volume control was designed to give normal volume when the moving contact is at or near the tap.

Removing Defective Controls. When you find an open or noisy potentiometer, leave it in the set until you have the correct replacement unit on hand. If wires were unsoldered for testing, put them back on their correct terminals while waiting for a new unit. This minimizes chances of making a mistake when changing potentiometers.

To remove a potentiometer, first unsolder all leads. If there is any danger of getting the leads mixed up, make a simple connection diagram first.

Next, loosen the nut that holds the potentiometer in place on the chassis. This nut can be loosened with a wrench or with ordinary pliers. Standard thread is always used, so turn the nut counterclockwise to loosen it.

Cutting the Shaft of a New Control. The shaft on a replacement control will usually be much longer than is needed. Before installing the control, cut off the excess length. Use the length of the old shaft as a guide for marking the shaft of the new control with a pencil, crayon, or file. Next, clamp the end of the shaft in a bench vise and cut off the excess length with a hacksaw. Filing the rough end of the shaft will give the job a professional look.

Another way of cutting the shaft is to file a small nick in the shaft over the cutoff mark. Now hold the control side of the shaft with a pair of pliers, and break the shaft at the nick by bending it away from the nick. This method is faster, but can be used only with an aluminum-alloy type of shaft that breaks easily.

Assembling Concentric Potentiometers. Exact-duplicate replacements for concentric potentiometers come factory-assembled as a single unit, with the shaft already cut to the correct length. Universal replacements are purchased as separate parts, but assembly instructions are always in-

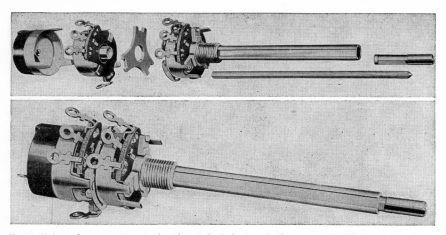

Fig. 6. Universal concentric control with switch, before and after assembly. (Mallory photos)

cluded. These instructions may seem complicated the first time you read them, but once you have assembled a concentric control you will agree that the job is not at all hard.

One important step in assembly is getting the potentiometers in their correct positions. Remember that the hollow outer shaft always gets the knob and potentiometer that are next to the front panel of the set. The solid inner shaft takes the smaller front knob and the farthest-back potentiometer.

Whenever a switch is used on a concentric control, it will be mounted on the farthest-back potentiometer and controlled by the front knob. An example showing assembly of a universal concentric control with switch is given in Fig. 6.

Mounting a New Potentiometer. Insert the shaft of the new assembled control through the hole in the chassis or the mounting bracket, working from the inside of the chassis so the new control is in the same position as was the old one. Place a lock washer and nut on the shaft, and tighten the nut with your fingers. Do not tighten the nut with a wrench until after all the wires have been connected to the control. This will allow you to turn the control slightly so as to be able to connect all the wires to the terminals without working in a crowded space.

In some sets an aligning hole is provided on the chassis for a control. New controls generally have a corresponding pin or projection on the shaft side of the control. When the control is rotated so that the pin fits into the hole on the chassis, the pin prevents the control from rotating when the shaft is turned. Ordinarily, this method of keeping the control from rotating is not needed because the nut on the shaft, when tightened properly, provides sufficient holding force.

If the new control has an aligning pin but there is no hole for it in the chassis, simply bend the pin out of the way or break or cut it off.

Connecting a New Potentiometer. After the new unit is in place on the chassis, attach all the wires to the terminals with permanent hook joints, by following the connection diagram that you previously made. Next, adjust the control to its final position, and tighten the nut with a wrench or pliers. It is important that this nut be tight. If it is not, the entire control will rotate when the shaft is rotated.

Finally, solder all the connections on the new control. Soldering is done last because movement of the control after soldering may break wires or loosen joints.

Checking Performance. With the new control installed and all connections soldered, turn the set on and try out the control. It should vary the volume of the sound (or whatever else it is controlling) smoothly and without noise or irregularity.

If the volume decreases instead of increasing as a volume control is rotated clockwise, the wires to the two outside terminals of the volume control are crossed. Unsoldering these two end terminals and reversing the wires will usually clear up the trouble.

If the new control has little or no effect on the volume of the radio, see if the old control had an internal ground connection between one of the terminals and the metal shaft of the control. If the ohmmeter test or the circuit diagram shows that a ground is required, connect a short piece of wire from the correct terminal to a nearby grounded terminal on the chassis.

Stocking Replacement Potentiometers. The best way to start your stock of universal replacement potentiometers is with a collection or kit of single units, shafts, and switches assembled by manufacturers especially for servicing requirements.

The control that you will replace most often in television and radio sets is the volume control. The following sizes will make a good starting stock of individual volume controls: 0.1 megohm; 0.25 megohm; 0.5 megohm; 2.0 megohms. The 0.5-meg size will be needed most, since it is used in about three out of every four sets. Each should of course have the correct taper for audio volume control (taper 1 for Mallory units and taper E for IRC units). All are carbon controls. Get a few attachable on-off switches for these controls also.

Your own parts jobber can give you good advice on controls to carry in stock to meet the needs of your particular locality at the start. Once you have been servicing sets for a half year or so, you will have acquired experience to guide you in reordering.

Types of Switches. Although there may seem to be many different types of switches in television and radio sets, they differ chiefly in external appearance. All have contacts inside that either come together or go apart when the switch is operated. These contacts usually serve either to open or close a single circuit, but can also be used to connect one circuit to any of two or more other circuits.

The switch you will encounter most often is that mounted at the rear of a volume control, where it serves as the on-off switch for the receiver. This switch is open when the knob is turned to its maximum counterclockwise position, and is closed when the knob is turned clockwise. Other switches are operated by pushing or pulling a knob, by pushing a button, by flipping a lever, or by rotating a knob that serves only for the switch.

A switch is perhaps the easiest part of all to check with an ohmmeter. Connect any range of the ohmmeter to the switch terminals and operate the switch. If the reading is infinity for one position of the switch and zero for the other position, the switch is good. If the reading is infinity for both positions, the switch is open or burned out and needs replacement. If the reading is zero for both positions, the switch contacts are fused together or otherwise shorted and the switch likewise needs replacing.

Since the on-off switch is by far the commonest, it will be covered in detail. The same general instructions will apply to all the other types of switches that you will find in television and radio receivers.

Combining Switches with Controls. The on-off switch for a television or radio set is usually mounted on the back of the volume-control potentiometer. This allows one front-panel knob to do two jobs. When the volume control is turned to its counterclockwise limit (turned all the way to the left), the volume will decrease to a minimum and the switch will click off. When the control is rotated clockwise from this position, the switch will click on. Further rotation clockwise will then increase the volume.

In television sets, on-off switches are often mounted at the rear of a concentric control and are then actuated by the knob for the rear control.

More and more television sets are using push-pull switches on a sliding volume-control shaft, or pushbutton switches having the button inside the volume-control knob. These permit turning the set on or off without changing the volume-control setting.

On-off switches for auto radios often have two pairs of contacts. These are called double-pole single-throw switches, abbreviated dpst switches. The contact ratings will be 5 amperes or 10 amperes, which is generally higher than for home receivers, so be sure you get the correct switch for an auto radio.

Repair of On-off Switches. If the switch on a potentiometer is not operating, the control should be taken out of the chassis and the switch cover or entire switch unit removed. To determine whether the trouble is electrical or mechanical, connect an ohmmeter to the switch terminals and operate the switch lever with a finger or screwdriver. For one switch position the ohmmeter should read zero. It should read infinity for the other position. If these readings are not obtained, the switch is electrically defective. A new switch is needed. This will usually mean buying a new potentiometer to fit the new switch, as a new switch will rarely fit on an old potentiometer.

If the ohmmeter shows that the switch is in good condition electrically, it is possible that the switch-operating arm at the rear end of the potentiometer shaft is bent out of position. A bent arm will not operate the switch toggle properly; hence the switch stays either open or closed all the time. In this case, try to bend the arm back into proper position with long-nose pliers.

Assembling a New Switch. Assembly instructions always come with replacement on-off switches. It may take you 10 minutes to read and follow the directions the first time you put one on. Once you have done it, however, you will be able to put additional controls together in a few minutes.

Assembly methods vary considerably among the different manufacturers; hence only a general procedure can be given here. Assume that the problem is to put together a volume-control and an on-off switch.

First, remove the back plate or cover of the volume control by bending up the clamping lugs.

Second, rotate the shaft of the volume control until the movable contact arm is in the middle position.

Third, snap the actuating arm of the switch to the on position. This will be the clockwise position as you look at the inside of the control. Remember that you turn a set off by turning the shaft knob counter-

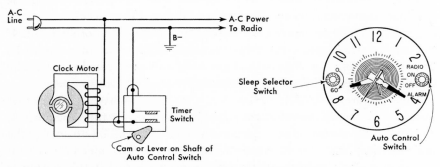

Fig. 7. How timer switch of clock radio does work of on-off switch

clockwise. The set is on when the contact arm is in mid-position, so turn the switch on before assembling it.

Fourth, set the switch in its closed position by checking with an ohmmeter. Place the switch on the back of the volume control. Hold it in place with your hands and turn the shaft of the volume control counterclockwise as far as it will go. The switch should click off. If the switch does not work, study the assembly instructions again.

Fifth, fasten the switch to the volume control permanently by bending the clamping lugs.

Combining Switches with Clocks. On many radios and even a few television sets, the on-off switch is in a built-in clock rather than on the volume control. A typical power-circuit arrangement for one of these clock radios is shown in Fig. 7. The motor winding of the clock is permanently connected across the line-cord wires, as required to make the clock run continuously. The radio receives power only when the timer switch in the clock is closed.

The timer switch in a clock is operated by a cam or lever mounted on a shaft. The shaft projects forward through the clock face, where it generally has a handle or knob for operating the switch manually to turn the radio on or off, as also shown in Fig. 7.

The timer switch can also be closed or opened at a desired time by the clock movement. The user chooses this time in much the same way that any alarm clock is set, by turning an alarm-set knob until the alarm pointer on the face of the clock points to the time at which automatic turn-on of the radio is desired.

Some clock radios also have a conventional on-off switch on the volume control. This permits turning the radio on and off manually when the automatic-alarm features are not desired. If these two switches are connected in parallel in the set, one loses control when the other is on. If the switches are in series, the on-off switch on the volume control must be turned on before the automatic-alarm switch features can be used. The circuit of the set will show how the switches are intended to operate.

Extra Features on Clock Radios. Some clock radios have a buzzer that sounds at the same time that the radio is turned on by the timer switch. This serves to awaken the sleeper even when the station to which the radio is tuned is not yet on the air in the morning. One added refinement is a snooze alarm button, usually on top of the radio. If this button is pressed when the alarm comes on, the buzzer is silenced for 5 to 10 minutes while the radio stays on, to permit a short extra nap.

Another clock-radio feature is a sleep switch. This turns off the radio at a desired interval of up to 60 minutes after the switch is set, to provide music for going to sleep.

Some clock radios have a power outlet at the rear. This is energized when the timer switch turns on the radio. It can be used to turn on a lamp automatically, turn up a furnace thermostat, turn on a coffee maker, or even turn on an electric window closer.

Clocks in Transistor Radios. A few transistor radios have been made with clocks and timer switches. The operation of the timer switch is essentially the same as in a-c radios. The only difference is in the clock motor, which in transistor radios is a tiny d-c motor designed to operate from a single 1.5-volt flashlight cell. The transistor radio must therefore contain one extra flashlight cell that is permanently connected to the clock motor. This flashlight cell must of course be replaced whenever the clock loses time or stops. It is a good idea to replace all dry cells when either those for the radio or that for the clock goes bad, because the cost

Controls, Switches, and Clock-radio Timers

of an extra cell is small compared with time needed to obtain and install a dry cell.

Repairing and Replacing Clock-radio Timers. Practically all the timers in clock radios are made by one of three manufacturers—Telechron, Westclox, and Sessions. Each has a number of different models of timers, with the model number generally stamped on the back of the faceplate, on the back of the mounting plate, or on a gummed label attached to the back of the clock.

Replacement parts, warranty work, and factory service on clock timers are generally handled by the clock manufacturer's organizations rather than by the radio manufacturer's distributors. The local radio distributor will of course be able to tell you exactly where to get what you need for clock repairs. Sometimes the radio distributor is also able to furnish new control knobs, crystals, faceplates, and clock hands.

Adjusting Clock-radio Timers. Although there are rarely any true service adjustments on clock radios, you may often be able to clear up a trouble by bending a lever. Thus, if the alarm buzzer is not working loud enough, you can fix this by carefully bending the buzzer arm. Look for this arm close to the magnetic field of the motor winding. The arm is normally held stationary by a damper. The clock mechanism moves the damper out of the way at the time for which the alarm is set, allowing the buzzer arm to vibrate as it is alternately attracted and released by the a-c magnetic field of the motor. This arrangement is much the same as in ordinary electric clocks.

You may occasionally encounter a jammed mechanism. This trouble can occur if the user turns one of the knobs on the clock too far. Look for some part that has jumped out of its pivot point or jumped out of the groove through which it normally travels. By loosening adjacent mounting screws, you can often slip the part back into position.

Some clocks have an adjusting screw that controls the point at which the cam or other actuating lever opens and closes the timer switch. If the radio does not come on at the correct time, simply turn this screw until it does.

If there is no alarm-adjustment screw, study the clock mechanism carefully. Sometimes you can bend a cam or lever to achieve the same adjustment that is provided by a screw. More often, however, wrong timing can be fixed by removing the crystal from the clock face, turning the alarm-set button at the rear until the radio comes on, then moving the alarm pointer to the time indicated by the clock hands. The pointer is pressed onto its shaft, and can be turned in this way without damage. After this adjustment is made, the radio will come on automatically at the time to which the alarm pointer is set. The clock switch must of course be set to the alarm position before making all these adjustments.

Clock-radio Switch Troubles. When the contacts in a timer switch become burned or broken, the switch no longer turns on the radio. An

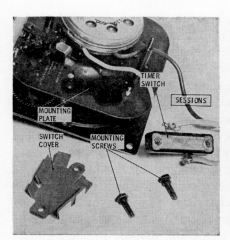 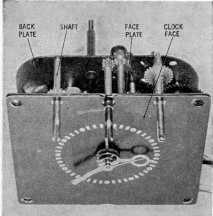

Fig. 8. In Sessions clock at left, the timer switch can be removed after taking out only two mounting screws. The clock timer at the right has three control shafts. (Figures 8 and 9 reprinted from July 1959 issue of *PF Reporter*, copyright 1960 by Howard W. Sams and Co., Inc.)

ohmmeter test will quickly verify that the switch stays open when its actuating button or arm is pressed.

When switch trouble occurs, it is generally best to remove the entire clock and take it to your authorized service station for that make of clock. These service stations have efficient facilities for removing riveted mounting plates, along with a complete stock of the spare parts needed. They will usually replace the actuating arm when replacing the switch, because any wear in this arm, even though not visible, can make the new switch operate improperly. Bring in only the clock, because clock service stations often charge several dollars extra if they have to take the clock out of the radio.

Of course, if the switch is easily replaced as a complete unit, as in the Sessions unit shown in Fig. 8, you may prefer to do the repair job yourself. It is recommended that you take on clock repairs yourself only when you

Controls, Switches, and Clock-radio Timers 395

have a bit of spare time; the first job on each make of clock will usually take you far longer than after you have acquired experience, and this time should be charged up to your own education rather than to the customer.

Replacing Broken Clock Shafts. One common trouble in clock timers is a broken control shaft. There may be three of these shafts, as in Fig. 8. Replacing these is a clock-service-station job because it generally requires removal of the riveted or staked clock face and faceplate.

Your parts jobber may have kits of slip-on shafts for making a repair. If the correct size of shaft is obtainable, it can usually be pushed over the broken shaft without even removing the chassis from the cabinet.

Fig. 9. Motor in Westclox unit at left can be removed as shown after taking out small screws in the two mounting studs. In Sessions unit at right, three mounting studs fit into slotted holes; turning the motor counterclockwise a small amount permits lifting out the entire motor, much as a bayonet-base lamp is removed from a socket. (Howard W. Sams photos)

Replacing Clock Motors. A new motor for a clock can usually be obtained from the appropriate clock-service station for around $3. Replacement is usually easy. In fact, in some Sessions clocks the motor is removed simply by turning it counterclockwise until the three mounting studs are out of the elongated holes in the motor-mounting plate. Other motors use mounting screws, as in Fig. 9.

Telechron motors consist of a core with its coil, entirely separate from a sealed rotor unit having a small drive gear on a shaft projecting from one end. The rotor unit may be replaced separately from the coil, and vice versa.

All clock motors have two leads or terminals, so only two soldered connections need be made when replacing a motor.

18

Testing and Replacing Capacitors

Capacitor Troubles. Next to tubes, fixed capacitors cause more trouble than any other part in television and radio receivers. Some capacitors are frequent troublemakers, while others rarely go bad. Once you are acquainted with the peculiarities of each type, you will know which to suspect first and how to test it.

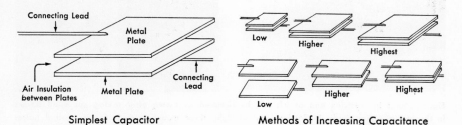

Fig. 1. Getting-acquainted facts about capacitors. Capacitors were formerly known as condensers, and many servicemen still use this term

Just as important as testing capacitors is knowing how to order and install the correct replacement capacitor. You will learn all these things in this chapter.

What Is a Capacitor? A capacitor is a device for storing an electric charge. A simple capacitor can be made from two metal plates separated by air or other insulating material, as shown in Fig. 1. The connecting leads are soldered or riveted to the metal plates.

The charge-storing ability or electrical size of a capacitor is called its capacitance. This ability depends on the total area of the capacitor plates, their spacing, and the insulation used between the plates.

Any two metal objects that are separated by insulation have capacitance. Even two insulated wires have capacitance when side by side or when twisted together. In some television and radio sets, short lengths of hookup wire twisted together are actually used to give extra capacitance between two points.

With capacitor plates of a given size, moving the plates closer together increases the capacitance.

The better the insulation is between the plates of a capacitor, the higher is the voltage that can be applied before a spark jumps through the insulation between the plates and causes a short-circuit.

The capacitance of a capacitor can be increased by using a better insulation than air between the plates. This insulation is technically called the dielectric. Examples of insulating materials used in capacitors are air, mica, paper, special chemicals, mineral oil, ceramics, thin sheet plastic material, and even glass.

The different types of capacitors are named according to the kind of dielectric between their plates. The commonest types of fixed capacitors used in modern television and radio sets are paper capacitors, electrolytic capacitors, and ceramic capacitors.

Paper Capacitors. In this commonest of all capacitors, thin strips of metal foil are used as plates, with one or more thicknesses of insulating paper between the plates. The foil and paper strips are wound into a tight roll to occupy minimum space. This roll is then inserted in a tubular cardboard or metal container, as in Fig. 2. Sometimes a plastic housing is molded around the roll. Because of this tubular shape, paper capacitors are often called tubular capacitors.

The capacitance of a tubular paper capacitor is made larger in any or all of three ways: (1) by using wider metal-foil and paper strips, which makes the capacitor longer; (2) by using longer paper and metal-foil strips, which makes the capacitor thicker; (3) by using thinner paper insulation between the layers of the metal foil.

Thinner insulation breaks down more easily when voltage is applied. A spark jumps through a flaw in thin insulation and burns out the paper, allowing the opposite metal-foil layers to touch and short out the capacitor. Thick paper insulation is therefore needed when capacitors must withstand high voltage.

Sometimes no metal foil is used in paper capacitors. Instead, a thin metallic coating is deposited on one side of each paper strip. These tubular capacitors are called metallized paper capacitors. Whereas ordinary paper capacitors stay shorted when a conductive flaw develops between the foil plates, metallized paper capacitors fix their own shorts. The metallic coating burns out around the flaw and isolates it, as shown in Fig. 2. This leaves the capacitor practically as good as new.

A thin film of Mylar or other good insulating plastic is used in place of paper in many tubular capacitors.

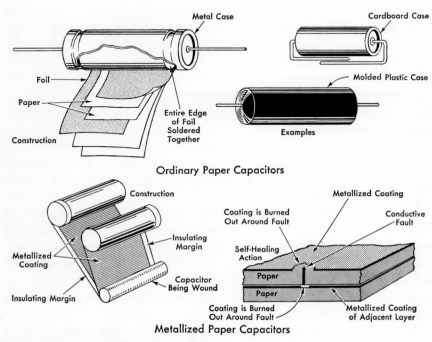

Fig. 2. Construction details of the two important types of paper capacitors

Electrolytic Capacitors. As Fig. 3 shows, an electrolytic capacitor has metal-foil plates rolled together much like paper capacitors, but uses a chemical paste between the plates. This paste or electrolyte is held in position by gauze or blotter paper. It produces on the positive plate an extremely thin chemical oxide film that acts as the insulating material or dielectric. This film gives a much higher capacitance than can be obtained with ordinary insulation.

Ceramic Capacitors. A special ceramic material serves as the insulation between the plates in ceramic capacitors, also shown in Fig. 3. The ceramic material is molded into a thin disk or cylinder and baked until it is hard as rock. Thin coatings of silver are then placed on each face of the ceramic

Capacitors **399**

to serve as plates. Connecting leads are soldered directly to the silver or are soldered to crimped-on end terminals that make contact with the silver. The entire unit is then covered with insulating material or molded in plastic to keep out moisture.

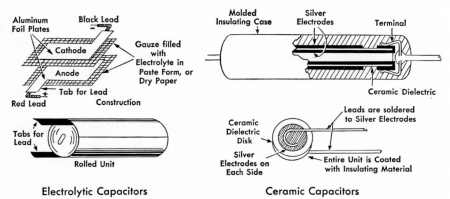

Fig. 3. Useful facts about electrolytic and ceramic capacitors

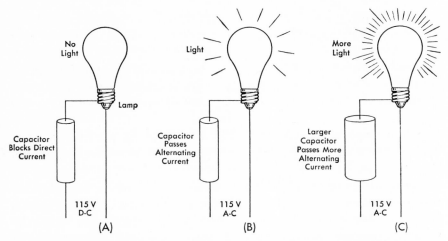

Fig. 4. The larger the capacitor in an a-c circuit, the more alternating current it passes. Low-wattage lamps and large capacitors are needed to actually demonstrate this effect

What Capacitors Do. Before starting on capacitor test and replacement procedures, you need to know just a little about what capacitors do, along with practical information on capacitor ratings and capacitor symbols.

If a capacitor and lamp are connected in series across a d-c voltage as in Fig. 4A, the lamp will not light. When the same capacitor and lamp are connected across an a-c power line as in Fig. 4B, however, the lamp lights up. This simple experiment shows that a capacitor passes alternating current but blocks the flow of direct current.

If the electrical size of the capacitor is increased, as in Fig. 4C, the capacitor passes more alternating current and the lamp glows brighter. The larger the electrical size or capacitance of a capacitor, the lower is its opposition to alternating current. This opposition is measured in ohms just as for resistors but is called *impedance* rather than resistance.

Capacitor Impedance. The opposition or impedance of a capacitor is not a fixed value like the opposition of a resistor, because the impedance depends on frequency as well as on the electrical size of a capacitor. Instead, the *impedance* of a capacitor *increases* when the *frequency* of the a-c voltage is *decreased*.

At zero frequency, which is d-c, the impedance of a capacitor is a maximum, and current flow is blocked. At extremely high frequencies, on the other hand, the impedance is practically zero ohms and the capacitor passes alternating current readily.

For a given frequency, a higher capacitance gives a lower impedance. Low impedance is needed in power-supply filter circuits to cut down hum; this is why you will find such large capacitance values in power-supply filter circuits.

In radio-frequency and video-frequency signal circuits, the frequencies are so high that only very small capacitance values are needed to get low impedance. This is why you see such small capacitors in the tuners and other signal circuits of television receivers.

How a Capacitor Passes A-C. The equivalent of alternating-current flow through a capacitor is produced by back-and-forth movements of electrons. The electrons pile up on one set of plates, leaving a shortage of electrons on the other set of plates. When polarity of the voltage is reversed, the electrons pile up on the other set of plates. The larger the capacitance value of the capacitor and the higher the voltage, the more electrons will be piled up.

When a capacitor is connected across a battery or other d-c voltage source, the electrons are pushed to one set of capacitor plates. When the available voltage can push no more electrons, the electron-transferring action stops, and the capacitor is fully charged. There is then no further movement of electrons in a d-c circuit.

If the battery is disconnected from a charged capacitor, the capacitor stays charged. A high-quality capacitor can hold its charge for several hours.

If initially charged to a dangerously high voltage (even 115 volts can be dangerous sometimes), a capacitor can give a deadly shock if accidentally touched by someone. This is why television servicing instructions so often call for discharging a high-voltage capacitor by holding a 50,000-ohm resistor across its terminals for a few seconds.

When a charged capacitor is discharged by short-circuiting it with a copper wire or a screwdriver, there is a popping sound and a spark as electrons rush through the shorting path to equalize the number of electrons on the plates. This action may damage the capacitor and burn the screwdriver.

When a capacitor is connected in an a-c circuit, the capacitor discharges and then charges up with opposite polarity each time the polarity of the a-c voltage source reverses. It is this back-and-forth movement of electrons during charge and discharge that produces the effect of alternating current passing through a capacitor. This is why we say that a capacitor blocks d-c but passes a-c.

Ratings of Capacitors. The amount of alternating current that a capacitor will pass depends on the ability of the capacitor to store electrons on its plates. The greater the capacitance of a capacitor, the more electrons it can store on its plates and the better it conducts alternating current. Capacitance is also called capacity.

Another important rating of a capacitor is its *voltage rating*. This is the maximum voltage that can safely be applied to the capacitor. A greater voltage may cause the capacitor to break down and conduct a current directly from one plate to the other. The breakdown occurs when a spark jumps between the plates through a tiny flaw in the insulation. The heat of the spark burns out the insulation and makes it conductive. Sometimes the heat even melts the metal foil of the plates, so the metal flows between the plates and causes a direct short-circuit.

The two important ratings of a capacitor are thus the capacitance rating and the voltage rating.

Capacitance Rating. Just as the resistance of a resistor is given as so many ohms, the capacitance of a capacitor is given as so many *microfarads*. The greater the capacitance of a capacitor, the more microfarads it has.

With many capacitors, the microfarad is much too large a *unit of capacitance*. Capacitance is therefore also expressed in millionths of a microfarad, called *micromicrofarads*. One microfarad is equal to one million micromicrofarads.

Capacitors used in television and radio sets range in capacitance from about 10 micromicrofarads for the smallest ceramic or mica unit to several thousand microfarads for the highest capacitance value used in transistor circuits.

Abbreviations for Microfarad. On circuit diagrams, in catalogs, and in books and magazines, the word microfarad is usually abbreviated as mfd, mf, or μf, used either with or without periods. The last version uses the Greek letter μ (pronounced mu). Thus, an 8-microfarad capacitor might be listed as 8 mfd, 8 mfd., 8 MFD, 8 MFD., 8 mf, 8 MF, 8 μf, 8 μF, or 8 μFD. The meaning of the abbreviations will always be clear, since there are no other similar abbreviations.

In asking for an 8-mfd capacitor at a jobber's, a serviceman would probably say, "Give me an 8-mike capacitor," because *mike* is the slang abbreviation for microfarad.

The abbreviation for micromicrofarad is mmfd or $\mu\mu$f. Thus a 500-micromicrofarad capacitor would be designated as 500 mmfd, 500 $\mu\mu$f, or 500 $\mu\mu$F.

Changing Microfarads to Micromicrofarads. Sometimes you may want to know the capacitance of a capacitor in micromicrofarads when its value is given in microfarads. To change from microfarads to micromicrofarads, multiply the microfarad value by one million. This can be done by moving the decimal point six places to the right, as follows:

$$0.0002 \text{ mfd} = 000200. \text{ mmfd} = 200 \text{ mmfd}$$
$$0.001 \text{ mfd} = 001000. \text{ mmfd} = 1{,}000 \text{ mmfd}$$

Changing Micromicrofarads to Microfarads. The value of a capacitor in micromicrofarads can be changed to its value in microfarads by dividing the micromicrofarad value by one million. This can be done by moving the decimal point six places to the left, as follows:

$$500 \text{ mmfd} = .000500 \text{ mfd} = 0.0005 \text{ mfd}$$
$$10{,}000 \text{ mmfd} = .010000 \text{ mfd} = 0.01 \text{ mfd}$$

The zero at the left of the decimal point in the final values 0.0005 mfd and 0.01 mfd has no particular meaning. It is the standard way of writing decimal numbers.

Voltage Rating of Capacitors. The voltage rating is printed on many capacitors along with the capacitance rating, somewhat as shown in Fig. 5. This voltage is often labeled DCWV, which means direct-current working voltage. Sometimes V.D.C., D.C.V, D.C.W., D.C.V.W., or other com-

Capacitors   403

Fig. 5. Typical methods of labeling paper capacitors with capacitance and voltage ratings

binations of the letters D, C, W, and V are used, but all have the same meaning as DCWV.

Some capacitors, particularly electrolytics, also have a peak voltage rating. This is the highest voltage the capacitor is rated to withstand momentarily. The rating is important in power-pack filter circuits and other circuits that may have voltage pulses along with a steady d-c voltage.

Common voltage ratings found on capacitors in television sets are 200, 400, and 600 volts for paper capacitors and 150, 350, and 450 volts for electrolytic capacitors. Ceramic capacitors generally have much higher voltage ratings.

Transistor circuits use low-voltage capacitors, having voltage ratings ranging from 3 to 25 volts.

Symbols for Capacitors. On circuit diagrams, fixed capacitors are represented by the standard symbol shown in Fig. 6, which has one straight line and one curved line. An older symbol using two straight lines is still widely used and means the same thing.

If the capacitor is variable, as for tuning capacitors, an arrow is drawn diagonally through the symbol just as for variable resistors, or an arrowhead is placed on the curved line.

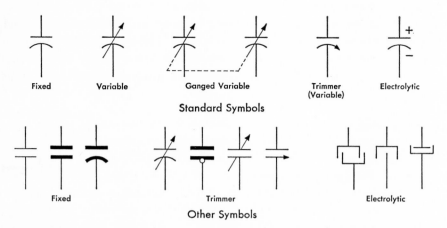

Fig. 6. Symbols used on circuit diagrams to represent different types of capacitors

Many variations of the capacitor symbol may be found, but all have two plates facing each other, just as in an actual capacitor. Examples of these are also shown in Fig. 6.

The capacitance value of the capacitor will usually be found alongside the symbol on a diagram. Only rarely, however, will the voltage rating of the capacitor be shown on a circuit diagram. Look in the parts list in a service manual to find the voltage rating of a capacitor.

Connecting Capacitors in Parallel. When capacitors are connected in parallel, their combined capacitance is the *sum* of their individual capacitance values. Thus, 0.02 mfd and 0.03 mfd in parallel give 0.05 mfd. Parallel connections are seldom used, however, because single capacitors are generally available in the required size or close enough to it.

Avoid connecting capacitors in series. With a series connection, one capacitor will usually get more voltage than the other because the leakage resistances of two capacitors are seldom the same. The capacitor with the higher resistance will get the greater share of the voltage and may break down.

Testing and Replacing Capacitors. With this general description of what capacitors look like, what they do, and how they are rated, you are ready to learn how to check and replace each type of capacitor in turn, starting with the most common type—the paper capacitor.

How Paper Capacitors Go Bad. Unlike most other parts, paper capacitors need not be abused to go bad. They can fail at any time, especially if they are of cheap make or if they have too low a voltage rating for a particular circuit.

Capacitor failure can occur in three ways. It can be an *open* or actual break in the leads inside the capacitor. It can be a direct *short* between the plates due to failure of insulation. Third, the capacitor can become *leaky*. This means that the insulation between plates deteriorates and drops

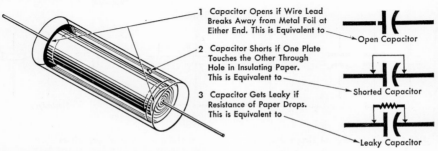

Fig. 7. Three common ways in which paper capacitors go bad

in resistance, allowing electrons to leak through from one plate to the other.

These three types of faults in paper capacitors, shown in Fig. 7, will now be considered separately.

How Paper Capacitors Open. Paper capacitors are said to be open when they break somewhere internally. Usually one of the leads breaks away from the metal foil inside. Open up an old paper capacitor and you will see exactly how this can happen.

When a paper capacitor becomes open, the effect is pretty much the same as if a lead to the capacitor itself were cut. In addition to blocking d-c, the capacitor now also blocks the a-c television or radio signal.

The best way to find an open paper capacitor in a receiver is to shunt each suspected capacitor in turn with a good capacitor, as described later under capacitor-substitution tests. An ohmmeter is not reliable for finding an open, because open capacitors give the same near-infinity resistance reading as good paper capacitors.

Sometimes the open in a capacitor is not permanent, but intermittent. When this happens the capacitor works properly part of the time and does not work at all the rest of the time. An intermittent open can also cause noisy operation.

An intermittently open capacitor often occurs because one of the leads of the capacitor pulls away from the metal foil when the capacitor is heated, cooled, or jarred. Thus a paper capacitor may work all right before the receiver has had a chance to warm up. Once the set has warmed up, the heat may move the wire lead of the capacitor just enough to make the set stop playing. Similarly, a paper capacitor may work for a while, then quit, but start operating properly again as soon as the set is jarred.

How Paper Capacitors Short. A paper capacitor is said to be shorted when a direct connection or a low-resistance connection develops from one plate of the capacitor to the other. This happens when the insulation between the capacitor plates breaks down.

The insulation may break down because the voltage across the capacitor is too high for its rating. The high voltage causes a spark to jump from one plate to the other, carbonizing the insulation and producing a low-resistance connection between the plates.

More often, a paper capacitor breaks down because its insulation deteriorates. Paper insulation tends to weaken after many years of use, especially if the capacitor becomes warm in the set or if the receiver is operated in a humid location and moisture gets inside the capacitor. Eventually

406 *Television and Radio Repairing*

the insulation may weaken sufficiently to allow a spark to jump between the plates and produce a low-resistance carbon path from one plate to the other.

How Paper Capacitors Become Leaky. Paper capacitors are said to be *leaky* when they pass an appreciable amount of current directly from one plate to the other. The term leakage does not mean dripping of wax or liquid. Rather, it refers to lowered electrical resistance that allows electrons to leak or flow through the capacitor.

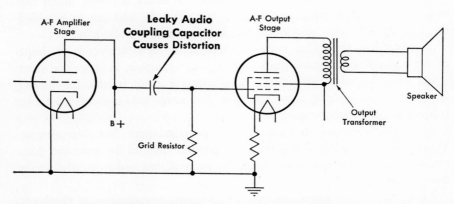

Fig. 8. If a receiver sounds badly distorted, the cause may be a leaky coupling capacitor. This will be a paper capacitor somewhere between 0.002 and 0.05 mfd, connected between the control grid of the a-f output tube and the plate of the a-f amplifier tube

Leakage of paper capacitors is usually specified as the leakage resistance measured by an ohmmeter connected to the leads of the capacitor. The leakage resistance of a good paper capacitor is very high, from 50 megohms up to 5,000 megohms. With continued use, however, the leakage resistance may become low enough to upset normal operation of the set.

Leakage is not ordinarily an important defect of paper capacitors. In most circuits using these capacitors, the leakage has no effect on performance until it gets bad enough to approach the low resistance of a short-circuit.

One place where a leaky capacitor can really make trouble is in audio coupling circuits. When the coupling capacitor in the circuit of Fig. 8 becomes leaky, the effect is that of a resistor connected across the capacitor. This resistor allows current to flow from B+ through the grid resistor to ground. The resulting voltage drop across the grid resistor acts as a positive grid bias that causes the tube to distort the audio signal badly.

Capacitors   407

This is why servicemen suspect a leaky audio coupling capacitor whenever the sound is badly distorted in a television or radio receiver.

If you measure less than 20 megohms leakage resistance for any suspected paper capacitor, disconnect it and try a new capacitor of the same size temporarily. In most cases, there will be no change in performance, but there are enough times when low leakage resistance is the trouble to justify trying a new one. Better yet, replace any capacitor that measures below 20 megohms, because it is likely to get much worse soon and cause trouble.

When to Suspect a Paper Capacitor. The majority of bad paper capacitors can be found by noting certain visible clues in the receiver.

Any capacitor that appears to have become hot during operation of the set should be suspected. Heat often causes capacitors to become open, shorted, or leaky. Heating is indicated when the wax at the ends of the capacitor bulges out. Sometimes the wax runs out of the ends of the capacitor. Sometimes all the wax inside the capacitor may disappear, so that the cardboard housing fits loosely over the rolled foil. The capacitor then feels hollow when pinched with the fingers.

To check a capacitor you suspect of being bad because of overheating, substitute a known good capacitor and see if this clears up the trouble.

Whenever you find a burned-out resistor or a smoking resistor in the radio, look for a shorted paper capacitor that could draw current through the resistor. All paper capacitors that connect to the resistor lead that is farther away from the power supply should be suspected. Check the resistance of each suspected capacitor with an ohmmeter.

If wiggling any capacitor or twisting the outer case slightly produces a noise in the loudspeaker while the set is on, suspect that capacitor as being bad. If noise or hum is produced continuously, wiggling the bad capacitor may cause additional noise or may cause the noise and hum to stop altogether.

Audio distortion that starts as soon as a set warms up can be caused by a leaky coupling capacitor connected to the control grid of the audio-output tube, or by a leaky plate bypass capacitor in the first a-f stage. Check the paper capacitors connected to these tubes, even if you do not yet understand circuit operation.

Checking Paper Capacitors with an Ohmmeter. When you suspect that a certain paper capacitor is shorted, connect an ohmmeter across the terminals of the capacitor, and measure its resistance. The capacitor does not ordinarily have to be disconnected to make this test, as a dead short will show up as zero ohms even if the capacitor is in parallel with other parts.

If the resistance as measured is zero or almost zero, chances are that the capacitor is shorted. To make sure of this, unsolder one lead of the capacitor and again measure the resistance of the capacitor. If the resistance measured is still zero, the capacitor is definitely shorted and should be replaced.

If the resistance of the capacitor measures infinity when disconnected, the capacitor is not shorted. A coil or resistor was in parallel with the capacitor in the circuit and was giving the low-resistance reading. Replace the capacitor connection and continue your hunt for the shorted capacitor.

Ohmmeter Kick Test for Paper Capacitors. The actual capacitance of a capacitor can be measured only with special test equipment. The approximate capacitance, however, can be determined by connecting the highest range of an ohmmeter to the capacitor terminals. The battery in the ohmmeter charges the capacitor, and the sudden but momentary charging current makes the meter pointer kick momentarily. The larger the capacitance of the capacitor, the greater is the kick. A good 1-mfd or larger paper capacitor will give a large kick of the meter pointer, while that of a 0.01-mfd capacitor will be barely visible. By comparing the kick with that of a good paper capacitor of known capacitance, you can roughly estimate the value of an unknown capacitance.

Measuring Leakage Resistance with an Ohmmeter. When connecting an ohmmeter to measure the leakage resistance of a paper capacitor, the meter pointer will first kick to the right. After the capacitor has charged, the pointer will come to rest. The amount of resistance indicated by the ohmmeter will then be the leakage resistance of the capacitor.

With a 20-megohm ohmmeter range and with a good capacitor, this leakage resistance reading will be infinity on the meter scale. Only with a very leaky capacitor will the reading be enough below 20 megohms so it can be read on the meter scale.

Checking Paper Capacitors by Substitution. When you suspect a paper capacitor of being open, intermittent, or leaky, the best way to check the capacitor is to substitute a known good capacitor of about the same size in its place. If the new capacitor clears up the trouble, then the suspected capacitor is bad and should be replaced. For testing, the value of the new capacitor is not important, but for the permanent replacement you will want to get the correct value.

To check paper capacitors by the substitution method, keep 0.0001, 0.0005, 0.001, 0.002, 0.005, 0.01, 0.02, 0.05, 0.1, and 0.25 mfd paper capacitors with 600-volt ratings on hand. It is a good idea to get three or four of each to provide a stock of spares, since these are the sizes of paper capacitors you will need most often. Some capacitor manufacturers have capacitor kits containing quantities of most of these values, in convenient boxes for bench and toolbox.

For a temporary check, unsolder one lead of the suspected capacitor and then hold the leads of the new capacitor on the terminals to which the old capacitor was connected. Be sure the leads of the new capacitor make good contact. If in doubt, solder them to the terminals with temporary lap joints for the check. With the new capacitor connected, turn on the set and check its performance. If the new capacitor has no effect on the set, remove it and connect the old capacitor back again.

When testing by substitution, do not shorten the leads of the good capacitor. Just be careful that these long bare leads do not touch any other bare leads or terminals.

In connecting the test capacitor, be sure that the lead at the end marked OUTSIDE FOIL or having a black band goes to the same terminal as did the corresponding lead of the original capacitor.

Hold the new capacitor by its insulated housing, not by a bare lead, when making a substitution test with the set turned on, to avoid getting a shock. Only when capacitor leads are insulated, as they often are in electrolytics, can you safely hold them while power is on.

Sometimes the presence of your hand on the substitute capacitor may cause hum or prevent normal operation. Whenever you suspect this condition, solder the substitute capacitor in position temporarily for the test.

Remember that one lead of the old capacitor must be disconnected before substituting a new capacitor to check for a leaky or short-circuited paper capacitor. Only when the faulty capacitor is open can you connect a good capacitor directly across a suspected capacitor, without bothering to disconnect one lead of the suspected capacitor.

Always leave the bad capacitor in the set until the new one is on hand. Since one of its leads is disconnected for the final ohmmeter test or for temporary substitution of a good capacitor, you can easily read printed values even on the underside of the bad capacitor.

It is a good idea to hook the disconnected lead of the bad capacitor back on its terminal when going for a replacement, so you will not connect it wrong later.

When printed values are on the underside of a connected paper capacitor in a set, there is some risk of breaking the leads by turning the body of the capacitor so the values can be read. You may prefer instead to keep a small mirror in your toolbox, to use with your flashlight for reading values underneath such parts.

Ordering Replacement Paper Capacitors. Only two things need be known to order a replacement paper capacitor: the capacitance and the voltage rating of the defective capacitor. If these ratings are not printed on the capacitor or if the printing is faded, the capacitance rating must be obtained from the parts list or the circuit diagram just as for resistors.

If the voltage rating of a paper capacitor is unknown or if it was below 600 volts, always get a 600-volt unit. It costs only a few pennies more than the 200- or 400-volt unit that may have been in the set, and will give much longer service. It is foolish to try to save a few pennies on a repair job that will bring you a good many dollars.

When the cheapest possible parts are used, the law of averages works against you because the part is more likely to fail during your guarantee period on the repair job. This means a repeat call, taking many dollars' worth of your time to fix the set over again. Widely advertised makes of capacitors are generally high in quality, so stick to these for replacements. Stay away from the bargain counters.

To find the value of the capacitor when a pictorial diagram or photograph of the chassis is part of the servicing data, locate the defective capacitor on the photograph and note its identifying number. Look up this number in the parts list to find the capacitance value.

If only the circuit diagram of the set is available, locate some signpost, such as a tube terminal, to which one end of the capacitor is connected in the set. Locate this same signpost on the circuit diagram and note the value of the capacitor connected to it. If more than one capacitor is connected to this point, note which parts are connected to the other end of the bad capacitor in the set, and look for these on the circuit diagram.

Capacitor values are generally assumed to be in *microfarads* when no unit is given after the value. Microfarad values will generally be decimal values or whole numbers up to 100, while micromicrofarad values will be whole numbers ranging from around 10 up to several thousand.

To reduce the number of different capacitance values that have to be ordered by manufacturers and kept in stock by servicemen for replacement, EIA has standardized capacitance values so their significant figures are 10, 12, 15, 18, 22, 27, 33, 39, 47, 56, 68, 82, and 100. Some manufacturers still use other values, however. Order the value nearest to that of the old capacitor, as capacitance values rarely need to be exact.

Be on the lookout for special paper capacitors having unusual markings or housings. Exact replacements are required for these. Some are oil-impregnated to give longer life, and others have special Mylar or other plastic dielectric films to give a high degree of stability.

Obtaining Replacements for Capacitors of Unknown Capacitance. Occasionally you will run into a set in which the capacitors are unmarked and there is no servicing information available. In such a case, you will have to make a guess at the size of replacement. It is helpful here to look at a few circuits of somewhat similar receivers of other makes, and note the value of the capacitor used in that same position in these sets.

The physical size of a bad capacitor is also an aid to guessing its capacitance value. Actual sizes of capacitors vary considerably depending on their make and voltage rating, but comparison with your stock of new capacitors will give you a rough guide to the capacitance value of the unmarked paper capacitor.

When guessing at a capacitor value, connect your first choice in temporarily and try the set. If the set works again, connect the capacitor in permanently. If the set does not operate quite right with the new capacitor, take out this capacitor and try one of different capacitance. Continue this testing and trying until a capacitor is found that restores the set to normal. In most cases, this will require only one or two quick substitutions.

If the original capacitor has a voltage rating above 600 volts, replace it with the same higher value. Buffer capacitors in auto-radio vibrator circuits usually have a voltage rating somewhere between 1,200 and 2,000 volts. Use a replacement capacitor with the same voltage rating as the original when above 600 volts.

Outside Foil. The designation OUTSIDE FOIL, OUTER FOIL, GROUND, or similar wording, usually with a heavy black ring or line, will usually be seen at one end of a tubular paper capacitor. The lead at that end is connected to the outer metal foil inside the capacitor. This outside-foil lead is connected to the circuit terminal which is at or near ground potential. The outer foil then serves as a shield for the capacitor. Use the same outside-foil connections as for the old capacitor and you will be correct.

On molded paper capacitors, a plastic or solder bead may be used around the lead at one end to indicate the outside-foil connection.

Color Code for Paper Capacitors. The capacitance value, voltage, and tolerance of tubular paper capacitors are sometimes indicated by colored bands, as in Fig. 9. The colors have the same values as for the standard EIA resistor color code; hence there is nothing new to memorize. Only the method of reading the colors is different.

First of all, the capacitance rating will always be in *micromicrofarads*

412 Television and Radio Repairing

when expressed by color-code markings. Remember that you multiply mfd values by 1,000,000 to change them to mmfd, so that 0.002 mfd becomes 2,000 mmfd and 0.27 mfd becomes 270,000 mmfd.

Color-code markings will always be crowded toward one end of a paper capacitor. This is the end you start reading from.

As Fig. 9 shows, the first color ring stands for the first figure in the capacitance value. The second color ring stands for the second figure in the capacitance value. The third color ring gives the value by which you multiply the first two figures. So far, this is exactly the same as for resistors.

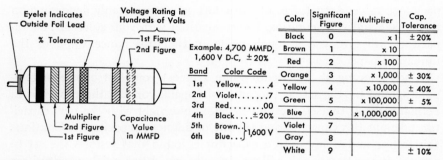

Fig. 9. Color code used by one manufacturer for tubular paper capacitors having molded plastic housings. One voltage band means the voltage rating is under 1,000 volts

The fourth color ring gives the capacitance tolerance in per cent, and here is where the system differs from resistors. For capacitors, black is 20 per cent, white is 10 per cent, and green is 5 per cent. There are also two new tolerance colors: orange for 30 per cent and yellow for 40 per cent.

Usually you will not have to bother to figure out the meaning of the tolerance color, because values of paper capacitors are rarely critical in radio or television circuits. Servicemen do not bother to specify tolerance when ordering replacement paper capacitors, because the units given them by jobbers have close enough tolerance for all practical replacement purposes.

After the fourth ring there will be either one or two color rings to specify the significant figures of the voltage rating. Two zeros are always added to the significant figures given by the rings. Thus, red is 200 volts, yellow is 400 volts, blue is 600 volts. Two rings are used only for ratings above 900 volts. Thus, brown and black means 1,000 volts, brown and yellow means 1,400 volts, and brown and blue means 1,600 volts.

Removing the Defective Paper Capacitor. Once a replacement paper capacitor of the same capacitance and an equal or greater voltage rating than the original has been obtained, make a simple sketch of the terminals to which the defective capacitor is connected. This sketch should identify the outside-foil terminal. You can then remove the old capacitor safely.

Installing the New Paper Capacitor. To install the new capacitor, first place it in the set so the lead marked OUTSIDE FOIL goes to the terminal to which the outside foil of the defective capacitor was connected. Now make permanent hook or wrap-around joints to the correct terminals, squeeze the joints with pliers, and solder them to complete the job.

Leave the capacitor leads long enough so they do not pull on the capacitor after installation. In auto radios or in other radios subject to a lot of vibration, this means placing the capacitor against the chassis or some other fairly solid part to prevent it from vibrating excessively.

When soldering, apply heat on each terminal only as long as necessary. Excess heat may melt the wax inside the capacitor or even loosen its soldered connection to the rolled foil inside.

If these precautions are not followed, the capacitor may open or become intermittent as a result of one of the leads inside the capacitor pulling away from the metal foil. Avoid introducing a new trouble in this manner, because you rarely suspect a new part.

Resonant Capacitors. A tubular paper capacitor having a coil wound around it is called a resonant capacitor. The coil is in series with the capacitor. The combination acts as a filter that passes signals at a particular frequency, usually the intermediate-frequency value, while offering high opposition to undesired signals above or below the desired i-f signal frequency.

When a resonant capacitor goes bad in a radio set, squealing will usually be heard from the loudspeaker as the receiver is tuned in to a radio station. The pitch of the squeal may change as the receiver is tuned. Install a new ordinary capacitor of the same size and transfer the coil to it. Be sure to make the connection exactly the same as for the old unit.

Electrolytic Capacitors. Just as important as paper capacitors in servicing are electrolytic capacitors. These units are generally called *electrolytics*. They come in a wide variety of sizes and shapes, as shown in Fig. 10. The can types are made especially for mounting above the chassis, while the tubular and rectangular cardboard types are made for mounting under the chassis. Some are single units having only two leads or terminals. Others have two or more capacitors in the same housing and hence have three or more leads or terminals.

The commonest type of electrolytic capacitor is the dry electrolytic. This has two foil plates rolling loosely together, with the plates separated from

414 *Television and Radio Repairing*

each other by paper soaked with a moist chemical called the electrolyte. Older types used a liquid electrolyte instead of a paste between the plates, and for this reason were called wet electrolytics.

Where Electrolytics Are Found. Electrolytics are used in every television and radio receiver, because this type of capacitor provides greater capacitance in a given space than does any other type of capacitor.

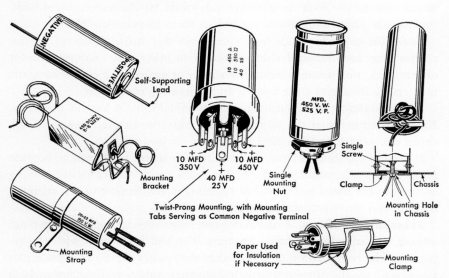

Fig. 10. Typical housings and mountings for electrolytic capacitors. On twist-prong types, outer prongs are twisted after they are inserted in the slots in the mounting wafers

Most electrolytics are used in power-supply circuits, where large values of capacitance are needed. Here the electrolytics help filter or smooth out the d-c voltage supplied by the power supply to the set, hence are often called filter capacitors. A power-supply filter has two or more electrolytics, usually combined in a single housing.

Another place where you will sometimes find an electrolytic is in the audio output stage. Here the electrolytic is connected across the cathode resistor which is between the cathode terminal of the output tube and ground (the chassis). The electrolytic provides a low-opposition path for audio signals around the cathode resistor.

Low-voltage electrolytics are used extensively in transistor portables, because they provide high capacitance values in minimum space.

You will be able to distinguish an electrolytic from other capacitors by its large physical size, by its high capacitance value (above 4 mfd), and by terminal polarity markings on the actual units and on circuit diagrams.

Polarity of Electrolytics. Unlike other capacitors, electrolytics have a definite polarity. The positive lead of the capacitor must always be connected to the positive side of the voltage source.

With single-unit tubular electrolytics, which look somewhat like large paper capacitors, the lead marked + should go to the positive terminal in the circuit. The other lead, either unmarked or marked —, is the negative lead of the capacitor and should go to the negative terminal in the circuit.

Checking Low-voltage Electrolytics. The miniature high-capacitance electrolytics used in transistor circuits have such low voltage ratings that they can be damaged if the battery inside the ohmmeter is applied to them with the wrong polarity.

The polarity markings on the panel of a multimeter do not always represent the polarity of the internal battery, because they are provided primarily for d-c voltage measurements. You will have to check this for yourself with your own multimeter.

Get a good ordinary 150-volt or higher-voltage d-c electrolytic capacitor and measure its resistance with your ohmmeter. Now reverse the ohmmeter leads at the capacitor and repeat the test. The higher of the two readings will be the one obtained with correct ohmmeter polarity. If this is obtained with the black common lead on the negative terminal of the capacitor, you know that the markings on your multimeter are correct for low-voltage electrolytics.

If the higher resistance reading is obtained with the common black leads on the positive terminal of the capacitor, write this note on a small piece of paper and fasten it to the back of your multimeter with clear scotch tape: BLACK OHMMETER LEAD GOES TO + TERMINAL OF LOW-VOLTAGE ELECTROLYTIC.

Colored-lead Codes for Cardboard Electrolytics. Tubular electrolytics often have different colors of insulation on the leads to indicate polarity. For a single unit (having two leads), the red lead is positive, and the black lead is negative.

With tubular electrolytics having two or more sections in one housing, each lead will usually be a different color, and the color identification code will be printed on the housing. Usually there will be only one negative lead, serving in common for all sections. Occasionally you will find units with separate negative leads for each section. These are for receivers where the negative terminals of the electrolytics do not all go to the same point.

No standard color code has been adopted for electrolytic capacitor leads; hence each manufacturer uses his own combination of colors. As a rule, *black* will be the *common negative lead*.

Terminal Identification on Can-type Electrolytics. With can-type electrolytics, the metal can itself is usually the negative terminal, and should go to the negative terminal in the circuit (most often the chassis). Convenient soldering tabs project from the bottom of the can for making the negative connection.

Sometimes there may be a separate lead or terminal coming out the bottom of the capacitor for the negative terminal. The metal can then serves only as a housing. Polarities of terminals or leads are usually marked on the housing.

With can-type electrolytics having two or more capacitor units, the terminal lugs are identified by stamped or punched-out squares, triangles, half-circles, and other geometric shapes. The meanings of these code markings are given on the side of the capacitor. There is no standard code for these, so always read the data on the side of the capacitor. An example is shown in Fig. 10.

If by mistake, or otherwise, the polarity of an electrolytic is reversed when connected into a receiver, gas forms inside the capacitor when voltage is applied. The pressure of this gas can get high enough to explode the housing of the capacitor and cause considerable damage to the receiver. For this reason, be sure to get the correct polarity of the leads when connecting a new electrolytic.

How Electrolytics Go Bad. Electrolytics do not have an indefinitely long life. Some may need replacement before the set is a year old, while others may last 5 or 10 years. Hardly ever does an electrolytic last as long as the set itself. The life depends on the quality of the original capacitor used and on the design of the set.

An electrolytic capacitor opens when one of the leads inside corrodes because of the presence of the chemical. Such a defect is rare, however, because most capacitors are designed to prevent it from happening.

An electrolytic also effectively becomes open when it loses its original capacitance or develops a high series resistance. Both these troubles are caused by drying out of the chemical paste. Heat hastens the process of drying out. Blocking of ventilating holes by operating a radio against a wall can increase the heat and cause failure of the electrolytics in the radio.

Electrolytics are said to be shorted when a low-resistance path develops between the plates. A shorted capacitor can be found by measuring its resistance with an ohmmeter. The resistance of a shorted electrolytic is very low, usually close to zero.

When the resistance is lower than normal (way below 50,000 ohms) but still nowhere near zero, the capacitor is said to be leaky. A leaky capacitor will usually have lost part of its capacitance also. For this reason, leakage and loss of capacitance will be considered together as the last major trouble encountered in electrolytics.

Leakage Current in Electrolytics. Unlike other types of capacitors, electrolytics always have quite an appreciable leakage current. If this leakage current were measured for a good electrolytic, the current indicated might be as high as 8 milliamperes under normal operating conditions. Servicemen rarely measure this, however, because of the difficulty of making current measurements.

Instead of measuring leakage current, the leakage resistance of an electrolytic can be measured with an ohmmeter. The resistance indicated will generally be above 300,000 ohms for a good electrolytic used as a power-supply filter capacitor, and can be as high as 10 megohms. Low-voltage electrolytics (50 or 25 volts) used as bypass capacitors in cathode circuits can have a resistance as low as 100,000 ohms.

An electrolytic has two different resistance values, depending on the polarity of the ohmmeter connections. If the ohmmeter is connected wrong, no harm is done, but the measured resistance is much lower. Instead of bothering to figure out polarities, most servicemen measure the resistance both ways and use the *higher* of the two readings. The lower reading has no meaning. Wait a few seconds for the ohmmeter pointer to stop moving before reading the resistance value of an electrolytic.

Do not rely on ohmmeter tests for anything more than finding open or shorted electrolytics. Leakage-resistance readings mean so little toward deciding whether an electrolytic is good that many servicemen do not even bother to make them. Performance of the electrolytic in the set is what really counts.

With age, the leakage of an electrolytic usually increases. This means that its leakage resistance decreases. Eventually the leakage resistance may become so low and the leakage so great that the capacitor no longer works right in the set. An annoying hum is then usually heard along with the program. Such a hum is an indication that the capacitor is leaky and needs **replacing.**

How to Figure Leakage Resistance. The permissible leakage current for low-voltage electrolytics (rated at 50 volts or less) is about 0.25 milliampere per microfarad at the rated working voltage.

The minimum permissible leakage resistance of a low-voltage electrolytic can be found by dividing the working-voltage rating by the permissible leakage current in amperes.

As an example, a 20-mfd cathode bypass capacitor rated at 25 volts would have a permissible leakage current of 20×0.25 milliampere, or 5 milliamperes. This is the same as 0.005 ampere. Dividing 25 volts by 0.005 gives 5,000 ohms as the permissible leakage resistance. Most units of this size will have much higher resistance than this when new.

Remember to measure the leakage resistance both ways and use the higher of the two readings. This will be obtained when the common negative terminal of the ohmmeter goes to the negative terminal of the electrolytic.

Danger of Relying on Ohmmeter Tests. Although a low ohmmeter reading (below 50,000 ohms for the higher of the two readings) is proof that an electrolytic is bad, a high reading does not clear the capacitor of suspicion. An electrolytic may show extremely high leakage resistance when subjected to the low test voltage of an ohmmeter but still leak badly when the normal operating voltage of the receiver is applied. Therefore, whenever a leaky electrolytic is suspected, replace it temporarily with a good electrolytic and note whether the trouble is cleared up.

Shelf Life of Electrolytics. When electrolytics are not used for a long period of time, they tend to deteriorate. This occurs even when they sit on the shelf, so beware of bargain sales on electrolytics. Old electrolytics can have high leakage current and often also very much less capacitance than their rated value.

If an electrolytic has stood on the shelf for several years, it probably should not be used at all. If you have to use an electrolytic that is a few years old, allow the set to operate for 15 to 30 minutes after completing the repair job. If the set still works O.K. and hum is not objectionably loud, you can assume that the capacitor has re-formed itself and is as good as new.

Watch the rectifier tube for the first few minutes when trying a questionable new electrolytic. Turn off the set instantly if the tube gets red-hot inside. A long-idle electrolytic can have such high leakage current that it acts almost as a dead short on the power supply, overloading the rectifier tube. Such an electrolytic should be thrown out.

Reactivating Electrolytics. Sometimes an electrolytic can be reactivated if leaky because it has been stored for a long time. This is done by applying d-c voltage gradually. Start with about one-third of the rated working voltage, and increase to the full working voltage over a period of about 2 minutes. This calls for a variable-voltage d-c power supply, which is rather expensive and should therefore be deferred until you are well established. Stay away from bargain electrolytics or another serviceman's old stock at the start. If the reactivated electrolytic still tests leaky, throw it out and get a new one.

When to Suspect an Electrolytic. If a receiver has a loud hum that can be heard when a regular program is on, and the hum is not affected by the volume control, suspect an electrolytic in the power supply as being bad. This hum will be just about the same loudness when the set is tuned to a station as when tuned between stations.

If a radio receiver squeals or whistles as you tune in the stations, suspect an electrolytic. This squeal will change its pitch as you tune in and out of the station, coming to zero frequency (no squeal) when you are right on the station. Oftentimes hum will also be noticeable in such a set, giving a double clue to a bad electrolytic in the power supply.

If a radio receiver motorboats (makes a put-put sound like an outboard motor), suspect one of the electrolytics in the set. Sometimes the motorboating will speed up and slow down as you tune the radio in and out of a station. The same trouble can occur in transistor radios and in the audio system of a television receiver.

If the set has been taking longer and longer to warm up, or if it hums only for a few minutes while the set is warming up, suspect an electrolytic filter capacitor.

If a can-type electrolytic mounted above the chassis has a deposit of white on the top, or one of the electrolytics under the chassis looks discolored and bulged out, suspect that electrolytic as being bad. The white foam alone on top of the can does not mean the capacitor is bad, though; suspect the capacitor only if the set also hums, squeals, motorboats, or has something else wrong. Foaming is normal for some older types of electrolytics in metal cans, because they have gas escape vents around the top.

If any electrolytic feels warm, suspect it. If the plates of the rectifier tube become red or the tube shows other signs of overheating while the set is in use, suspect one of the electrolytics in the power supply as being shorted. More often than not, however, some other part is at fault when you observe this condition.

Checking Electrolytics by Substitution. There is only one sure way of telling whether a given electrolytic in a receiver is good or bad. Substitute a known good electrolytic of the same capacitance and voltage rating in place of the one you suspect. If this capacitor clears up the trouble, you know that the old electrolytic is defective.

Since you will want to use this substitution method for checking electrolytics, it is a good idea to make up a few electrolytics with small alligator clips on their leads, as shown in Fig. 11, to speed up testing by substitution. Some types of alligator clips will have to be soldered to the capacitor leads,

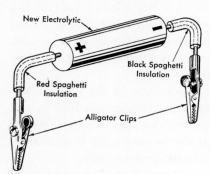

Fig. 11. Method of placing alligator clips on the leads of an electrolytic so it can more easily be connected into a circuit temporarily to check a suspected unit

while others will have screw terminals for the leads. Either type will do. Slip spaghetti insulation over each capacitor lead before putting on the clips. Do not shorten the capacitor leads.

The sizes of electrolytics to make up in this manner should include 80-mfd 150-volt, 50-mfd 150-volt, 20-mfd 450-volt, and 10-mfd 450-volt units. These will take care of most of the substitutions you will have to make when checking for a bad electrolytic.

If you are testing by substitution to determine whether an electrolytic in the set is good or bad, the test capacitance value need be only approximately the same as that in the set. If none of the test capacitors agrees in value with that of the original capacitor, use the next higher capacitance. Thus, to check a 16-mfd electrolytic, use the 20-mfd test electrolytic.

Use the test units for replacements about once a year and buy new ones for test purposes, so as to have only fresh electrolytics on hand.

Checking Single-section Electrolytics. To check an electrolytic capacitor when it is the only one in the capacitor housing, first try clipping a test capacitor of about the same size across the suspected unit. Be sure the clips grip the terminals when the set is turned on. If the capacitor in the set is open, has lost capacitance, or has developed high series resistance, the good capacitor will improve the performance of the set. This proves that the capacitor in the set is bad. Always *pull out the line-cord plug* before changing an electrolytic, as electrolytics have *dangerously high voltages* when the set is on.

Some servicemen connect their multimeter to measure the B+ voltage (across the output filter capacitor) while shunting each filter capacitor in turn with a good unit. If the voltage goes up more than about 3 volts when the new unit is shunted across, or if hum is noticeably reduced, the old filter capacitor in the set should be replaced.

Shunting a good capacitor in this way does no good if the suspected capacitor is shorted or leaky. The capacitor in the set can therefore still be shorted or leaky if the shunt test gives no change in performance. You then have to disconnect one lead of the suspected capacitor so you can connect the test capacitor in its place. Repeated burning out of rectifier tubes and blowing of receiver fuses are signs of a shorted electrolytic.

If the capacitor in the set has leads, unsolder one of them to check for shorts or leakage. If it has terminals, unsolder all the leads from one of the terminals. Keep these leads connected together but away from the terminal.

Many of the older can-type electrolytics have only a single terminal. The other connection is made to the chassis automatically by the capacitor mounting. Here the metal can is the negative terminal and is directly connected to the mounting arrangement.

After disconnecting the suspected electrolytic, clip the test capacitor in its place. The positive lead of the test capacitor should be clipped to the terminal or bunch of wires to which the positive of the original capacitor was connected. The negative lead should be clipped to the negative side of the old capacitor, which will often go to the chassis.

If the test capacitor restores normal operation, you have found the trouble. Remove the test capacitor and proceed to replace the original bad capacitor.

Checking Multiple-section Electrolytics. To check a multiple-unit electrolytic, made up of two, three, or four electrolytic capacitors in a single housing, check each section separately, starting with the highest-capacitance section.

To find the highest-capacitance section, note which lead or terminal is marked or coded as having the highest capacitance. Ordinarily only the positive lead or terminal for each section of the capacitor is marked, because the negative leads or terminals of all the capacitor sections are connected internally to the single common negative terminal. This common lead will be marked COMMON NEGATIVE on tubular and rectangular cardboard-housing electrolytics and will be the metal can or a lead coded as COMMON NEGATIVE for can-type capacitors.

First, try shunting a test capacitor across the highest-capacitance section, just as when testing a single-unit electrolytic. If this gives no change in performance, disconnect the suspected section and clip a test capacitor of the same capacitance in its place with the correct polarity, just as for testing a single capacitor.

If substitution of the test capacitor clears up the trouble, that section of the electrolytic in the set is bad. Usually the highest-capacitance section is first to go bad; if it tests good, however, repeat the substitution test for each other section in turn.

If only one section of a multiple-section electrolytic capacitor is shorted, you can connect a single-section replacement in place of the defective section after cutting off the positive lead of the bad section.

If only one section is open, leaky, has lost capacitance, or has too high an internal resistance, it is best to replace the entire unit. These troubles are caused by natural aging and drying out, so the other sections will probably go bad soon also. This would mean repair work that you would usually have to do free, because the customer expects his set to work for a reasonably long time after you have repaired it.

Checking Electrolytics of Unknown Capacitance. If the capacitance of a suspected electrolytic capacitor is not known, a rough check of its condition can still be made. To do this, disconnect the suspected capacitor, and use your judgment as to which of the test capacitors to clip in first. For an a-c receiver use 450-volt test capacitors, and for an ordinary universal a-c/d-c set, use 150-volt test capacitors. Some a-c/d-c sets use a voltage-doubling rectifier circuit that gives more than 150 volts, so measure the d-c voltage across the old electrolytic first if there is a chance that it is higher. Use the 450-volt test capacitors for anything above 150 volts.

If the capacitor is of the single-unit type, start off by using a 10-mfd test capacitor and progress to larger capacitors. If one of these clears up the trouble, chances are that the original electrolytic is bad.

If the capacitor is one section of a multiple-unit electrolytic in a universal a-c/d-c set, start with a 30-mfd test capacitor.

Ordering Replacement Electrolytics. Practically all electrolytics have capacitance values and voltage ratings marked directly on the units, as a guide for ordering a replacement. If the defective electrolytic is one of the rare units that are not marked, look up the capacitor in the servicing data for the receiver just as you would for any other unknown part.

Capacitance values for electrolytics are not critical, but it is good practice to get values that are within a few mfd of those originally in the set. Using too small an electrolytic may cause the set to sound distorted, to hum, to squeal, or to motorboat. Too big an electrolytic, on the other hand, costs you more money unnecessarily. A capacitor many times larger than the one originally in the set may also have some undesirable effect on receiver performance.

When the voltage rating of an electrolytic capacitor is unknown, use a 450-volt replacement electrolytic if it is for a modern a-c set. (Only a few old sets used 500-volt electrolytics.) Use a 150-volt unit if it is for an ordinary universal a-c/d-c set. Use a 450-volt unit for universal sets having a voltage-doubler rectifier circuit.

Another way of finding the required voltage rating is to measure the highest voltage that will be across the replacement capacitor. Connect the positive lead of the voltmeter to the positive lead of the test electrolytic that you have clipped into the set. Connect the negative lead of the voltmeter to the negative lead of the test electrolytic. Use a high d-c voltage range (around 500 volts) initially. Turn the set on and note the highest voltage reading on the voltmeter as the set warms up.

The replacement capacitor must have a working voltage at least 25 volts greater than the voltage measured. Thus, if the voltage measured is 125 volts, a 150-volt replacement is satisfactory. If the voltage measured is 270 volts, a 350-volt replacement should be used.

A replacement electrolytic can have the same working voltage rating as the original unit or a slightly higher voltage rating, but never more than about twice the voltage rating of the original. This is the opposite of what is recommended for paper capacitors, where a higher voltage rating means longer life. With electrolytics, rated capacitance is obtained only at rated working voltage.

Surge Voltage Rating. The maximum voltage that an electrolytic capacitor can withstand for short periods of time without breaking down (without shorting internally) is called the *surge* or *peak voltage rating*. This rating is important because, just after a receiver is turned on, the voltage applied to the electrolytics is greater than the voltage after the set has warmed up. You can neglect the surge voltage rating when ordering a replacement electrolytic, however, because this rating is of interest chiefly to the designer of the set. Get a new unit having the same or higher d-c working voltage and you will be all right.

Type of Replacement to Get. In general, get a tubular electrolytic to replace a tubular or rectangular paper-housing electrolytic. Get a can-type electrolytic to replace a can-type electrolytic. Do not worry if the new unit is much smaller in physical size than the old; modern capacitors are all smaller than those made 5 or 10 years ago.

If an exact replacement is not available for a multiple-unit tubular electrolytic, get one with the closest higher set of capacitances, provided

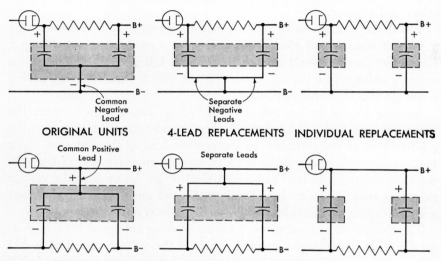

Fig. 12. Methods of replacing multiple-section electrolytics having common leads

that no one capacitance is more than one and a half times the value it should be. Otherwise, use separate tubulars, or use one dual electrolytic and one or more single electrolytics to replace the sections of the defective capacitor.

In obtaining replacements for can-type electrolytics, always try to get a replacement using the same type of mounting as the defective capacitor. If this is impossible, be sure to order the hardware you will need to mount the new capacitor.

One other thing to note when ordering can-type electrolytics is whether or not the metal can of the capacitor serves as the negative terminal. If a separate negative lead or terminal is used, you will save yourself work by getting a replacement with a similar separate negative lead or terminal.

Examples of replacements for multiple-section electrolytics are shown in Fig. 12. The original capacitor blocks in the left-hand column contain the same capacitors, and each has three leads, yet the blocks could not be interchanged. The upper block has a common negative lead for both capacitors, while the lower block has a common positive lead for both capacitors. A two-capacitor block with separate positive and negative leads can be used to replace the capacitors, as shown, or separate capacitors can be used if an exact-duplicate replacement is not conveniently available.

Removing the Defective Electrolytic. Once you have the correct replacement electrolytic, make a rough sketch that identifies the leads on the old

Fig. 13. Using pliers to break off the mounting strap of an old tubular electrolytic. The rivet remaining in the chassis can then be cut off with diagonal cutters

unit. Mark the polarity of each lead on your sketch. If the capacitor is of the can type, mounted above the chassis, make a note as to whether the can is insulated from the chassis with a fiber washer or not. Only after doing all this should you unsolder all the leads of the capacitor from the set.

With all the leads disconnected, the defective electrolytic can be removed from the set. With tubular or square types, this involves merely lifting out the capacitor or first removing a single screw or rivet on a clamp or bracket and then lifting out the capacitor.

When the mounting strap of an electrolytic is riveted to the chassis, it can often be twisted off with pliers as in Fig. 13. The rivet can then be cut or sawed off so it will not rattle. The strap of the new unit can be bolted to the chassis by using the same hole, or soldered to the chassis.

Mounting the Replacement Electrolytic. Ordinarily you will be able to get the required replacement for any electrolytic to fit into the space made available by removing the defective capacitor.

To mount a tubular electrolytic, merely connect the wires themselves to the proper terminals in the set. The larger units often have a metal mounting strap that can be bolted or soldered to the chassis. Smaller units are light enough to be supported by their own leads, hence require no mounting hardware.

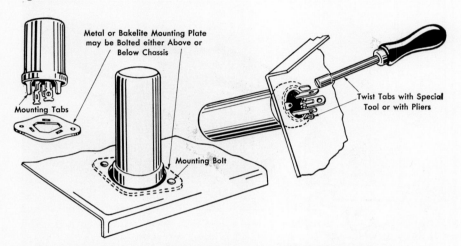

Fig. 14. Common method of mounting replacement can-type electrolytic capacitor

A common mounting arrangement found on replacement can-type electrolytics is that shown in Fig. 14. This fits into the hole left by removal of the old can-type electrolytic. The metal or plastic mounting plate, shaped much like a wafer-type tube socket, is bolted to the chassis first. This may fit the holes left by the old mounting plate; if not, set the plate in position on top of the chassis, mark the chassis with a pencil or a sharp tool where holes are needed, and drill holes of the correct size for the mounting bolts you will use. Always place lock washers under the nuts when mounting parts.

After bolting the plate in position either above or under the chassis, insert the electrolytic from the top so that mounting tabs project downward through the slots in the plate. With the chassis on its side, hold the electrolytic in position with one hand on top of the chassis, and twist each mounting prong underneath with pliers. One of these prongs is used also as a terminal when the can is the common negative, so one or more of the prongs may have soldering holes.

An under-chassis spring clamp mounting is used for can-type electrolytics in some television receivers. The old unit is pulled out of the clamp and the new unit pushed in position, so that the whole mounting job takes only a few seconds.

Connecting the Replacement Electrolytic. Using the sketch you drew of the connections to the original capacitor, connect the leads and ter-

Fig. 15. Connecting leads of replacement tubular electrolytic. Sketch at upper left was made before removing the old unit. New unit is bolted to chassis after being connected

minals of the new capacitor as illustrated in Fig. 15. In doing this, pay particular attention to polarity. With multiple-section electrolytics, be especially careful to get each section of the new capacitor connected correctly according to your diagram.

With can-type electrolytics, if the metal can of the original unit was insulated from the chassis by a fiber washer or some other insulating means, mount and connect the new unit the same way.

As with all other repairs, double-check the new installation against your diagram before trying out the set. Even experienced servicemen occasionally make a mistake.

Replacing Can-type Electrolytics with Tubular Electrolytics. There may be jobs where you want to replace a can-type electrolytic with a tubular paper-housing electrolytic. One reason is that tubulars are usually on hand, whereas can-type units must be ordered specially from your jobber. Such replacements can be made if a few precautions are observed.

Electrolytic capacitors are seriously affected by heat. Replacement tubular units should therefore be kept well away from voltage-divider resistors, filament voltage-dropping resistors, and other parts which give off considerable heat.

In general, it is not safe to use a tubular electrolytic capacitor under the chassis as a replacement for a can-type unit mounted vertically above the chassis. The can-type units have metal housings that can dissipate substantially more heat. A tubular unit placed in a much warmer location under the chassis may fail prematurely even though it has the same capacitance and voltage rating as the original can-type unit.

If you decide to place the new tubular capacitor above the chassis, install a mounting lug near the defective electrolytic, move the wires from the terminal on the bad electrolytic to this lug, and connect one lead of the new tubular capacitor to this lug. Connect the other lead of the tubular electrolytic to the chassis if the can-type electrolytic was connected to the chassis, or to the lead going to the can if the can was insulated from the chassis.

Testing the Set. With the new electrolytic connected, turn the set on. If everything has been done properly, the set should work like new again.

Sometimes, however, the receiver will not work right. Look at all the connections—are they good? Look for pieces of solder that may have dropped into the set and are shorting some terminal to the chassis. Check connections again to make sure the replacement electrolytic is connected properly. If the leads happen to be reversed, the capacitor will not work. It will become very warm in just a short time, and will have to be replaced again even if it does not explode. If the receiver worked all right with the test electrolytic, something must be wrong with your permanent installation of the new electrolytic.

Recommended Stock of Electrolytics. In addition to a supply of test capacitors with attached clips, as previously described, you will want to keep on hand a small supply of new electrolytics for replacement use. Your requirements will depend on the type and number of receivers you service per week; hence only a minimum starting stock can be given. Experience will tell which sizes and types require a larger stock.

Practically all 150-volt and lower-voltage units are tubular or rectangular cardboard-housing types for mounting under the chassis; hence low-voltage can-type units need not ordinarily be carried in stock. Both can and tubular units will be needed in the higher voltage ratings, however. A few multiple-section units of the types most used should also be kept on hand, because they are much quicker to install than separate capacitors. Table 1 gives the recommended minimum stock.

Table 1. Recommended Starting Stock of Replacement Tubular Electrolytic Capacitors

Voltage, volts	Capacitance, mfd	No. in Stock	Voltage, volts	Capacitance, mfd	No. in Stock
10	25	1	450	8	1
50	25	1	450	16	2
50	150	1	450	20	1
150	30	2	450	30	1
150	40	2	450	40	1
150	40–40	1	450	16–16	1
150	50–50	1	450	30–30	1

Ceramic Capacitors. Small values of capacitance, as required in i-f and r-f circuits, are today obtained with ceramic capacitors. These have a hard tubular or flat ceramic insulating material, with a silver coating fired onto the opposite faces to serve as the plates. Connecting leads are soldered to the silver coatings.

Tubular Ceramics. In a tubular ceramic, the insulating material is a ceramic tube. A silver coating on the outside of the tube serves as one plate of the capacitor, and a silver coating on the inside of the tube serves as the other plate. Wire connecting leads are soldered to the coatings and brought out either axially or from the sides as in Fig. 16, just as for resistors. These units therefore look much like resistors, but are recognized easily because of their special color-code markings. Another recognition clue is the white, light-brown, or dark-purple body color commonly used.

In receivers the frequency of each resonant circuit will change slightly with changes in temperature. This is due to a great many small changes in resistance, capacitance, or inductance in coils, tube bases, tube sockets, resistors, variable capacitors, and wiring. It is impractical to correct each of these characteristics by itself at a reasonable cost. Instead, a single temperature-compensating tubular ceramic capacitor is used in the oscillator circuit to correct automatically for the over-all frequency change of the rest of the receiver with temperature. Usually this compensating capacitor will have a negative temperature coefficient of capacitance. This means that an increase in temperature reduces the capacitance of the capacitor. A negative coefficient of 1,400 parts per million per degree centigrade, for example, means the capacitance will drop 8.4 per cent when the temperature is raised from 25 to 85°C.

Since tubular ceramics are generally used only in critical circuits, re-

430 Television and Radio Repairing

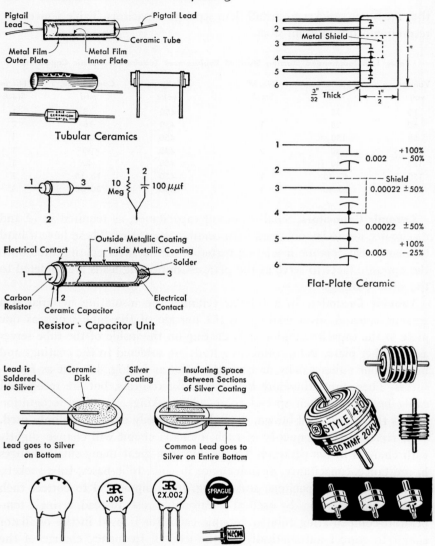

Fig. 16. Examples of different types of ceramic capacitors. Many of these have printed values or type numbers instead of color codes; when ordering a replacement, copy all the printed data or bring in the old unit

placements for them should have exactly the same capacitance, temperature, and tolerance ratings as the original.

A few receivers are using tubular ceramic capacitors having a carbon resistor inside, as in Fig. 16. Sometimes the ceramic capacitor is in two sections, to give two capacitors and a resistor. If one element goes bad, try to get a replacement for the entire unit. If this is not available, replace with standard individual parts having the values specified in the service manual for the set.

Disk Ceramics. These coin-size units have a thin disk of ceramic material, with the silver coating fired onto opposite faces. Wire leads are soldered to the coating and brought out as in Fig. 16. By dividing the conductive coating on one face and running separate leads, two capacitors can easily be obtained from a single disk. Disk ceramics are coated with a plastic material that serves to insulate the capacitor from other parts and protect it from effects of high humidity.

The characteristics of the insulating material and the methods of manufacture are such that production of disk ceramic units with close capacitance tolerance is not practicable. Disk units will therefore have high tolerances, sometimes 100 per cent.

For coupling and bypassing applications, close capacitance tolerance is usually not important. For these applications the disk ceramic capacitor is widely used because it is cheaper than a comparable mica capacitor.

The voltage rating of a disk- or plate-type ceramic capacitor is usually 500 volts.

Multiple Ceramic Capacitors. To save space and simplify wiring, the capacitors associated with one part of a circuit are often all placed on one rectangular ceramic plate. An example of this is shown in Fig. 16. This multiple-unit capacitor was designed for use in the second-detector circuit of a universal a-c/d-c radio receiver. The metal shield that separates the first capacitor unit from the others helps to keep signals in their proper paths, so that i-f signals cannot get into the audio stages and so the audio output signal cannot get back to the first audio stage. Other combinations of capacitance values will be found on ceramic plate units used in television receivers.

Ceramic plates are tested just as if they were individual units mounted side by side. When a defective section is found, replace the entire plate if an exact-duplicate replacement can be obtained. These units do not often go bad, but when they do, other sections of the same unit are likely to fail also in the near future.

If the correct replacement plate is not obtained, get individual disk ceramic capacitors of the correct sizes and use the circuit diagram of the receiver as your guide for connecting these properly in place of the flat-plate unit.

High-voltage Ceramic Capacitors. By stacking silver-coated ceramic disks so the disks are in series, extremely high voltage ratings can be obtained. For high-voltage power-supply filters in television receivers, practically all units have a value of 500 mmfd. The voltage rating may be 15,000 or 20,000 volts. The molded plastic version shown in Fig. 16 is widely used. Instead of being soldered, this unit fits into special clips or screw connectors that have smoothly rounded surfaces. Any sharp metal corners in a high-voltage circuit can be the starting point for a high-voltage discharge; hence even the ends of the terminal rods of these capacitors are rounded.

The high-voltage filter capacitor filters the voltage delivered by the high-voltage rectifier, so as to provide pure direct current for the cathode-ray tube. One capacitor is used in most sets employing a single high-voltage rectifier. Several are generally used in sets having a high-voltage multiplier circuit using two or more rectifier tubes.

Testing Ceramic Capacitors. All types of ceramic capacitors will have extremely high leakage resistance when good, comparable to that of mica capacitors. An ohmmeter test will therefore read infinity for a good ceramic. Any reading below 50 megohms is therefore an indication of trouble, particularly in high-voltage ceramics. If substitution of a new unit having somewhere near the same ratings restores performance of the set, you have confirmation of the ohmmeter indication.

When a ceramic capacitor is suspected of being open or leaky, check it by substituting a known good ceramic capacitor of about the same rating in its place. Since ceramic capacitors hardly ever go bad, though, do not suspect one as being defective unless you have good reason to.

Ceramic-capacitor Color Code. To replace a ceramic capacitor, you need to know the ratings of the defective capacitor. Sometimes the capacitance value will be printed right on the capacitor itself. You can always look up the capacitance value of the capacitor in the parts list or on the circuit diagram for the receiver. Most often, however, you will read the color code on the capacitor to get its value and other ratings.

The standard EIA color code used on ceramic capacitors of all types is shown in Fig. 17. Always start reading from the farthest counterclockwise dot on disk ceramics.

Capacitors 433

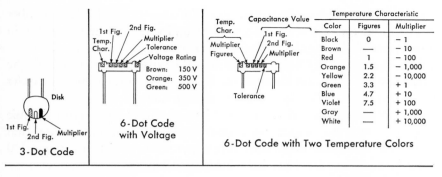

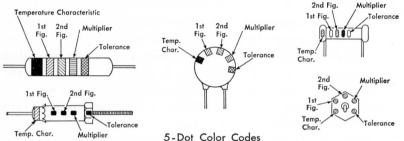

Color	Temperature Characteristic in Parts per Million per Degree C	Printed Temp. Char.	1st Figure	2nd Figure	Multiplier	Capacitance Tolerance	
						10 mmfd or Less	Above 10 mmfd
Black	0	NP0	0	0	1	± 2.0 mmfd	± 20%
Brown	− 30	N030	1	1	10	± 0.1 mmfd	± 1%
Red	− 80	N080	2	2	100	—	± 2%
Orange	− 150	N150	3	3	1,000	—	± 3% or 2.5%
Yellow	− 220	N220	4	4	10,000	—	—
Green	− 330	N330	5	5	—	± 0.5 mmfd	± 5%
Blue	− 470	N470	6	6	—	—	—
Violet	− 750	N750	7	7	—	—	—
Gray	+ 30	P030	8	8	0.01	± 0.25 mmfd	—
White	Any	—	9	9	0.1	± 1 mmfd	± 10%
Gold	+ 100	—	—	—	—	—	—
Silver	Bypass & Coupling Only	—	—	—	—	—	—

Fig. 17. Color codes for ceramic capacitors. If other code markings are encountered, take the old unit to your jobber and ask for an exact replacement

The color of the first band or dot indicates the temperature coefficient of the capacitor. This tells how much the capacitance changes for every degree change in temperature. A negative sign means that the capacitance decreases when the temperature goes up. A positive sign means that the capacitance goes up with the temperature.

The next two dots or bands of color give the first and second figures of the capacitance value. The fourth dot or band of color gives the number of zeros that follow the first two figures to give the capacitance value in micromicrofarads.

The fifth dot or band of color indicates the capacitance tolerance of the capacitor. This tells how much the actual capacitance value can differ from the value specified by the color-code markings.

As an example, suppose that the colors as read in order on a ceramic capacitor are green for the wide-band end, then gray, red, brown, and black. This capacitor would have a temperature coefficient of minus 330 parts in a million for each degree centigrade rise in temperature, a capacitance of 820 mmfd, and a capacitance tolerance of ± 20 per cent.

Replacing Ceramic Capacitors. Disk ceramic capacitors used as plate or screen bypass capacitors are not critical. The size of the replacement may vary somewhat from the size of the original. For example, in a set that uses 1,000-mmfd ceramic capacitors as screen bypass capacitors on the video i-f stages, the replacement can be a 1,500-mmfd unit. You might even be able to use 1,000-mmfd replacements in a set that originally had 1,500-mmfd bypass capacitors.

Since ceramic plates are simply combinations of disk ceramics, capacitance values are likewise not critical when replacing a plate with individual disk units.

Values of tubular ceramics are usually critical, but in television receivers particularly the trend is toward using general-purpose tubulars in place of mica capacitors in noncritical circuits. When the color-code marking indicates that the unit may have any temperature characteristic, the replacement unit can have any other temperature color and can generally be somewhat different in capacitance without affecting set performance.

Mica Capacitors. Before ceramic capacitors were developed, small capacitance values were obtained with mica capacitors. Today you will rarely find a mica capacitor in modern television and radio sets, but there will be many of them in older sets. These units are made of alternating thin sheets of metal foil and mica, as shown in Fig. 18. Usually they are no bigger than a postage stamp and less than a quarter of an inch thick. Capacitance values range from about 5 to about 10,000 mmfd.

Most mica capacitors have brown or black molded plastic housings on which capacitance values are stamped directly or indicated by a color code of from three to six colored dots.

In silvered mica capacitors, the foil layers are omitted and a silver coating is deposited directly on each mica sheet to serve as a plate. Button-type silvered mica capacitors use round mica sheets.

Mica capacitors are tested and replaced in the same way as ceramic capacitors. An old mica unit can generally be replaced with a ceramic unit having the same capacitance value and voltage rating.

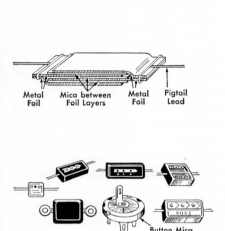

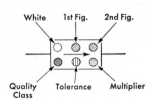

Color	Significant Figure	Multiplier	Tolerance ± %
Black	0	X1	20
Brown	1	X 10	—
Red	2	X 100	2
Orange	3	X 1,000	3
Yellow	4	X10,000	—
Green	5	—	5
Blue	6	—	—
Violet	7	—	—
Gray	8	—	—
White	9	—	—
Gold	—	0.1	—
Silver	—	0.01	10

Fig. 18. Construction, appearance, and EIA color code for mica capacitors. The first dot is always white, and is ignored. Capacitance value is given in mmfd. When the two rows of dots are on different faces, rotate the capacitor about the axis of its leads after reading the first row, then read the second row from right to left. The quality class can be ignored except in television oscillator circuits; here the new unit should have the same quality color

Mica Capacitor Color Code. Some mica units have values stamped or printed directly on their plastic housing, while others have color-coded values. The latest standard EIA color code for mica capacitors is given in Fig. 18.

Whenever a mica capacitor has two rows of dots, the dots are read in clockwise sequence when the capacitor is held so printing on it can be read. Some capacitors also have arrows to indicate the direction in which the dots are read. Reading clockwise means reading the top row from left to right, then going down and reading the bottom row backward, from right to left.

In rare cases you may encounter a mica capacitor having an older color code. When in doubt about code markings, the safe procedure is to refer to service data on the set.

Mica capacitors are often in critical signal circuits where even the movement of a lead out of position can cause trouble. Never move leads around unnecessarily.

Combining Capacitors in Parallel. Sometimes you may be faced with the problem of obtaining a paper, electrolytic, mica, or ceramic capacitor whose electrical size may be quite critical. When you do not have the time to obtain the correct replacement, or the correct replacement is not available, you can use two capacitors in parallel to give the desired electrical capacitance. For example, suppose you need a 0.15-mfd capacitor. Here you could connect a 0.1-mfd capacitor in parallel with a 0.05-mfd capacitor to obtain a total capacitance of 0.15 mfd, since *capacitance values add when capacitors are in parallel.*

Discharging Capacitors. A capacitor stores energy, and can give an unpleasant or even dangerous shock if charged to a high voltage. The best way to discharge a charged capacitor is by holding a 100,000-ohm or 500,000-ohm resistor across its terminals for a few seconds. Repeat this two or three times. Do not use a bare wire, as this may damage the capacitor. Do not use a screwdriver across the terminals, as the resulting arc may burn the screwdriver as well as damage the capacitor.

19

Testing, Repairing, and Replacing Coils and Transformers

Why Coils and Transformers Are Important. Television and radio receivers have quite a number of different coils. Some of these, called *choke coils*, work individually to keep signals in their proper paths. Some coils act with capacitors for tuning a circuit to a desired signal frequency and blocking undesired frequencies.

Two or more coils working together form a transformer that transfers signals from one circuit to another without wire connections. Transformers also serve to change a-c voltages to higher or lower values as required by receiver circuits.

In television receivers, four coils work together in the deflection yoke on the picture tube to make the electron beam move back and forth on the screen.

Most of the coils in a receiver are not visible because they are mounted in metal housings or shields. Once you have worked on a few sets, you will have no trouble recognizing the shields that contain coils. A quick way to get acquainted with coils and transformers is by looking at those in a typical radio set, as shown in Fig. 1.

If a ferrite-rod antenna coil fails (opens), the signal path from the antenna to the converter is broken. The set then goes dead or gets very weak.

If the oscillator transformer fails, there is no oscillator signal for changing the incoming radio signal to the fixed 455-kc value which alone can get through the i-f amplifier. Again the set goes dead.

If either i-f transformer fails, the i-f signal path is blocked, and the set goes dead.

If the output transformer fails, the path for audio signals from the output tube to the loudspeaker is broken, and no sound is heard.

If the filter choke coil goes bad, the receiving tubes get no plate voltages, no signals get through, and the set is dead.

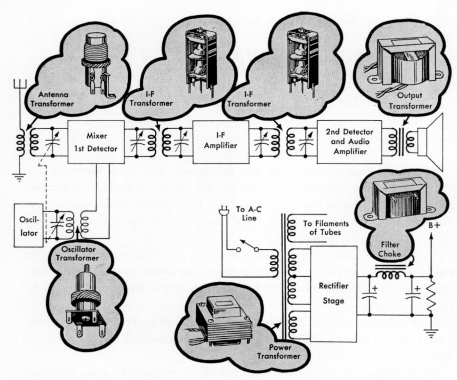

Fig. 1. Transformers in a typical a-c superheterodyne radio receiver, and their symbols. One side of each i-f transformer has been cut away to show the coils inside. Auto radios still use antenna transformers, but other broadcast-band radios have ferrite-rod antennas that replace this transformer. A resistor is often used in place of a filter choke in the power supply

If the power transformer goes bad, tubes get neither filament nor plate voltages, and the set is again silent.

Still other troubles develop if any of these coils or transformers become shorted instead of open.

Coil Repair and Replacement. Even though coils and transformers are important to the operation of a set, they are fortunately not too much of a servicing problem. Coils do not go bad as often as tubes, capacitors, and resistors. When they do go bad, you either resolder a joint, splice a broken wire, or install a new part. If factory parts are no longer available, you can often get universal replacements.

To fix the majority of receivers, you need to have only a general idea of how coils and transformers work. This will be given first, because it will help you to locate bad coils.

How Coils Are Made. A coil is nothing more than a length of copper wire wound to give one or more turns. The wire may be self-supporting as in television tuners, or may be wound around some sort of wood, paper, or plastic form.

A coil may have a single layer of turns or many layers of turns. It can

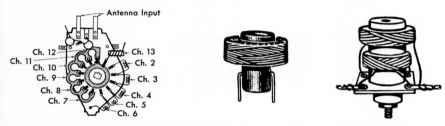

Fig. 2. Examples of coils used in vhf television tuner. At right are crisscross windings on r-f chokes used in television sets; these are also known as universal-wound chokes

even have only a fraction of a turn, as for the high-band channels in the vhf television tuner section shown in Fig. 2. Coils for some r-f circuits have crisscross windings, but most coils are wound in layers with the turns of wire side by side. A coil will have one lead for the start of the winding and one lead for the finish. The insulation on the wire may be enamel, cotton, silk, a plastic such as nylon, or Formvar.

The resistance of a coil as measured with an ohmmeter will always be the resistance of the length of copper wire used in winding the coil. Since copper wire has low resistance, a coil offers little opposition to direct current.

A break anywhere in the wire used for making a coil creates an open circuit, which has very high resistance. An ohmmeter is therefore the instrument you use for testing coils. A low reading means the coil is probably good, while an infinite reading means the coil is open.

Coil Impedance and Inductance. Although coils have very little effect on the flow of direct current, they provide an excellent control over alternating current. This is because coils oppose *any change* in the current flowing through them. The more often the current is changing, the greater is the opposition. This opposition of a coil is called *impedance,* and is expressed in ohms the same as for capacitors and resistors. At higher frequencies, current is changing faster; the coil then offers more opposition or impedance to the change in current.

A coil opposes a change in current by producing in itself an a-c voltage that opposes the applied a-c voltage. This opposing voltage is called a *self-induced voltage*.

The coil rating that determines how much self-induced voltage there will be for a given change in current is called *inductance*. The inductance rating is the *electrical size* of the coil, just as capacitance is the electrical size of a capacitor and resistance is the electrical size of a resistor. The higher the inductance rating, the greater will be the impedance of a coil for alternating current. The impedance of a coil thus increases when the inductance is increased and when the frequency is increased.

The unit of inductance is the *henry*, abbreviated h or H. A smaller unit is the *millihenry*, abbreviated mh or MH. One millihenry is equal to a thousandth of a henry. A still-smaller unit is the *microhenry*, which is a millionth of a henry.

Inductance values of coils in the power supplies of television and radio sets can be as high as 1,000 henrys. Inductance values of coils in tuning circuits are much smaller, ranging from a few microhenrys to several hundred millihenrys.

How Inductance Is Changed. The inductance value of a coil depends on three things: the number of turns of wire in the coil, the shape or dimensions of the coil, and the material used in the center or core of the coil. Winding more turns on a coil makes its inductance go up. Removing turns makes its inductance go down. Spreading out a coil, so that the same turns take more space, also reduces the inductance.

Changing the core material is the easiest way of changing the inductance of a coil. The three common types of core materials—air, powdered iron, and solid iron—will now be taken up one by one.

Air-core Coils. When alternating current is sent through a coil, a magnetic field is produced around the coil. This can be represented on diagrams as loops threading through the center of the coil, as shown in Fig. 3A for a coil that has only air and insulating material inside.

The inside of a coil is called the *core*. A coil with only air inside is called an *air-core coil*. Since insulating material is nonmagnetic, it is considered to be the same as air. Coils wound on fiber, paper, solid wood, or plastic are therefore also called air-core coils. A coil has its lowest possible inductance when the core is air.

Powdered-iron-core Coils. A given alternating current through a given air-core coil produces a definite number of magnetic lines of force. If a powdered-iron core is placed in the coil under the same conditions, many more lines of force are obtained, as shown in Fig. 3B. The tiny iron particles in this core provide much better paths than air for the magnetic field inside the coil, so a given current through the coil makes more lines of force. This means that the inductance of the coil increases when a better core than air is put in.

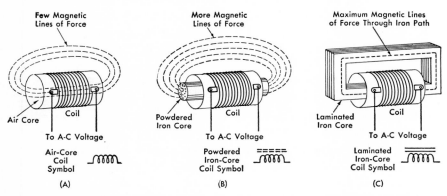

Fig. 3. Three types of coils, all wound on much the same cardboard form but having different cores. Corresponding coil symbols used in circuits are shown below

Powdered-iron cores and their newer ferrite versions are widely used because they need less copper wire to get the same amount of opposition to a-c. In addition, it is easy to adjust the inductance of the coil by moving the core in or out. The inductance is highest with the core all the way in the coil, and the coil is then tuned to the lowest frequency at which it can be used. The inductance is lowest with the core all the way out, and the coil is then tuned to its highest frequency.

Iron-core Coils. When a complete path through sheets of iron is provided for magnetic lines of force, as in Fig. 3C, the maximum possible number of magnetic lines of force is obtained. The coil then has maximum inductance, and is known as an *iron-core coil*. The sheets of iron are called laminations. The opposition that an iron-core coil offers to alternating current is extremely high.

Coils in Series and Parallel. In television receivers and in all-wave radio receivers, two or more coils are sometimes combined to get extra tuning bands. In a television set, the station-selector switch connects coils together automatically as required. In all-wave radio sets the band-changing switch does the job.

When coils are connected together in series, their inductances add just as for resistors. The combined inductance is therefore greater than that of any of the individual coils, so coils in series tune to a lower frequency than any of the coils alone. Many television tuners depend on this.

When coils are connected together in parallel, the combined inductance is less than that of the smallest inductance in the group. Coils in parallel tune to a higher frequency than does the smallest coil by itself.

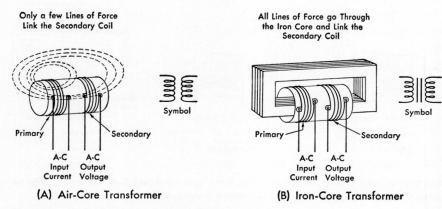

Fig. 4. Two types of transformers and their circuit symbols

How Transformers Work. Whenever the alternating current reverses in a coil, the magnetic lines of force reverse their direction also. These continually changing or reversing lines of force can transfer electricity from one coil through space to a second coil. The first coil is then called the *primary coil,* and the second coil is called the *secondary coil.* This action is indicated in Fig. 4A for a simple air-core transformer. The alternating voltage that is produced in the secondary coil has the same frequency as that of the current flowing through the primary coil.

Increasing the number of lines of force that are changing in the secondary coil increases the induced voltage. The maximum number of lines of force is obtained when a complete iron core passes through both coils as in Fig. 4B. An iron-core transformer thus gives the highest possible secondary voltage for a given pair of coils.

The secondary voltage of a transformer depends also on the ratio of secondary turns to primary turns. If the primary and secondary have the same number of turns, the ratio is 1 to 1, and the secondary voltage is the same as the primary voltage. An example of this is an isolation transformer.

When the secondary has more turns of wire than the primary, the secondary voltage is higher than the primary voltage. The high-voltage secondary winding of a power transformer is an example.

When the secondary has fewer turns than the primary, the secondary voltage is less than the primary voltage. An example of this is a filament winding on a power transformer, which gives 6 volts for tube filaments when the primary voltage is 115 volts a-c.

How Coils Go Bad. The commonest trouble you will encounter in an individual coil or in one of the windings of a transformer is a break in the wire, called an open circuit. Less common is a short-circuit between adjacent turns due to failure of the insulation on the wire. Still rarer is a change in the electrical value of the coil when in a tuned circuit.

The first two troubles apply to all types of coils and transformers, and for this reason will be taken up in detail now, before considering testing and replacement of individual parts. The third type of trouble, involving a change in value, is fixed by adjusting or realigning the receiver.

Open Circuits in Coils. A break in the wire of a coil is generally caused by a chemical action known as electrolysis, which is similar to corrosion and eats right through the copper wire. This action generally occurs near or at one of the terminals to which the ends of the coil are connected.

Ordinary vibration can also break the fine wire of a coil right at the terminal. Slipping of a sharp tool or careless handling of an exposed coil can nick the wire anywhere so that it eventually breaks. Sometimes the coil form expands enough with temperature on a hot day to stretch and break the coil wire at a weak point. A poorly soldered joint at a coil terminal can also be the cause of an open coil.

Testing Coils with an Ohmmeter. When a coil of any type is suspected of being open, the quickest way to check it is by measuring its resistance with the lowest ohmmeter range. Try this first without unsoldering any coil connections, unless the circuit diagram shows low-resistance parts connected across the coil. The correct resistance of each coil is generally given on circuit diagrams in service manuals. Examples of resistance values for radio and television coils are given in Fig. 5.

If an ohmmeter test between the two leads of a coil gives an infinity reading, inspect the terminals carefully. If no damage is visible, wiggle

444 *Television and Radio Repairing*

each soldered connection on the coil while watching the ohmmeter, to see if you can find an easily repaired external break.

If the measured resistance of a coil is much lower than the specified value, unsolder one lead of the coil and measure again. The low reading may be due to some other part in parallel with the coil. If the reading is still way low, some of the turns in the coil may be shorted together.

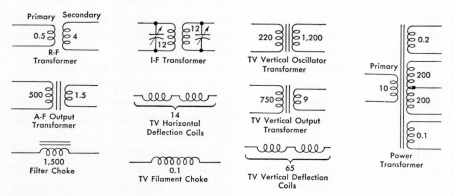

Fig. 5. Examples of coil resistance values in ohms. These values differ greatly in almost every set, so always refer to the service manual when the exact value is needed

Repairing Breaks in Coils. With patience and the right technique you can repair a break in even the finest wire used in coils, provided it is at the end of a winding so you can get at it. Clean the insulation off the broken end of the coil wire with a strip of fine sandpaper, as shown in Fig. 6. Use very gentle pressure because sandpaper can cut right through fine wire.

Assuming the break is near a terminal or external lead, solder a short length of extra wire to this terminal or lead, and splice its other end to the cleaned end of the coil wire. A simple hook joint is adequate. Leave a

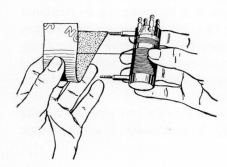

Fig. 6. Using small piece of sandpaper to remove enamel insulation from coil wire

little slack in the repaired connection, and anchor the repaired joint in position with electrical Scotch tape or with a few drops of speaker cement. Soldering the new piece of wire to the terminal first minimizes chances for breaking the delicate wire of the coil while working.

The windings of iron-core transformers are usually covered with insulating paper. Insulated leads will come out from between the outer layers of paper. If the ohmmeter indicates an open, carefully cut the outer layer of paper to expose the coil joints for examination. Once the paper is cut, avoid moving the leads because some of the wires that they connect to may be extremely fine. If the break is found to be here, it can be repaired and the insulating paper fastened back in position with electrical Scotch tape.

Shorted Turns in Coils. When the insulation between two or more adjacent turns in a coil breaks down, the effect is the same as adding a short-circuited secondary winding to the coil. This is a serious trouble that will affect performance or cause overheating. Repairs are not practical, so replacement is required when a coil or transformer has shorted turns.

It is generally not possible to detect shorted turns by measuring the resistance of a coil with an ohmmeter. Most coils have a large number of turns, so shorting out a few of the turns will not appreciably change the resistance value. The normal manufacturing tolerances in coils cause far greater variations in measured resistance values.

Coils and transformers must be judged by their performance in a receiver. If an analysis of symptoms could mean shorted turns in a certain stage, eliminate other possibilities of trouble first. Try a new coil or transformer only as a last resort. If the new unit clears up the trouble, the old one very likely had shorted turns.

Shorted turns in r-f or i-f transformers upset the tuning action so much that adjustments cannot cure the trouble. In iron-core transformers shorted turns make the coil get hot and may eventually burn it out. This heat is accompanied by a pungent smell of burning insulation, which is easily recognized. Short-circuits in capacitors and other parts of a set can also make power transformers and filter chokes get hot, so check this possibility first before replacing a hot transformer or choke.

Servicing Procedures for Each Type of Coil and Transformer. In the remainder of this chapter, each type of coil and transformer will be taken up in turn from a servicing standpoint. Detailed instructions will be given for testing suspected units, for repairing them where possible, and for ordering and installing replacement units.

Replacing R-F Chokes. Single air-core or powdered-iron-core coils are called r-f chokes. They have two leads or terminals, as shown in Fig. 7. The resistance of a good r-f choke coil will usually be less than 10 ohms. Inductance values can be found on circuit diagrams. Replacement is just as easy as for resistors.

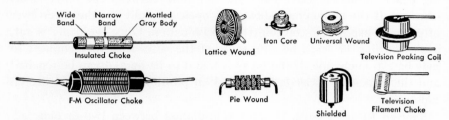

Fig. 7. Examples of r-f choke coils. Each has two terminals.

Molded R-F Chokes. Some small r-f chokes look much like carbon resistors. The circuit diagram will always show at a glance whether the part is a coil or a resistor. The values of these chokes range from 0.47 to 10 microhenrys, and the resistance as measured with an ohmmeter will be under 10 ohms even for the largest size. The inductance value is in microhenrys, often indicated by the standard EIA color code, in which the bands of colors have the same values as for resistors. The wide-color band gives the figure to the left of the decimal point, and the narrow bands give the figures to the right of the decimal point. Thus, a wide black band and a narrow yellow band mean 0.4 microhenry; a wide red band and a narrow violet band mean 2.7 microhenrys.

Video Peaking Coils. Single coils are often used in the second-detector and video-amplifier stages of television receivers to improve the high-frequency response. These are called video peaking coils or simply peaking coils.

An open peaking coil can affect picture quality seriously or cause complete loss of the picture, because the video signal travels through each peaking coil. The coils can be checked for opens with an ohmmeter without disconnecting them, after turning off the set. Open peaking coils should be replaced, unless the break is visible and can be repaired. If a new coil is not available, an emergency repair can be made by shorting out the coil. This will give a small loss in picture quality, but the effect may not even be noticeable.

Shorted turns in peaking coils have practically no effect, so do not worry about them. This is one of the very few examples where shorted turns in a coil do not cause serious trouble.

Occasionally you will find a damping resistor connected across a peaking coil. The value will be somewhere under 25,000 ohms. In rare cases an open damping resistor can cause ringing. This is an effect that produces multiple images on all stations, uniformly spaced and progressively weaker. (Ghost images due to reflections of signals from buildings or hills are not uniformly spaced, are not progressively weaker, and are seldom identical on all stations.) One end of a shunt damping resistor must be disconnected before checking its value with an ohmmeter to detect an open.

A resistor is sometimes used as a convenient support for a peaking coil. Here the resistor will be 1 megohm or more and will have no effect on performance.

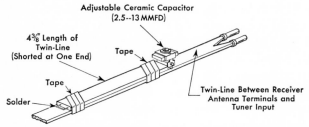

Fig. 8. Wave trap consisting of a half-turn coil (twin-line shorted at one end) and a trimmer capacitor. It can be made up as needed, or purchased from an RCA jobber

Wave Traps. An r-f choke coil and an adjustable capacitor are often used together as a wave trap. This serves to block out a strong local station that is interfering with reception of other stations.

Wave traps come with various tuning ranges. Order one whose range includes the interfering frequency. Full instructions come with the unit for making the necessary tuning adjustment and making correct connections to the antenna and ground terminals of the receiver.

The interference trap shown in Fig. 8 can be taped to the twin-line of a television receiver, as close to the tuner as possible, and adjusted to suppress any frequency from 40 to 170 mc. Its chief use is in blocking out an interfering f-m station in the f-m band between 88 and 108 mc. Adjust the capacitor slowly while watching the interference on the picture, because the trap tunes very sharply.

You can easily make up traps like this yourself by shorting one end of a 4⅜-inch length of twin-line and connecting to the other end an adjustable ceramic capacitor having a range of 2.5 to 13 mmfd.

Width and Linearity Coils. The horizontal section of a television receiver often has coils with adjustable ferrite cores, as shown in Fig. 9. When such a coil is used to permit adjusting the width of the picture on the screen, it is called a width coil. When used to adjust the horizontal linearity of the picture, it is called a linearity coil.

A width coil may have an extra winding used as an agc (automatic gain control) or afc (automatic frequency control) winding. A linearity coil

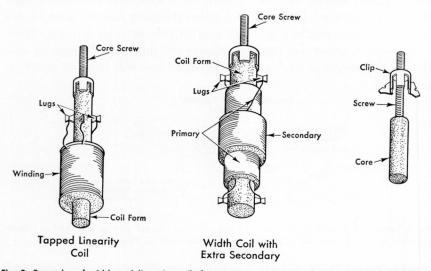

Fig. 9. Examples of width and linearity coils for antenna receivers and construction of adjustable ferrite core used in each

may have a tap on its winding. A glance at the service manual will show how continuity can be checked on any of these coils, regardless of the number of coils or taps they have. When one goes bad, an exact replacement is generally required.

When the core in the width coil is moved all the way out from the center of the coil, there should be a noticeable change in picture width. If adjustment of the core has little or no effect, disconnect one end of it and check the resistance with an ohmmeter. Similarly, check the resistance of the linearity coil if movement of its core does not have a normal effect on horizontal linearity.

Signs of overheating in the width coil, along with a narrow picture that cannot be widened by adjusting the width coil, indicate shorted turns. The width coil should be replaced. The low-value high-voltage capacitor across the width coil will generally need replacement also when the coil goes bad.

R-F Transformers. In older radio sets having loop antennas, you will usually find an antenna coil, an r-f coil, and an oscillator coil. These should really be called r-f transformers since they usually have two coils, but they are called coils in most catalogs and parts lists.

The antenna coil will almost always be above the chassis. It may or may not be shielded. One of its terminals will be connected to the antenna of the set, either directly or through a small fixed capacitor. Another terminal will be connected to one stator section of the gang tuning capacitor.

The oscillator coil is usually under the chassis, unshielded. One of its leads goes to a stator section of the gang tuning capacitor, but none of its leads goes to the antenna. If the sections of the gang tuning capacitor have different sizes of rotor plates, the oscillator lead will go to the section having the smaller rotor. Some oscillator coils have a single coil with a tap. About half of all coil troubles in a radio will be in the oscillator coil.

In radio sets that have a three-gang tuning capacitor and an r-f amplifier stage, there will be an r-f coil also. One of its terminals will go to the plate of the r-f amplifier tube. Another terminal will go to the additional stator section on the gang tuning capacitor.

In modern sets the ferrite-rod antenna serves in place of the antenna coil, and an r-f coil is seldom used. A typical new radio thus has only an oscillator coil.

Quality of R-F Transformers. The performance of an r-f transformer is greatly reduced by a poor soldered connection, by insulation that is weakened by dampness or natural aging, and by corrosion of the wire itself. All these defects lower the quality rating of the coils, commonly called the Q.

When the Q of an r-f transformer drops in a radio, sensitivity and selectivity both drop. Stations on adjacent frequencies then interfere with each other, the volume is less than normal, and distant stations cannot be received.

One rather common cause of low Q is a poor soldered connection on a coil. For this reason, it pays to apply a hot soldering gun and fresh rosin-core solder to each coil terminal. The rosin flux helps to remove corrosion at the joint.

Replacement is the best procedure when corrosion exists or insulation is bad. Fortunately, coils in modern sets seldom develop these troubles. It is best to leave coils alone until you have eliminated tubes and all other parts as possibilities.

Gimmicks. Occasionally you will find on an antenna coil a short length of bare or insulated wire that is connected to a terminal at one end only. This wire is called a *gimmick*. It provides a small value of capacitance between the primary and secondary windings, to make the performance of the radio set more nearly uniform for all stations in the broadcast band.

Ordering New R-F Coils. When a new antenna, oscillator, or r-f coil is required for a high-quality radio receiver, try first to obtain an exact-

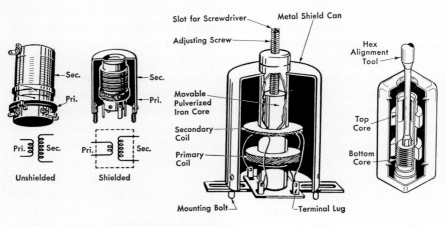

Fig. 10. Practical information on r-f transformers for a-m radio sets, and example of i-f transformer with top adjustments. Hexagonal-head alignment tool is used first to adjust top core, then pushed through it to adjust bottom core

duplicate unit from the local distributor of that set or directly from the factory. For ordering, give the make and model number of the receiver and get the part number from the service manual. An exact duplicate will also be needed for units having taps or extra windings.

For ordinary table and portable radios you can more easily obtain universal antenna, oscillator, and r-f coils from jobbers. Examples are shown in Fig. 10. If the original coil was unshielded, order an unshielded new coil.

Replacing R-F Coils. When you have an exact-duplicate replacement, the terminals, leads, and mounting facilities of the new transformer will be exactly like those of the old, and your job will be simple. Even so, it is easy to get careless and make mistakes in connections. To play safe, always make a rough sketch that shows all connections accurately, before removing the old transformer.

When replacing any parts in r-f and i-f circuits, get the habit of making lengths and positions of leads essentially the same as for the old part. This arrangement or dress of wiring is critical in a few broadcast sets, and becomes extremely important in f-m and television sets. Training yourself now to observe and to record on sketches the positions of wires is excellent preparation for television work.

When a universal replacement coil is used, the terminals and mounting facilities may be different from those of the old unit. Here you may have to drill new holes in the chassis. Use nuts and bolts for mounting, with a lock washer under each nut.

Connecting R-F Coils. To connect a universal replacement coil, locate the terminals of this smaller winding on the old coil, number these terminals A and B on your sketch, then locate the terminals for the smaller winding on the new coil and mark them A and B. The remaining two terminals on the old and new coils can then be marked C and D. It is now a simple matter to make connections to the new coil, even though its arrangement of terminal lugs is entirely different.

For antenna and r-f coils, it does not matter which of the two terminals for the smaller coil gets marked A. For the oscillator coil, you have a 50-50 chance of getting connections right the first time. If the set does not work after the new oscillator coil is installed, reverse the leads to terminals A and B. If more convenient, you can reverse the leads to terminals C and D and get the same result.

In some two-coil transformers, one lead of each coil may go to the same point in the circuit. Transformer manufacturers save the price of one terminal lug by connecting both coil leads to one terminal lug. The r-f transformer then will have three terminals rather than four. On the universal replacement transformer there is a separate terminal for each lead of each winding. Locate the correct separate terminals and connect them together with a short length of bare or insulated wire to duplicate the connections of the original coil.

Tracking of r-f coils means much the same as follow the leader in games. Tracking is good if all stations come in at their correct tuning-dial setting.

In small radios, where tuning-dial settings cannot be read closer than 50 or 100 kc, you can forget about tracking problems. The customer will never notice that stations come in at slightly different points after a new coil is put in. When you take up receiver alignment procedures, you will learn how to make coils track the tuning dial in larger and more expensive radio sets.

Identifying I-F Transformers. One of the easiest parts to identify in a radio set is the i-f transformer, because it is always in a square or cylindrical shield can on top of the chassis and always has two adjustments somewhere. Sometimes these adjustments are trimmer-capacitor screws that are accessible through holes in the top of the unit. In newer sets you will usu-

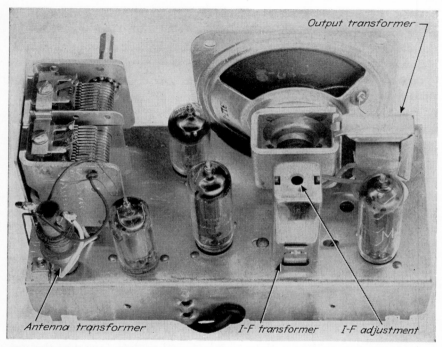

Fig. 11. Location of i-f transformer on older radio set. Antenna transformer is unshielded. Output transformer is mounted on loudspeaker. (Howard W. Sams photo)

ally find threaded-shaft adjustments for powdered-iron or ferrite slugs, accessible through one top hole or through separate top and bottom holes. One version is shown in Fig. 10.

The leads or terminals of i-f transformers always come out under the chassis through a large hole. The shield can will be riveted, bolted, snap-fastened, or otherwise mounted on the chassis. Spring clips can be released by squeezing them with pliers, one at a time, while pulling up on the shield. The parts inside will in turn be fastened to the shield with a screw or other means. Sometimes it is possible to remove the shield without disturbing the coils and their connections, for checking parts inside and repairing broken coil leads.

Coils and Transformers

What I-F Transformers Do. You will recall that in a superheterodyne receiver the oscillator and the mixer–first detector act together to change the frequency of the incoming station signal to a new value called the intermediate frequency, which for modern home radios has been standardized at 455 kc. This 455-kc signal is fed into the i-f amplifier where it

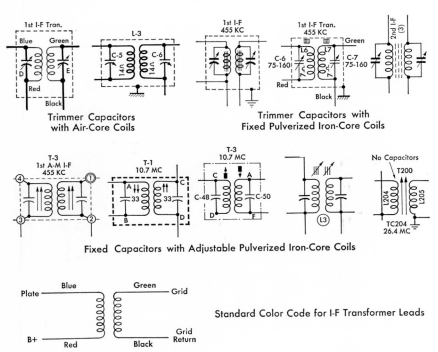

Fig. 12. Symbols used to represent i-f transformers on circuit diagrams. The dashed-line boxes that represent the metal shield can are not always shown, even though practically all i-f transformers are shielded. Color code helps identify leads

gets tremendous amplification. The i-f transformers serve as the tuned circuits for doing this. Their tuning adjustments are all set to 455 kc at the factory, and need to be readjusted only occasionally by servicemen.

One i-f transformer is usually used ahead of and after each i-f amplifier tube. Small table-model sets generally have only one i-f amplifier tube, so look for only two i-f transformers on these. An exception is the set shown in Fig. 11, which has only one i-f transformer. On larger sets having two i-f amplifier tubes, look for three i-f transformers.

The symbols most often used on circuit diagrams to represent i-f transformers are shown in Fig. 12. Solid lines with arrows, drawn between or alongside the coil symbols, indicate that the powdered-iron or ferrite core is adjustable.

I-F Transformer Color Code. Many i-f transformers have terminals or leads that are identified according to the standard EIA color code, as indicated in Fig. 13. This is a help in making connections to a new i-f transformer.

Fig. 13. Tuner used in vhf television receiver, showing coils having only a few turns of heavy wire for the antenna, r-f, and oscillator coils. (Howard W. Sams photo)

Common Troubles in I-F Transformers. Broken leads are perhaps the commonest defect you will encounter in i-f transformers, yet even these are rare. You may not have to replace a transformer for months at a time even when operating a full-time servicing business.

A simple break in a lead can generally be repaired by soldering in a short section of extra wire, just as for any other coil. If the receiver has been operated for a long time in a salt atmosphere near the seashore or in a damp basement location, so that there are corrosion spots on all exposed leads of the windings and mold on the insulation of the windings, replace rather than repair.

When an i-f transformer is close to a tube that gives off a lot of heat, the heat may melt the wax out of the transformer windings and change the tuning. Replacement is the only remedy here.

Mechanical failures in the adjustments of an i-f transformer are likewise a reason for replacement.

Ordering and Replacing I-F Transformers. Exact-duplicate replacements are rarely needed for i-f transformers. The important thing is to get a new unit having the same i-f value as the old. For broadcast-band radios this will be 455 kc. On auto radios it can be either 262 kc or 455 kc, and in f-m radios it will be 10.75 mc.

You will also have to specify whether the transformer is input, interstage, or output. The input i-f transformer is the first one, coming right after the mixer–first detector. The output i-f transformer is the last one, coming just ahead of the second detector. Any i-f transformers in between these two, if present, are called interstage i-f transformers.

Mounting and connecting procedures for i-f transformers are the same as for shielded r-f coils.

Replacement i-f transformers are generally adjusted only approximately to the correct frequency at the factory. The set may work quite well when the new part is connected, but usually realignment is necessary. This job should be turned over to another service shop until such time as you have learned alignment techniques and obtained the necessary signal generator.

Television I-F Transformers. The i-f transformers used in television receivers require exact-duplicate replacements. These can be obtained from the distributors for various makes of sets. For many of the more common makes of sets, your jobber will have equivalent replacement units made by transformer manufacturers. These are quite satisfactory as a rule and can save you a trip to the distributor, so try your own jobber first.

Practically all replacement i-f transformers for television receivers require critical adjustments after installation. This adjustment calls for special test equipment, and should not be attempted until you are familiar with television-receiver alignment procedures.

The crystal-diode second detector and a few associated small parts are often located inside the shield of the final i-f transformer of a television set. Shielding of these parts is thus achieved without the expense of a separate shield. If the top of the shield is not removable, you will have to cut open the top of the shield to get at these parts. This can be done easily with a sharp knife or with side-cutting pliers since the shield is made of soft aluminum. Make four cuts diagonally, from the center to each corner, so you can bend the pieces back to restore adequate shielding after completing your tests or repairs.

Television Tuner Coil Troubles. Television receivers have antenna, r-f, and oscillator coils just as do radio sets. The individual coils are small and have only a few turns. There is usually a separate set of coils for each of the 12 vhf channels, in a separate small tuner housing having the tuning mechanism and the tubes or transistors associated with station tuning (the r-f amplifier, mixer, and oscillator). A typical tuner is shown in Fig. 13.

If one of the coils in the tuner of a television receiver is damaged, it is usually impossible to obtain an individual replacement. Even if the replacement is available, costly special equipment is usually needed to realign the tuner after a repair.

Many distributors will exchange a defective tuner for one in good operating condition. The cost of the exchange is far less than the cost of a new tuner. If you run into tuner trouble, call the distributor for that make of receiver to see if he has this exchange service, before buying an entire new tuner. Tuner exchange services can also be obtained by mail, from tuner repair shops advertising in servicing magazines.

Horizontal Output Transformer. The television transformer that is connected between the horizontal output stage and the deflection yoke is called a horizontal output transformer, horizontal sweep transformer, or flyback transformer. It drives the horizontal deflection coils of the yoke and the high-voltage rectifier tube that provides the second-anode voltage for the picture tube, as indicated in Fig. 14. This transformer is always in the high-voltage compartment, and will have several windings along with a number of taps.

The horizontal output transformer is a costly part and rather difficult to replace. Eliminate other possible causes of trouble in the horizontal deflection section or in the high-voltage section before blaming the transformer.

If you can see burned or charred spots on one of the windings of the transformer, you can be pretty sure it needs replacement. Failure of some other part may have ruined the transformer, however. Check the associated circuits and tubes as carefully as possible before installing the new transformer. After completing the replacement, leave the high-voltage cage open so you can watch the transformer as you turn on the set. If the transformer begins arcing or smoking, turn off the set and continue looking for the part that is overloading the transformer.

A fuse is sometimes used in the circuit of the horizontal output transformer. If this fuse opens but there are no signs of overheating on the transformer, look for trouble elsewhere. Try a new horizontal output tube first. When you replace the fuse, watch the horizontal output transformer closely when you first turn on the set. If it starts to smoke or arc over,

Coils and Transformers 457

turn off the set and again check all other parts in the horizontal output circuit. If they are all good, you can be pretty sure that the horizontal output transformer is defective and needs replacement.

A study of the circuit diagram of a television receiver will show that most of the windings of the horizontal output transformer can be checked with an ohmmeter without disconnecting leads. Only the winding that drives the horizontal deflection coil will have to be disconnected; unplugging the deflection yoke is the easiest way of doing this.

The following symptoms point to possible trouble in the horizontal

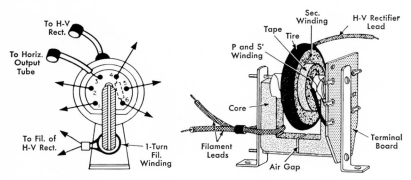

Fig. 14. Typical horizontal output transformer arrangement for television receiver

output transformer: no raster; a very narrow picture or simply a vertical line; blooming, which means poor focusing and low brightness combined with excessive width and height; ringing, in which equally spaced multiple images, gradually decreasing in strength, appear on the screen; vertical white lines anywhere on the raster; white horizontal streaks on the screen; popping sounds from the speaker, due to arcing or corona on the transformer.

Replacing Horizontal Output Transformers. Try to obtain the manufacturer's instructions for replacing the horizontal output transformer, because special procedures are often required. Sometimes the socket of the high-voltage rectifier tube must be loosened and lifted out of its plastic cup to get at its filament terminals, because they connect to a single-turn winding that goes around the core of the transformer. In any event, take off the high-voltage cage if you can, to get room to work. Ask for a copy of the service manual for the set when you order the new transformer from the manufacturer's distributor.

When an exact-duplicate replacement transformer has been obtained

458 *Television and Radio Repairing*

and installed, connect each lead first with a good mechanical joint, then solder it smoothly. Do not leave any strands of wire or points of solder sticking out, because high voltage will cause corona at any sharp point.

It is a good idea to apply an insulating coating to newly soldered terminals in the high-voltage compartment. You can get a bottle of this, with a brush attached to the cap, from your jobber. The coating prevents future corona and arcing troubles due to dust and moisture.

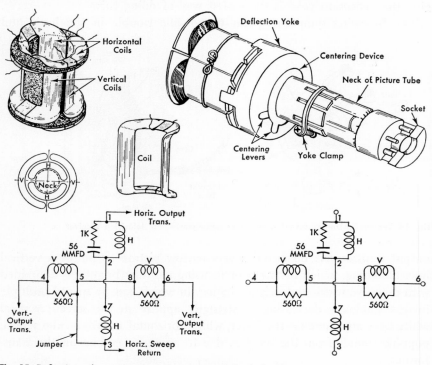

Fig. 15. Deflection-yoke construction, mounting, and circuit details

Deflection Yokes. The beam of electrons in a picture tube is moved back and forth horizontally and vertically by an arrangement of four shaped coils that are together called a deflection yoke. These coils are arranged in pairs on opposite sides of the neck of the picture tube, as close as possible to the funnel, as shown in Fig. 15.

In newer sets, the yoke is held in place on the neck of the picture tube by a clamp that goes around the neck just back of the two magnetic centering devices. In older sets the yoke is supported by a bracket mounted on the chassis.

Coils and Transformers 459

The horizontal coils, mounted at the top and bottom of the neck, are connected to the horizontal output transformer. The vertical coils, mounted outside the horizontal coils on opposite sides of the neck, are connected to the vertical output transformer. Two or more molded ferrite or powdered-iron core segments are clamped together around the vertical coils to complete the deflection-yoke assembly.

Early picture tubes used 70-degree deflection yokes. Here the maximum total angle through which the electron beam could be deflected was 70 degrees. As yoke design improved, this deflection angle was increased, so

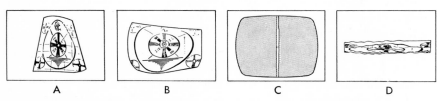

Fig. 16. Examples of picture defects that can be produced by a defective deflection yoke. A—shorted turns in one of the horizontal deflection coils; B—shorted turns in one of the vertical deflection coils, causing foldover at both top and bottom; C—open in horizontal deflection coil or its circuit; D—open in vertical deflection coil. A single horizontal line indicates an open in the damping resistor as well as in the coil, or a break in the connections to the vertical coils

that today picture tubes in new sets generally have 110-degree or larger angles. Each increase in deflection angle permits shortening the picture tube. As a result, many portable television sets are almost as thin as suitcases.

The circuit arrangement of a typical deflection yoke is also given in Fig. 15. Many other connecting arrangements are used. Sometimes the horizontal damping resistor and its coupling capacitor are omitted entirely. The two coils of each pair may be connected either in series or in parallel. The circuit diagram should always be checked to see what leads must be disconnected to check one of the deflection coils.

Some of the symptoms that point to deflection-yoke troubles are shown in Fig. 16. Other deflection-circuit parts can cause some of these distortions, so check them also.

Removing Deflection Yokes. A yoke can become frozen to the neck of a picture tube if the insulating varnish on the horizontal coils softens during operation of the set on a hot day, then hardens and sticks to the glass. A procedure for heating the yoke to soften the varnish again is given in the chapter dealing with picture tubes.

Here is another procedure, used chiefly when the yoke is bad and will be thrown out. Tip the chassis so the picture tube faces upward, then drip varnish solvent on the neck of the picture tube. Apply enough solvent all around above the yoke so it runs down through the yoke and softens the insulation. This method is not recommended for good yokes, because the insulating varnish may soften on the coils and cause shorted turns.

It is often wiser to pass up jobs involving old sets on which the picture tube has never been removed. Just point out politely to the customer that the repair may cost more than the set is worth, if you have to install an entire new yoke along with a new picture tube. If you do tackle the job, try tilting the yoke before you give an estimate, to see if the yoke is frozen.

Replacing Deflection Yokes. When you order a replacement for a deflection yoke, it may come with or without the associated damping resistors. Your job is to connect these parts correctly by using the old yoke and the circuit diagram as your guides.

There are no standard terminal numbering systems or lead color codes for deflection yokes. When installing a new yoke, connect its windings exactly the same as the windings were connected on the original yoke, regardless of terminal markings.

In newer sets, be on the lookout for a yoke that has an internal connection between the center junction of the vertical coils and one of the horizontal coils. This jumper must be installed in the new yoke to get vertical sweep.

If a new yoke gives a reversed picture, with signs reading from right to left, reverse the leads going to the horizontal coils. If the picture is upside down, reverse the leads going to the vertical coils in the yoke.

After installing a new deflection yoke, a number of adjustments are needed. These include the ion trap, tilting, centering, focus, width, horizontal drive, horizontal linearity, height, vertical linearity, and pincushion magnets. All these adjustments are described elsewhere in this book.

Deflection-yoke Thermistors. You will often find temperature-compensating thermistors mounted on 110-degree and larger-angle deflection yokes. These generally range from 20 to 30 ohms when cold and 3 to 4 ohms when the yoke reaches normal operating temperature. Use the circuit diagram as your guide for troubleshooting here.

Width-control Sleeve. On some picture tubes you will see a metal band that surrounds the neck of the tube and slides in and out of the yoke. This is a width control. It acts as a shorted turn for the horizontal coils of the yoke. Pushing the sleeve further into the yoke makes it absorb more energy from the horizontal coils, thereby reducing picture width.

What Output Transformers Do. The iron-core unit that gives the most trouble in receivers is the audio output transformer. This transformer transfers the audio signal from the output stage of the receiver to the voice coil of the speaker.

The primary winding of an output transformer is always the one that has the most turns and the finest wire. The primary winding always has one direct connection to an output tube or transistor and one direct connection to a B+ terminal in the set. The secondary winding has just a

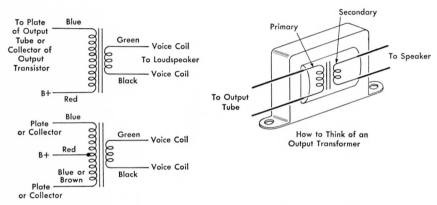

Fig. 17. Practical information on audio output transformers for tube and transistor circuits

few turns of heavier wire and connects directly to the two terminals of the voice coil on the loudspeaker, as in Fig. 17.

In large sets having two output tubes or transistors (said to be connected in push-pull), the primary winding will have a center tap that connects to B+. Each end of this primary winding will go to an output tube or transistor, as also shown in Fig. 17. Standard EIA color codes for output-transformer leads are indicated on this diagram.

The chief job of the output transformer is matching the characteristics of the output stage to those of the loudspeaker. Each type of tube or transistor requires a particular size of load for best operation. The voice coil of the average speaker, on the other hand, has an a-c impedance somewhere between 4 and 16 ohms.

Ratings of Output Transformers. Catalogs generally specify the power output of an output transformer in watts, along with the primary and secondary impedance values. For average table-model radios this transformer rating will be 3 watts or less. The average console will have an

8-watt output transformer, and only more expensive sets will go 30 watts or higher.

Common Troubles in Output Transformers. Just as with other coil devices, a broken lead or wire is the commonest trouble in output transformers. This break is generally due to corrosion.

Short-circuits may occur between adjacent turns or between layers of turns if insulation is damaged by overheating. Shorted turns cause weak volume and distortion, but these symptoms cannot be blamed on the output transformer immediately because several other receiver defects give the same symptoms. Substitution of a good universal output transformer is the only practical test for shorted turns.

Testing Output Transformers. An open circuit in an output transformer primary winding can be detected with an ohmmeter. It is best to disconnect one lead of the winding before making the test. There is often a capacitor across the primary winding. If leaky or shorted, this capacitor can make the ohmmeter indicate continuity even when there is a break in the primary winding.

The voice coil is seldom open because it is wound from heavier wire than the primary.

Do not expect to obtain ohmmeter readings corresponding to rated impedance values. The d-c resistance of a winding, as measured with an ohmmeter, is always much less than the a-c impedance value. A reading of infinity for any winding indicates an open circuit.

The windings of an output transformer are insulated from the iron core. One secondary lead is often grounded, as also is the core. The ohmmeter reading between the primary winding and the core should be higher than about 1 megohm. If appreciably lower even when all leads to the primary winding have been disconnected, and if the tone quality of the set is unsatisfactory, you are justified in replacing the transformer.

Shorted turns cannot ordinarily be detected with an ohmmeter, since they have so little effect on the total resistance. Poor tone quality and signs of overheating are symptoms pointing to shorted turns. There will also be a disagreeable odor of burned insulation coming from the output transformer. Replacement is the only remedy.

Ordering New Output Transformers. It is rarely necessary to order an exact-duplicate replacement output transformer from the manufacturer or distributor of the receiver. Replacements units that will work satisfactorily can generally be obtained from your jobber, oftentimes with special universal mounting brackets that eliminate drilling new mounting holes.

The power rating required for the new transformer is the same as the audio output rating in watts for the entire set. This value is usually given in the service manual for the set. When in doubt, use a higher power rating. Transformer size goes up with power rating, so be sure there is space for the larger unit in the set. Another advantage you get from a higher power rating is improved tone. The higher the power rating, the greater the cost of the new unit.

You can usually go to a jobber and get a satisfactory replacement simply by naming the output-tube or transistor type number. The great majority of sets have more or less standard voice-coil impedances of 3.2 to 4 ohms. Salesmen know pretty well by memory the impedance required by different output tubes or transistors. Be sure to specify whether you have a single or push-pull output stage. Catalogs often abbreviate push-pull as PP and abbreviate voice coil as VC.

Keep on hand at least a few universal output transformers. These have extra windings and leads so that they can be used in a great variety of sets merely by choosing the proper leads and snipping off the others. Instruction sheets showing how to do this are generally furnished with the transformers.

You may have to charge the customer a dollar or two more to cover the extra cost of the universal unit. This extra charge is justified because you will be able to finish the repair immediately, without making a special trip to the jobber or waiting for his next delivery.

Installing New Output Transformers. Output transformers are generally riveted to the chassis or speaker during manufacture. Removal of the old unit therefore involves drilling out rivets first. Use nuts and bolts for mounting the new unit. When drilling out rivets on the speaker, pack cloth or tissue paper back of the cone to catch the drillings.

The leads of new output transformers are generally color-coded. It is still a good idea to make a rough sketch of connections before disconnecting the old unit. Connecting the new unit is merely routine soldering procedure. Leads will generally be longer than necessary, so do not be afraid to shorten them first.

Vertical Output Transformer. Between the vertical output stage and the vertical deflection coils of a television set you will find the vertical output transformer. This looks very much like an audio output transformer, but will have much better insulation because it must withstand a peak voltage of about 2,500 volts during each vertical retrace interval.

A vertical output transformer may have three leads going to a single winding with a tap, or four leads going to two separate windings. The circuit diagram will show which type is used and give the resistance of each winding. Troubleshooting and replacement procedures are the same as for audio output transformers.

What Power Transformers Do. The job of the power transformer in an a-c receiver is to change the 117-volt a-c line voltage to lower and higher

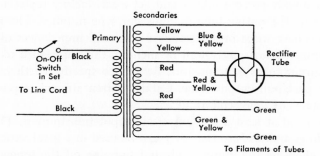

Fig. 18. EIA color code for power-transformer leads. If there is one more filament winding, it has brown outer leads and a brown-and-yellow center tap

values. The commonest type of power transformer is that shown in Fig. 18 in symbol form. This has a primary winding that connects to the line cord of the set through the on-off switch. It has one center-tapped high-voltage secondary winding that provides about 350 volts a-c between the center lead and each outer lead, or 700 volts between the two outer leads. It has one separate filament winding for the rectifier tube, usually providing 5 volts a-c. Finally, it has a filament winding serving all the other tubes in the set, usually providing 6.3 volts a-c. These are the approximate voltages that you should measure with an a-c voltmeter when an a-c receiver is turned on, if everything is operating normally.

The voltages in a power transformer can be dangerous to your meter as well as to you. Most dangerous is the voltage across the entire high-voltage secondary winding. To protect yourself and your meter, make a-c voltage measurements across only half the high-voltage winding at a time.

In very large radio sets and in television sets there may be more than two filament windings on the power transformer. Also, some sets may have one or more taps on the primary winding to take care of different line voltages in different communities.

Common Troubles in Power Transformers. Only two defects can ordinarily occur that make replacement of a power transformer necessary. One is an open winding, and the other is a partially or completely shorted winding.

The commonest of these troubles is the shorted winding. This is easy to recognize by its characteristic pungent and disagreeable odor of burned insulation. Do not blame the power transformer immediately when you smell this, however. A shorted electrolytic filter capacitor, shorted bypass capacitor, or some other short-circuit across the power supply may overload the transformer and make it heat up enough so the insulation begins smoking. To check for this, remove all tubes from the set, disconnect one lead of each power transformer secondary winding, then turn the set on again. For television sets, remove the socket from the picture tube also. If the transformer still smokes, or is too hot to touch after it has reached its maximum temperature, it is defective and should be replaced.

If the transformer does not overheat when disconnected, reconnect one of the filament windings and try the set again. If there is no overheating, plug in one by one the tubes that are operated from this filament winding. Use the circuit diagram to tell you which tubes should be plugged in.

If the transformer does not overheat with one circuit connected, repeat for each other winding in turn. Look for trouble in the first circuit or stage that makes the transformer overheat again.

Look also for signs of arcing on the insulating wafer of the rectifier tube socket. It will show as a charred path between terminals. If the rectifier has a wafer socket, the arcing may be occurring between the two wafers, where it cannot be seen. A charred socket should generally be replaced, but an emergency repair can be made by sawing or drilling through the charred area.

Testing Power Transformers with an Ohmmeter. An open power-transformer primary winding will make the receiver completely dead, with no tubes or pilot lamps lighted. To verify this, pull the line-cord plug out of the wall outlet, connect the ohmmeter to the prongs of the plug, and turn the set on. The reading should be around 100 ohms or less for a good primary, and infinity if the primary is open. Verify that the open is in the transformer by repeating the measurement directly across the primary winding. This must be done because the break can just as well be in the line cord, the on-off switch, or elsewhere in the primary circuit.

If one of the filament windings opens, the tube or tubes connected to that winding will be cold. If the rectifier filament is cold and the tube itself is good, check the rectifier filament winding by measuring between the filament terminals on the rectifier socket with an ohmmeter. The reading should be essentially zero ohms for a good reading.

If the rectifier is lit but all other tubes are cold, remove all these tubes and make a similar ohmmeter check between filament terminals on any one of the amplifier sockets. Again a low reading means good, and infinity means a break in the filament winding or in the connections to it. Measure directly across any suspected filament winding with the ohmmeter to verify the break.

To check for an open in the high-voltage secondary, measure with an ohmmeter between the two plate terminals of the rectifier tube socket. The reading should be somewhere under 1,000 ohms for a good winding. This gives a check on both paths of the secondary.

Testing Power Transformers with an A-C Voltmeter. If you prefer, you can check a power transformer by measuring the a-c voltage across each secondary winding in turn while the set is on. Measure each half of the high-voltage secondary separately, by measuring between the chassis and each rectifier-tube plate terminal in turn. This is possible because the center tap of the high-voltage secondary is always grounded to the chassis either directly or through a very small resistance.

Ordering New Power Transformers. To order a replacement power transformer, you need only specify the voltage and current ratings of each secondary winding. These values automatically ensure getting the correct primary winding and the correct size of core. Try to get a unit that has the same general shape and mounting arrangement as the old unit, to simplify fastening it to the chassis. An example of a good replacement power-transformer installation is shown in Fig. 19.

The biggest problem in ordering a replacement power transformer is determining the rating for the high-voltage secondary winding, because this is rarely given in service manuals. For this reason, try to get an exact-duplicate replacement power transformer from your jobber or from the distributor for that make of set. All connecting leads and lugs will then be in the same places as the old unit, simplifying changeover of the many wires involved. More important yet, the new transformer will fit in the available space and holes.

With auto-radio transformers an exact duplicate is best because the transformer usually is a tight fit in a shielded compartment. The transformer in an auto radio has only one secondary winding, for the high voltage, as all the tube filaments connect to the 6-volt auto storage battery.

Even with an exact duplicate, it is still highly desirable to make a rough sketch of connections before removing the defective power transformer. Some servicemen fasten marked slips of paper to each lead, because it is hard to avoid mistakes when there are so many leads to be changed.

Fig. 19. Good installation of replacement power transformer in hole for old unit at left rear corner of chassis. (Chicago Standard Transformer Corp. photo)

With power transformers, it does not matter whether connections to the ends of a winding are reversed or not.

If the replacement transformer has leads for an extra secondary winding that is not needed or a filament center tap that is not needed, cut the unused leads short and tape the ends to prevent the wires from touching anything.

Silencing Vibrating Laminations. If an annoying hum is heard from a receiver even when the loudspeaker is disconnected, the iron laminations in the power transformer may be loose and vibrating. First try tightening the bolts that hold the transformer together. If this does not help, loosen the bolts, apply shellac or service cement to the edges of the laminations and let it harden, then tighten the bolts. Sometimes a razor blade can be driven between loose laminations to stop the hum.

Testing and Replacing Filter Chokes. Some sets use a single iron-core coil called a filter choke to work with the electrolytic capacitors in suppressing a-c hum in the power supply. This choke generally has only two leads.

You can suspect the filter choke if the set is dead and there is no d-c voltage between one of the choke terminals and the chassis. An open circuit in the choke is the commonest trouble here since it has fine wire in its winding. Check with an ohmmeter directly across the terminals, with the set disconnected from the power line. The reading should be somewhere between 50 and 2,000 ohms if the choke is good.

In radio receivers using a power transformer, the filter choke will usually be rated at 8 or 10 henrys, with a current rating somewhere between 60 and 100 ma. If no service data are at hand, you are pretty safe if you order a 60- or 80-ma unit for sets having fewer than seven tubes and an 80- or 100-ma unit for sets having seven or more tubes.

Replacement of filter chokes is essentially the same as for output transformers. You will have to remove the rivets used for mounting the original unit, and mount the new unit with nuts and bolts. Connections are simple, and no special precautions are necessary.

Loudspeaker field coils in older sets were often connected to serve also as filter chokes. If the field coil opens, it is usually necessary to replace the entire electrodynamic loudspeaker.

20

Adjusting and Repairing Tuning Devices and Remote Controls

Why Tuning Is Needed. A good modern radio can pick up dozens of different stations in the broadcast band. A television receiver can likewise pick up several different stations in most localities. Tuning is the process of selecting the signal of the one desired station while rejecting all the other signals.

The actual job of tuning is done with coils and capacitors connected together to form tuned circuits. The tuned circuit accepts and passes a signal at its resonant frequency, but presents opposition to signals that are above or below this frequency.

When there is only one tuned circuit, the receiver will accept a desired signal but may let a couple of undesired signals on nearby frequencies get through also. This is why receivers have at least two adjustable tuned circuits in the station-selecting section and additional fixed-frequency tuned circuits in the i-f amplifier. The more tuned circuits there are, the more selective is the receiver.

Tuning Methods. To tune in different stations, either the capacitor or the coil in the tuning section must be changed in value. An adjustable tuning capacitor can be used with a fixed coil, or a fixed capacitor can be used with an adjustable coil. Both methods are used in modern a-m and f-m radios and in uhf television tuners.

Another way of tuning in different stations is to change the coils or capacitors in the tuned circuits by means of switches. This method is used in vhf television tuners.

Tuning adjustments are always ganged together to give single-knob tuning. The various ways of obtaining gang-tuning action are taken up in this chapter, along with troubleshooting, repair, and replacement of the tuning devices used in television receivers.

Gang Tuning Capacitors. The two-gang tuning capacitor shown in Fig. 1 is the commonest type you will find in radio sets. When the shaft is rotated, metal plates mounted on the shaft move between fixed metal plates. This varies the capacitance between the two sets of plates. The movable plates are called the rotor. The stationary plates are called the stator.

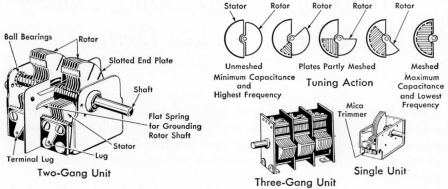

Fig. 1. Tuning-capacitor information. Two-gang units are the commonest

A two-gang unit has two rotors mounted on the same shaft, so they turn together and tune two circuits simultaneously. The rotors are grounded to the frame of the capacitor by a flat spring that presses against the rotor shaft. The two stator sections are mounted on insulating strips, and have separate terminal lugs.

A receiver is tuned to the highest frequency in a band when the tuning-capacitor plates are entirely apart or unmeshed, because this gives minimum capacitance. Fully meshed plates tune to the lowest frequency and have maximum capacitance, as indicated in Fig. 1. This fact is worth remembering when you have no tuning dial for guidance. In many sets the dial is printed on the cabinet, and does not come out when you remove the chassis.

Each section of a gang tuning capacitor usually has a trimmer capacitor connected between the stator and the frame. The trimmer is thus in parallel with the tuning-capacitor section. The trimmer usually consists of two metal plates with a sheet of mica in between for insulation. An adjusting screw permits moving these two plates together or apart to make the very small change in capacitance needed for matching the tuning-capacitor sections during alignment. Never touch trimmer screws except when actually aligning a receiver.

Some transistor radios have a dual tuning knob to provide coarse and fine tuning, much as on television sets. The knobs are rotated together for tuning in a station, after which the fine-tuning knob is adjusted for maximum volume and clarity. Each of the knobs controls one section of the gang tuning capacitor, eliminating the need for trimmers.

How Tuning Capacitors Are Mounted. In most radios the tuning-capacitor mounting bolts provide the connection between the rotors and the chassis of the set. The nuts that are on these bolts under the chassis should be tight and should have lock washers that bite into the chassis to ensure good contact.

In more expensive sets the tuning capacitor is often insulated from the chassis by vibration-absorbing rubber washers. One or more of the capacitor mounting bolts are then connected to the chassis with flexible wire that does not transmit vibration. With a floating rubber mounting, the rotor plates will not vibrate and cause howling in the loudspeaker each time the set is bumped.

Slotted Plates. On some gang tuning capacitors the two outer plates of one or more rotor sections will have deep slots. This can be seen in Fig. 1. The slots permit bending sections of the outer plate in or out individually during alignment, to make minor adjustments in capacitance so stations will come in at their correct dial settings. Do not try to straighten out slotted plates, because they are intentionally bent. Their settings can be just as critical as those of trimmer capacitors.

Tuning-capacitor Troubles. A symptom pointing to tuning-capacitor trouble is noise heard whenever the tuning knob is touched or when the set is tuned. The noise may occur at only one point on the tuning dial or at several or all points. It can be due to rotor plates rubbing together or to a poor connection between the rotor and the tuning-capacitor frame. Intermittent troubles, particularly those that can be reduced or cleared up by slapping the cabinet, are often due to poor rotor connections.

The flat spring that serves to make contact with the rotor shaft is usually easy to remove for cleaning. When you get the spring out, clean it on both sides with fine sandpaper, scrape out the groove in the rotor shaft, and bend the spring to get more contact pressure if necessary.

If the rotor spring is not readily removable, run a small strip of fine sandpaper between the spring contacts and the rotor shaft to improve the wiping contact. Any gummy dirt here should be removed with cleaning fluid and a small brush.

Dirt in the rotor shaft bearings can make a tuning capacitor turn hard. Apply a few drops of cigarette-lighter fluid, carbon tetrachloride, or any other cleaning fluid to each bearing, then turn the tuning capacitor back and forth about a dozen times to work out the dirt. Wipe off surplus fluid and all dirt, then apply fresh oil or petroleum jelly to the bearings to complete the job.

Shorted Tuning Capacitor. Whenever rotor plates touch stator plates, that section of a gang tuning capacitor is shorted. The condition can be recognized by a scraping sound heard when tuning the set.

Flaky conductive particles, such as peeled metal plating, can get between the plates and cause momentary or partial shorts. Such particles can usually be blown out or pushed out with a strip of paper.

When one or two inside plates of a rotor are bent out of line, they can usually be straightened with a thin razor blade or knife. When all are off center, look for an end-play adjustment, as the entire rotor shaft has shifted out of position.

An end-play adjusting screw presses against the end of the rotor shaft and is held in position by a locknut. Loosen the locknut, adjust the screw with a small screwdriver until the rotor plates are centered and there is no end play, then tighten the locknut without disturbing the setting of the screw.

Tuning-capacitor Grounding Troubles. With rubber-mounted tuning capacitors, poor soldered connections on the flexible grounding leads going from the capacitor frame or mounting bolts to the chassis are a common cause of trouble. The connection to the chassis should be suspected even though it looks all right, because rosin under the lump of solder may provide the insulating film that is causing trouble. Whenever in doubt, resolder these connections with a large soldering gun or iron, applying rosin-core solder to get fresh flux.

If the grounding leads go to soldering lugs that are riveted to the chassis, the rivet itself may be the cause of a poor connection. Solder the lug to the chassis to eliminate this possibility, because rivets can make bad contact even though they feel tight.

Tuning Knobs. The simplest way of tuning a radio is by rotating a knob mounted on the shaft of the gang tuning capacitor, as in Fig. 2. Sometimes the circular tuning dial will be printed directly on the cabinet, and the knob will have a pointer or other indicating mark. In other sets the mark will be on the cabinet, and the knob itself will have the dial scale. In television sets likewise, the channel numbers may be either on the front of the set or on the knob for the station-selector switch.

Replacements for loose or broken knobs are obtained from jobbers when needed. The knobs are held in place on the shaft with a screw, a spring, or simply by friction of corrugations inside the knob and on the shaft, just as for volume controls.

Fig. 2. Transistor radio in which tuning knob is mounted directly on the shaft of the miniature two-gang tuning capacitor. Note that the special 9-volt battery used in the set is no larger than a standard flashlight cell. (Ray-O-Vac photo)

To remove the chassis from the cabinet of most sets, you must first remove the tuning knob. If this knob has markings, it must be put back in its original position to make the tuning dial read correctly. For an a-m broadcast set, turn the tuning-capacitor plates until they are completely unmeshed or apart, then put the knob back on so that it reads at the farthest mark on the high-frequency end of the dial, usually just beyond the 1,600-kc mark. Check your work by tuning to a station whose frequency you know. Readjust the knob if necessary, to make the station come in at its correct point on the dial.

Dial-cord Tuning Systems. A dial-cord drive makes it possible to place the tuning knob and tuning dial anywhere on a cabinet while leaving the tuning capacitor at its best location on the chassis.

The simplest dial-cord drive is that shown in Fig. 3. Here the dial cord connects a separate tuning shaft to a large pulley mounted on the shaft of the tuning capacitor. A pointer is mounted on the capacitor shaft, in front of a round tuning dial, to indicate the frequency setting.

In another popular dial-cord drive, shown in Fig. 4, the pointer is attached to the dial cord and moves over a straight dial scale. This type is often called a slide-rule dial.

A surprisingly large number of service calls to a customer's home result

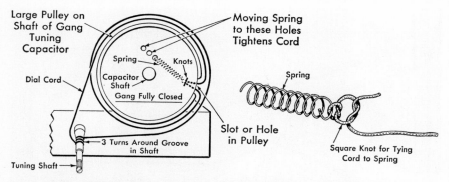

Fig. 3. Simplest dial-cord drive for a gang tuning capacitor, and method of tying square knot. When both ends of dial cord go to spring, slipping can often be stopped by unhooking cord ends from spring, twisting them together several times, then replacing

from failure of the dial-drive system. The tuning knob then spins freely but nothing happens, and the customer has no way of tuning the set. Knowing how to restring dial cords is therefore just as important as being able to replace a burned-out resistor.

Perhaps the greatest source of trouble in dial-cord drive systems is slipping rather than breakage. Slipping can be due to too smooth a surface on the cord, to insufficient tension, or to some part not turning or moving freely.

Tightening Dial Cords. When a dial cord slips but has no slack, its friction can often be restored temporarily by applying a friction-producing compound to the cord. This is available at jobbers in powder, stick, or liquid form. The liquid is easily applied with a dropper or small brush to the entire cord. Treated cords rarely work well for long, however. It is best to replace rather than repair troublemaking cords whenever possible.

A loose cord can be tightened in several ways. Tuning-capacitor drums often have several tabs or holes over which the free end of the spring can be hooked. If the spring is in the closest hole, moving it to stretch the spring will tighten the cord, as indicated in Fig. 3.

If it is possible to untie one of the dial-cord knots, pull up the string and tie a tighter knot. This may call for resetting the dial pointer, as covered later under restringing of cords. If the knot cannot be untied but there is sufficient slack, cut the cord close to the knot and retie it with more tension.

When slipping is due to binding of one of the pulleys or shafts in the tuning system, straightening or bending a pulley bracket may fix this. If oil

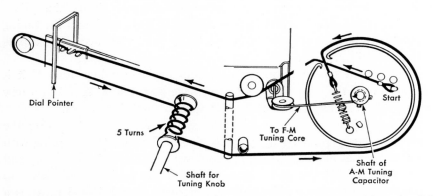

Fig. 4. Simple slide-rule dial-cord drive for making a pointer move over a straight tuning-dial scale

appears to be needed, use only a few drops and apply it carefully so that none gets on the dial cord.

A worn pulley may turn freely when loose but bind when pressure is applied to it by the cord, so inspect all pulleys carefully. If pulleys are gummed from old oil or grease, wash this away with carbon tetrachloride before applying new oil.

Slipping may occasionally be due to too much end pressure on the tuning-capacitor shaft. For this, loosen the end-pressure screw at the rear of the gang tuning capacitor, apply a drop of oil to the ball bearing inside, tighten the screw just enough so there is no end play in the rotor shaft, then tighten the locknut.

Choosing a Replacement Dial Cord. A broken or worn dial cord should be replaced with glass-core nylon-braid fishing line or equivalent special dial cord. The cord must have minimum stretch, high surface friction, ability to withstand bending over small shafts, and a breaking strength of at least 20 pounds.

Three sizes of dial cords are available for general replacement purposes: (1) extra-thin; (2) standard; (3) medium-thick. When you get into actual service work, keep a spool of each size on hand, and use the size closest to that of the old cord. If the cord is too thick, it will slip on a small drive shaft and jump out of idler pulleys.

Restringing Dial Cords. Before removing the old cord on a dial-cord system, make a rough sketch showing how the cord runs around pulleys if it is still on any of them. Do not rely on your memory; this sketch will save a bit of figuring when putting on the new cord. Next, remove the old cord. Do not bother untying knots; cut them with side-cutting pliers.

Before putting on new cord, check the tuning-knob shaft and the tuning capacitor to be sure they rotate freely. Apply light oil if necessary, wiping off surplus oil carefully.

Examine the cord tension spring inside the large pulley next. If it is stretched or if there are signs that turns have been removed, put in a new spring. You can buy an assortment of springs for this purpose.

On most jobs you can restring the system without first cutting the cord from the spool. Start by tying the end of the cord to the spring with a square knot as in Fig. 3. Bring the cord out of the hole in the pulley and around the top of the pulley, then down to the tuning-knob shaft. Go around this shaft the required number of times (use $2\frac{1}{2}$ or 3 turns as a trial number when in doubt), go around any idler pulleys in the system, then bring the cord back around the other side of the large pulley and down into the pulley slot. Pull the cord tight enough so the spring is exerting tension, then anchor the end temporarily to the same point at which the old cord was anchored. The cord tension should be enough to provide tuning without slipping, yet not so tight that it twangs like a violin string when tapped.

Extra turns around the tuning-knob shaft give additional friction to prevent slippage. If the shaft is grooved for the cord, the size of the groove will determine how many turns you can get on without having them pile up on each other and jam. If the shaft is not grooved, additional turns give no trouble, so put on three or more.

Crossing the cord in a drive reverses the direction of rotation. On most sets, the dial pointer is supposed to turn clockwise when the tuning knob is turned clockwise, but there are a few exceptions. When the cord is broken, there is no easy way of telling whether or not it was crossed. Restring such a cord so the pointer and knob turn in the same direction.

After tying the last knot, turn the tuning capacitor so its plates are completely meshed together, and push the pointer onto the tuning-capacitor shaft or dial cord at about 540 on the dial scale. Check tuning action over the entire dial now, to make sure the pointer moves to 1,600 when the plates are fully unmeshed. Check for binding and rubbing of any moving parts. If everything works all right, tie a permanent knot to complete the restringing job.

As a final check, turn on the set and tune in a station that you can recognize positively. Readjust the pointer to the frequency reading of this station if necessary.

Most pointers for slide-rule dial systems have spring clips that grip the cord. Others may require crimping or a drop of service cement. Finally, apply a dab of speaker cement or clear nail polish to each knot to ensure against loosening.

If the new cord must be cut to length before stringing, allow a few inches extra so you can tie the knots easily. It is far better to waste a few inches of cord than to do the job over because you did not leave enough to tie a secure knot easily.

Place the anchoring end of the spring in the hole closest to the opening in the tuning-capacitor pulley if there are several anchor points. This permits later shifting of the spring to tighten the cord if necessary.

With rotating-pointer dials the position of the tuning capacitor during restringing does not ordinarily matter. The pointer is attached to the tuning-capacitor shaft and will therefore read correctly after the job is done. Only with slide-rule dials is it necessary to give attention to the position of this capacitor during or after stringing.

In the slide-rule drive of Fig. 4 the sliding pointer has spring tabs that grip the cord tightly yet permit sliding the pointer along the cord for adjustment after restringing. The pointer is easily pulled off the cord and replaced on it.

The end of the cord that goes directly to a sliding pointer is usually tied to a rigid tab inside the drum, rather than to the spring. This ensures that stretching and contraction of the spring during use do not affect the position of the pointer on the tuning dial.

Other ways of making a pointer move over a straight tuning dial are shown in Fig. 5. The chief differences are in the locations and numbers of idler pulleys that serve to change the direction of travel of the cord.

In a few systems, the ends of the dial cord are anchored to the tuning

478 *Television and Radio Repairing*

shaft or to a spring that moves with the dial cord. These are just as easy to restring as conventional systems because the ends of the cord will still be in their correct positions even if the cord breaks.

Many f-m/a-m receivers have a separate gang tuning capacitor for f-m. These sets usually have two separate dial-cord systems. They may look complicated at first glance, but actually are just two entirely separate simple systems mounted close together. Usually only one cord breaks at

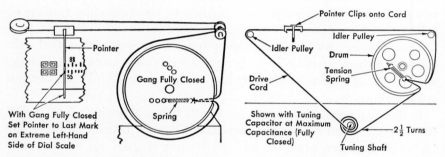

Fig. 5. Variations of slide-rule dial-cord systems, with restringing data

a time. After replacing this, examine the other cord and replace it if there are signs of wear.

Dial-cord Stringing Guides. Dial-cord restringing diagrams are usually included in service manuals. Handy books containing only dial-cord diagrams, indexed by receiver make and model number, are available from jobbers.

Cord-restringing Hints. When restringing a system having a number of idler pulleys, pieces of Scotch tape or adhesive tape can be placed over the cord at each pulley and shaft during the stringing operation, so that the cord cannot come off the pulleys if you accidentally drop or loosen it during the final tying operation.

A 6-inch piece of stiff wire with a loop or eye at one end makes an excellent threading needle. Use it to lead the end of the cord around the various pulleys and bends.

When working on a complicated restringing job without a diagram, use cheap wrapping cord for experimenting. When you are sure you have it right, use this cord as a guide for cutting the exact length of dial cord needed.

Replacements for broken dial pointers are often available at jobbers. Special pointers and special dial scales must usually be obtained from receiver distributors, however. The new pointer is simply pushed over the exposed end of the shaft, as friction holds it in place.

Metal or plastic bead chains are used in some television sets in place of long dial cords, to obtain positive drive of a channel-indicating dial. An example is shown in Fig. 6. The endless chain runs through sprocket wheels at each end, somewhat like a bicycle chain in reverse, and cannot slip.

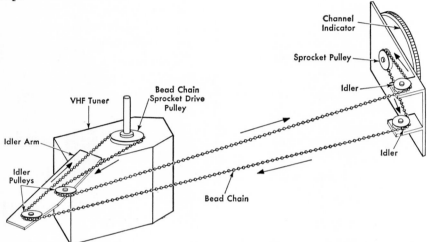

Fig. 6. Use of bead chain in place of dial cord to change television channel indicator when the shaft of the tuner is rotated by hand or by a remote-control tuning motor

Permeability Tuning. In some a-m and f-m radio sets, coils rather than capacitors are adjusted by the tuning knob. Powdered-iron slugs or cylinders are moved in and out of the coils to vary the inductance, giving what is known as permeability tuning. A dial-cord system is generally used to move the slugs in and out of the coils in unison when the tuning knob is turned.

The inductance of each coil is a maximum when its slug is entirely inside the coil, and a minimum when the slug is entirely outside the coil. Maximum inductance corresponds to the lowest frequency. Set both slugs entirely inside the coils after restringing a permeability-tuning job, then place the pointer at the lowest frequency on the dial for the initial check. Final adjustments sometimes require special aligning procedures as given in service data for the set.

Trimmers. Tuned circuits are adjusted during manufacture or servicing of receivers by means of trimmers. These are small adjustable capacitors or coils having screwdriver or wrench adjustments rather than knobs. Some have intricate shafts requiring special aligning tools to discourage tampering. Many set owners have in the past thrown receivers completely out of alignment by changing the screwdriver adjustments without knowing what they were doing.

Trimmer capacitors are named according to the type of insulation between their plates. Construction features of the commonest types are shown in Fig. 7.

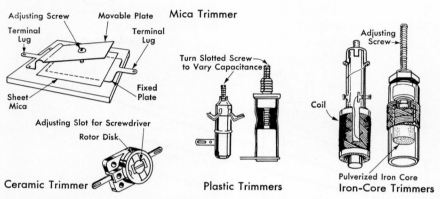

Fig. 7. Examples of trimmers used in receivers for adjusting tuned circuits

In a mica trimmer the adjusting screw moves a spring-brass upper plate toward or away from a fixed metal plate. When the two plates are closest together, the capacitance is a maximum since there is only a mica sheet between the plates. The farther apart the plates are, the lower is the capacitance. When a larger variation in capacitance is needed, two or more pairs of metal plates are used, all separated from each other by mica sheets.

Air trimmers are concentric meshing metal cylinders of slightly different diameters, separated from each other by air. The adjusting screw moves one cylinder in or out of the other. Sometimes a plastic material is used between the plates to increase the capacitance.

Ceramic trimmers use half-silvered ceramic disks. Turning the adjusting screw rotates one disk, changing the area of overlap of the silvered semicircles and thereby changing the capacitance. The action is very much like that of the rotor and stator in a tuning capacitor.

Some ceramic trimmers also change in value in a definite way with temperature. This characteristic is used to compensate for other changes with temperature in receivers. A zero temperature coefficient means that there is no change in value with temperature. A negative temperature coefficient means that capacitance goes down when temperature goes up.

Permeability-tuned trimmers are simply adjustable coils. Turning the adjusting screw moves a ferrite core in or out of a coil to vary the inductance. (Technically we are varying the permeability of the magnetic circuit of the coil, but this theoretical information is not needed to fix or adjust receivers.) In other versions the core is slid in or out simply by moving the bent end of the core shaft, which projects out of the rear of the chassis through a slot. The core is often called a slug.

Adjusting screws of permeability-tuned trimmers are sometimes anchored with cement to prevent loosening. A drop of acetone will generally loosen the cement in a few minutes and permit readjustment.

Trimmer Troubles. When you suspect that a trimmer capacitor is defective, measure between its terminals with an ohmmeter after first disconnecting all leads temporarily from one of the terminals. The reading should be infinity if the trimmer is good. A lower reading very likely means defective insulation. A zero or near-zero reading indicates a short-circuit. Once you know what the trouble is, you can either fix it yourself or put in a new trimmer of the same size and type.

Permeability-tuned trimmers are tested the same as coils. An ohmmeter reading should therefore be near zero for a good unit and infinity for an open unit. Repair and replacement procedures are the same as for television peaking coils.

Preset Fine Tuning. A number of television sets have means for adjusting the fine tuning control once for each channel at the time of installation. After this is done, a preset cam moves the fine tuning control to the correct new position each time the channel selector is changed, or a preset trimmer is switched into the circuit for each channel.

The preset fine tuning may require readjustment when tubes or other tuner parts age. Instructions for doing this are usually given in the operating instructions for the set, so many set owners know how to do this themselves. You will of course want to check these adjustments after changing tubes in the tuner. Always let the set warm up for about 15 minutes before readjusting.

The preset fine tuning adjustment may be a knob that is pushed in and rotated for clearest picture at each station position of the channel selector. Other sets have individual screwdriver adjustments that are accessible

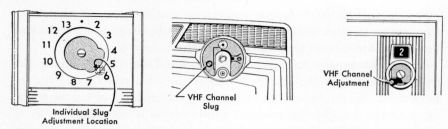

Fig. 8. Examples of individual fine tuning adjustments for each vhf channel on television receivers. The adjustments are usually accessible when the large channel-selector knob is removed. Adjust each for best picture, not for loudest sound, using a nonmetallic screwdriver or alignment tool

when the tuning knob is removed. Representative examples of these adjustments are shown in Fig. 8.

Automatic Tuning. Millions of auto-radio sets now in use have automatic tuning systems that tune in a station accurately and almost instantly when a button is pressed. There may be anywhere from four to twelve pushbuttons, each with an identifying tab giving the call letters of the station that it brings in. The chief advantage of pushbutton tuning is that it allows the driver to change stations without taking his eyes off the road.

A few home radios and television sets have pushbutton tuning as a convenience for users. Most television sets have switch tuning, however; this takes a little longer because the channel-selector knob often has to be rotated past unwanted settings.

A pushbutton or station-selector knob produces the required tuning action by rotating the regular tuning capacitor to the correct setting for a desired station by means of levers, gears, or cams, or by switching in preadjusted coils or capacitors.

Mechanical Pushbutton Tuning. In mechanical automatic tuning systems, the gang tuning capacitor is turned to the position for reception of a desired station by pushing a preset button all the way in, as shown in Fig. 9 for a typical mechanism. The listener does the manual work of moving the tuning capacitor and dial-cord system, so the buttons may therefore require considerable pushing force. The tuning mechanism merely serves to convert the straight pushing motion to rotary motion and stop the motion at the correct point.

Mechanical pushbutton tuning systems generally have a separate locking device for each pushbutton. It may be a screw located above, below, or behind each button, but more often it is simply a clamp that is released by pulling the button out about half an inch, sometimes with a sideways motion. A few older sets had round buttons that were turned counterclockwise to release the clamps.

Setting Up Buttons for Mechanical Tuning. To set up a station on a button, hold that button in firmly all the way, loosen its lock, manually tune in the station desired for that button, tighten the lock, and release the button. Repeat for each other button in turn to complete the job.

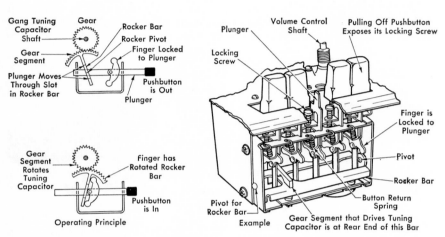

Fig. 9. Operating principles and example of typical mechanical automatic tuning system now used chiefly in auto sets. The dial pointer moves to a new position each time a different button is pushed

With mechanical tuning, any station in the broadcast band can be assigned to any button. It is logical to arrange the stations on the buttons in order of frequency just as they are on the tuning dial.

Station call-letter tabs go inside or alongside the buttons. They can be put in either before or after the adjusting procedure. Sheets of call-letter tabs come with new sets, and extra sheets can usually be obtained from jobbers.

Servicing Tips. When mechanical pushbutton tuning systems work too hard, try oiling the bearings of the tuning-capacitor shaft and the bearings for the shaft extension into the pushbutton unit. Inspect the dial-cord system carefully, to see if it is placing too heavy a load on the buttons because of binding or friction.

If moving parts and bearings are covered with dust, clean by brushing thoroughly with carbon tetrachloride. Pipe cleaners and old toothbrushes are handy for applying the cleaning fluid and getting off old grease. Apply a very thin coat of Lubriplate or an equivalent high-quality grease to each bearing or sliding part, but do not overlubricate. Too much grease serves only as a base for more troublemaking dust.

If buttons refuse to come out after being pushed, and cleaning does not help, look for weak, broken, or missing springs. Correct replacements can usually be found in assortments of dial-cord springs.

Electrical Pushbutton Tuning. In this older switching system, each pushbutton usually operates switches that insert preadjusted trimmer capaci-

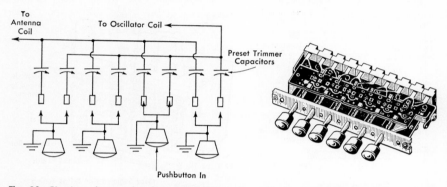

Fig. 10. Circuit and example of electrical automatic pushbutton tuning system found in older radios

tors in the two tuned circuits of the receiver, as in Fig. 10. The switching is usually done with a latching type of switch. Pushing in any button lifts a latch first to release whichever other button is in, then lowers the latch to hold in the button that was pushed. Permeability-tuned trimmers are sometimes used in place of trimmer capacitors.

Some older auto radios have a floorboard pushbutton that controls a ratchet-type switch. This changes stations each time the floor button is pressed, in a definite sequence that repeats itself.

Setting Up Buttons for Electrical Tuning. Turn on the receiver and let it warm up for at least 10 minutes while you assign stations to buttons. Give preference to the strongest of the customer's favorite stations in the frequency range covered by a button. Weak or distant stations require more critical tuning, so even slight drifting of trimmer settings would be immediately noticed.

The tuning ranges of the trimmers will be printed near the adjustments or given in the service manual. For each button, tune in the assigned station manually, note the program, push in the button, adjust the oscillator trimmer for that button until the same station is heard, then adjust the antenna trimmer for maximum volume. Readjust the oscillator trimmer again for maximum volume, then do the same for each other button. Use a nonmetallic screwdriver for trimmer adjustments.

Switch Troubles. Poor, dirty, or corroded contacts come first in the list of actual troubles with switch-type tuning. Loose contacts can often be tightened by bending carefully with tweezers or long-nose pliers. Dirt on contacts can be removed by brushing with a special contact-cleaning fluid or with carbon tetrachloride. Broken springs in the latching bar mechanism can usually be replaced from dial-cord spring assortments.

If only one button gives trouble and cannot be fixed, put a black station tab on that button and reassign its station to an adjacent button, giving up the station used least by the customer.

Pushbuttons are generally held in place by friction between the flat moving shaft and a slot molded into the plastic button. Sometimes there are springs inside the buttons also, much like those in certain types of control knobs. Looseness can generally be cured by placing a small piece of Scotch tape over the end of the shaft before pushing on the button.

If the looseness is due to a missing spring, you can generally find a satisfactory replacement spring in the spring assortment sold for control knobs. Missing buttons must usually be obtained from the manufacturer of the set.

Signal-seeking in Auto Radios. Many de luxe auto radios contain signal-seeking circuits. When a pushbutton on the floorboard or a bar at the front of the radio is pressed, the radio automatically changes to the station next higher in frequency that can be received satisfactorily. If the bar or foot button is held down, the tuning motor continues moving the tuning mechanism toward the high-frequency end of the broadcast band. When the high end is reached, the motor reverses at high speed to get back to the low-frequency end, then reverses again and slows down to start the signal searching over again.

In another signal-seeking system a stretched spring is used in place of the motor. When the spring pulls the tuning mechanism to the high end of the dial under control of the pushbutton, battery voltage is automatically applied to a solenoid to stretch the spring again and pull the tuning mechanism quickly back to the low end of the dial.

Special circuits are needed in an auto radio to stop the tuning action accurately for received signals that differ greatly in strength. These circuits contain familiar tubes or transistors working with resistors, coils, and capacitors, so there will be no new servicing problems. You will need the manufacturer's service manual, however, because the adjustment procedures are generally different for each set.

Signal-seeking auto radios will usually have a sensitivity selector. At one position the signal-seeking system will stop only on strong local stations. Other positions give a choice of weaker signal strengths, to permit tuning in weak stations automatically in rural areas.

Television Tuners. Some form of instantaneous switch tuning is used in the tuners of modern vhf television sets. In one type, called a turret tuner,

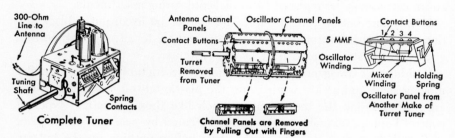

Fig. 11. Example of turret tuner for television sets, showing arrangement of removable panels containing coils for each channel. Turning the tuning shaft moves a different pair of panels up to fixed spring contacts that connect to receiver circuits

the desired coils and capacitors are actually moved into the circuit. The turret tuner has 12 sets of coils and capacitors, with each set mounted on an insulating panel that is held in position by its own spring contacts as in Fig. 11. The channel-selector shaft rotates the entire turret disk or drum to place the desired set in the receiver circuits.

Television receivers having vhf turret tuners can be adapted for uhf reception by removing a panel strip for an unused vhf channel and inserting in its place a special uhf channel strip obtained from the distributor for the set. It will then be necessary to add a uhf antenna input assembly and usually also a uhf antenna.

Another type of switching system for tuning a television receiver to one of the 12 vhf channels is shown in Fig. 12. In television receivers having three tuned circuits, there will be three sets of coils and capacitors (r-f, oscillator, and mixer) for each channel. When only two sets of coils and capacitors are used, the r-f stage will be omitted.

The coil and capacitor switched in at each setting may act independently to provide the total inductance and capacitance required. Here, adjustments can be made for one channel without affecting the other 11 channels, since there is a separate coil for each channel. In other switch-tuning arrangements, each switch setting serves to add or remove inductance or capacitance, so a given coil or capacitor serves for more than one channel.

At the high frequencies involved in television reception, coils in tuned circuits have only a few turns. For the highest-frequency channel (No. 13), there is just a fraction of a single turn. This means that the slightest bending of one of these coils, accidentally or intentionally, will disturb the alignment of the tuning system.

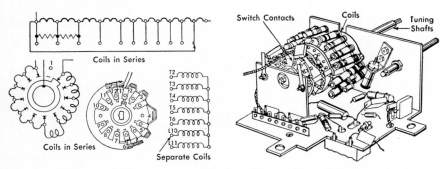

Fig. 12. Switch-type vhf television tuner, and coil-switching systems used in these tuners

A few older television receivers had continuous tuning. This most often used a variable inductance such as the Du Mont Inductuner. A slider moved along the turns of the coil for each tuning circuit to change the amount of inductance and thereby tune to a different channel.

Tuner Switch Troubles. Poor, dirty, or corroded contacts are a major problem in television tuners. If they occur only on one channel, look for trouble in the parts associated with that channel. When a poor contact is due to loss in spring tension, careful bending of the contact arm or the contact with long-nose pliers will usually cure the trouble. Erratic jumping of the picture, with or without associated noise from the speaker, is one symptom of a poor contact in the tuning circuits. Sudden drifting slightly out of tune is another.

Dirty contacts can often be cleaned with carbon tetrachloride applied with a small brush. All work on contacts must be done very carefully, so as not to disturb other parts.

When a tuner shaft turns too hard, try cleaning the shaft bearings with carbon tetrachloride. Wipe carefully, then apply fine machine oil and work it in by turning the switch.

Except for tube troubles and the simple contact and shaft troubles just described, television tuner repairs are best turned over to specialists. A tuner can sometimes be turned in to the manufacturer's distributor for another one on an exchange basis at a nominal price, to eliminate waiting for repairs.

UHF Tuners. Television receivers for uhf reception are not normally made separately. Instead, a uhf tuner is added at the factory to certain vhf sets in which mounting room and the required circuit connections have been provided.

One common type of uhf tuner tunes continuously over the entire uhf band between 470 and 890 mc. The tuner converts the uhf carrier signal directly to the 10.7-mc intermediate frequency of the set. The output of the uhf tuner is then fed directly to the input of the i-f amplifier.

A uhf tuner will generally have an oscillator tube and a mixer crystal diode. The tube and diode can usually be replaced without realigning the tuner. When parts inside the tuner are replaced, however, realignment is usually necessary. This calls for special uhf signal generators, so the job is best turned over to a shop that is equipped for such work. Fortunately, the parts inside a tuner are so small and so simple that they seldom go bad.

The tuning device in a uhf tuner is often a combination of a small variable capacitor and a variable coil. It is usually operated by a separate uhf tuning knob, but on a few sets the vhf fine tuning knob serves as the main tuning control for uhf.

Television Remote Controls. Many television receivers have provisions for operating the main tuning, volume, on-off, and picture controls from normal viewing positions. The command signals travel to the set in a variety of different ways, depending on the system. Ultrasonic waves are used in the majority of the remote-control systems on new sets today. In other systems the commands travel over wires, as radio waves, as carrier waves through house power lines, over a light beam, and even through a tiny plastic air hose.

The remote-control unit that sends out the command signals is called the transmitter. It usually has from one to four pushbuttons, each providing one of the receiver control functions.

Remote controls can be used only on receivers designed for this purpose. A special chassis receives and amplifies the control signals. The chassis also contains relay circuits that turn on tuning motors or solenoid stepper switches attached to the receiver controls. This remote-control receiver chassis is essentially the same for all types of control systems, even though its input device may be an ultrasonic microphone, a ferrite-rod antenna, a photocell, an air bellows, or a jack into which is plugged a cable going to the transmitter.

Remote-control systems come with all degrees of refinement, depending on how much the customer wants to pay. Some can turn the set on or off, while others lose control when they turn the set off. A few sets have

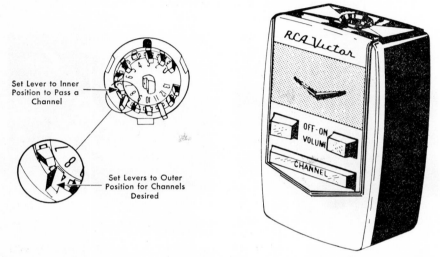

Fig. 13. Channel-setting lever arrangement at back of one vhf tuner. At right is a typical three-button remote-control transmitter

entirely automatic tuning, with no channel-selector knob. Pushing a tuning button either on the receiver or on the remote-control unit advances the set to the next selected channel. The service manual is particularly essential when working on tuning mechanisms like this or when making the initial adjustments for the desired channels.

One method of setting up a channel selector to stop only on preset channels is shown in Fig. 13. Here there are 13 levers arranged in a circle, for the 12 channels and an off position. Pushing a lever outward makes the tuner stop at that channel when the tuning motor approaches the channel position in response to orders from the remote-control unit.

Ultrasonic Remote Controls. A sound wave too high in frequency to be heard by human ears is called an ultrasonic wave. It travels through air much like ordinary sound waves, and can easily be picked up by a special microphone. These facts are the basis for ultrasonic remote controls.

490 *Television and Radio Repairing*

An ultrasonic wave can be produced in three different ways: (1) by tapping a short aluminum rod or tube with a hammer to make it vibrate; (2) by squeezing a bellows to send air through an ultrasonic whistle; (3) by using a transistor oscillator to generate an ultrasonic frequency that is fed to a tiny ultrasonic loudspeaker. In each case, the ultrasonic frequency is somewhere around 40 kc. In more elaborate remote-control systems there may be two or more different ultrasonic frequencies, each serving to operate a different receiver control.

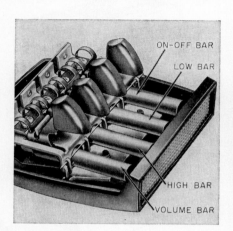 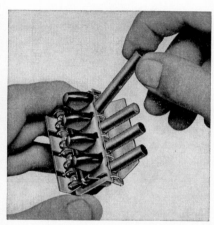

Fig. 14. Gong-type ultrasonic transmitter with cover removed, and method of replacing solid aluminum gong. (Admiral photos)

In a television receiver that is equipped for ultrasonic remote control there will be a microphone at the front of the cabinet to pick up the ultrasonic waves. The microphone signal is amplified enough to operate a relay which in turn actuates a solenoid or small motor that changes the setting of the appropriate receiver control. If two or more different ultrasonic frequencies are used, there will be discriminators that separate the frequencies and feed each to a different relay control system.

In a one-button system, each push of the button advances the channel selector to the next higher preset channel on which there is a station in the locality. With one button, the set is turned off by going past channel 13, but power for the remote-control receiver stays on. This is necessary so the receiver can be turned on again at some later time from the remote-control unit.

On some sets, the holder for the remote-control transmitter has a switch like that on telephones that cuts off power to the microphone circuit.

Lifting the transmitter out of its holder then turns on the remote-control receiver, in readiness for ultrasonic commands.

An ultrasonic remote-control unit works best when pointed at the television receiver, with no obstructions in the way. It will not usually operate through doors, walls, glass, or furniture, because ultrasonic waves are absorbed by all heavy materials.

If the ultrasonic receiver unit has a range control, set it at the lowest setting that gives reliable control action at the required operating distance.

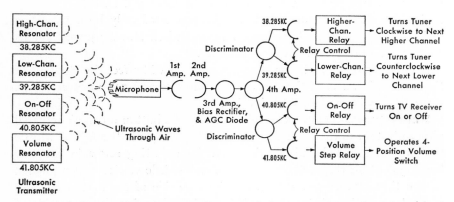

Fig. 15. Block diagram of four-frequency ultrasonic-control system used by Admiral for some of its television receivers

Higher settings make the control system respond to random sounds such as the jangling of keys, a chain on a dog's collar, a bell on a cat, or the ringing of a doorbell or telephone bell. If the telephone bell triggers the control system even at the lowest sensitivity setting, suggest that the customer ask his telephone company to change the tone of the bell slightly.

Gong-type Ultrasonic Controls. The ultrasonic transmitter shown in Fig. 14 has four different lengths of aluminum rods, each producing a different ultrasonic frequency when tapped by its hammer. When a button is pushed, a spring-mounted hammer is cocked and released. Button 1 advances the tuner to the next higher channel. Button 2 makes the tuning motor go in the opposite direction, to the next lower channel. Button 3 turns the television receiver on or off immediately. Button 4 advances the volume in successive sequences through off, low volume, medium volume, and full volume. The block diagram in Fig. 15 gives the frequencies produced by the transmitter and shows how they travel through the remote-control receiver in the television set.

On some ultrasonic transmitters the tuning button can be held down to make the receiver tune continuously through preset channels, to bring in more rapidly a channel at the other end of the band. When the button is held down, the rod continues vibrating and sending an ultrasonic signal to the receiver to make it continue tuning. When the button is released, however, a spring presses against the vibrating rod and damps it, stopping the signal. Tuning then stops at the next preset channel.

Dropping a gong-type ultrasonic transmitter can knock a rod out of its support. Just open the case and put the loose rod back in the same position as the other rods. It should be supported only by a mounting spring at its center, so it is free to vibrate when struck squarely on its end by the pushbutton-operated hammer.

A film of grease or dirt on a hammer or rod will affect the strength of the ultrasonic wave. If the areas where hammer and rod meet are dirty, clean them with carbon tetrachloride.

Whistle-type Ultrasonic Controls. Pressing the button on an air-actuated ultrasonic transmitter squeezes a bellows, forcing air through a whistle that produces a 37-kc ultrasonic wave. An adjusting screw in the tiny housing controls the exact frequency. A two-whistle version uses a pivoted button that controls two valves. Pressing one side of the button operates one valve, sending air through a 35-kc whistle. Pressing the other side of the button operates the other valve, sending air through a 22-kc whistle.

If an ultrasonic whistle does not seem to be working, it may be clogged. Blow into its opening to force out dust particles. If the bellows is leaking air, it may have to be reglued to the housing. Ask the manufacturer for instructions covering this and other repairs that may be needed on these ultrasonic whistles. An easy way to check a whistle is to try it on another set of the same type that is working properly.

Speaker-type Ultrasonic Controls. A single transistor operating from a battery generates the ultrasonic frequency in another television remote-control system. In a three-button model, the buttons switch in different tuning-circuit parts to give three different control frequencies.

The output of the transistor oscillator is fed to a type of high-frequency loudspeaker that is technically known as a magnetostriction transducer. Since these transducers are tuned to a particular frequency, one is needed for each button.

To tell whether a transducer is working, insert a wire through the grille on the transmitter and hold it lightly against the circular aluminum disk that serves as the speaker diaphragm. When you hold down the button for that transducer, an audible sound should be heard. For a more dramatic test, take the transmitter out of its housing, hold it upward, place a penny on the aluminum disk, and hold down the corresponding button. The penny should produce a loud squeaking sound if the transmitter is working properly. The other transducers can be tested in the same way by moving the penny and holding down another button.

As in all transistor equipment, a weak battery is the commonest cause of trouble. Always try a new battery first, and try a new transistor last.

Radio-type Remote Controls. Another type of remote-control system uses a single transistor to generate an r-f carrier signal that can be modulated with one of four different audio tone frequencies by pushing the appropriate button. Each button inserts a different ferrite-core coil in the oscillator circuit and applies power to the transistor. The modulated signal is radiated as a radio wave to the antenna of the remote-control receiver in the television set. Here the control actions are the same as in a four-button ultrasonic system.

Carrier-type Remote Controls. Instead of radiating a modulated carrier signal through space to an antenna in the television set, it is sent through the power-line wiring of the home in carrier-type systems. The remote-control transmitter in this system must be plugged into a wall outlet, and operates from a-c power. The carrier signal enters the television receiver through its line cord, and from there is routed to the remote-control receiver.

Photoelectric Remote Controls. Light-beam systems of remote control have not been popular, probably because of the difficulty of aiming the beam accurately with unsteady hands. Undesired triggering by sunlight or nearby lamps is another problem. The remote-control unit is simply a form of flashlight. At the front of the television receiver is a photocell that responds to the light beam. The remainder of the system in the television set is much the same as for other control systems.

Cable-type Remote Controls. Simple and reliable control action is obtained when direct wire connections are used between the transmitter and the television set. Continuous control of volume can then be achieved. The set can be turned on or off from the transmitter without leaving the receiver control chassis on 24 hours a day.

The cable running across the room to the viewing position is the chief drawback of a cable system. It is particularly good for sets used by invalids and elderly people in rooms where no one is likely to trip over the cable.

Pneumatic Remote Controls. A length of clear plastic tubing replaces wires in a pneumatic control system that has been used on portable television sets as well as on consoles. The home owner simply turns on the television set, picks up a rubber squeeze bulb that is attached to 15 feet of clear plastic tubing, and takes it to his chair. When he wants to change channels, he squeezes the bulb once for each channel position. The resulting air pressure in the tubing moves an aluminum piston that rotates the tuner shaft to the next channel position. The piston does this by acting on a stepper wheel at the rear end of the tuner shaft.

In one pneumatic tuning system the tuner has 13 positions, with channel 1 being REMOTE OFF. In another version, channel 1 is used for uhf reception and there are provisions for using either channel 8 or channel 9 for REMOTE OFF. If the television set is left on the REMOTE OFF position, it can be turned on remotely at any time by squeezing the bulb.

Remote-control Troubleshooting. Isolation of the trouble speeds repairing in all types of remote-control systems. When you press a button on the remote-control unit, listen for a relay to click in the set. If the relay closes but the correct control action does not occur, you know that the trouble is between the relay and that control on the set. The tuning motor or solenoid for that television control is then checked. If no click is heard, the trouble can be anywhere in the path taken by that command signal from the transmitter button through the remote-control receiver to the relay. Try all buttons; if some work, you know that the sections through which all the command signals pass must be good. Look for trouble only in stages or parts serving the command signals that do not work.

Checking of tubes is a logical starting point after isolating the trouble as much as possible. Voltage measurements can be made next, just as in any other receivers.

The commonest troubles encountered in remote-control systems are broken leads, defective contacts, bad tubes, and mechanical wear. Many of these troubles will be obvious and easily fixed. When the repair involves taking apart the tuning mechanism, however, you will definitely need to have on hand the manufacturer's instructions.

F-M Tuning Systems. The frequency-modulation radio band covers 88 to 108 mc, between channels 6 and 7 in the vhf television band. The tuning systems of f-m radios therefore use a combination of familiar television and a-m radio techniques. Once you are familiar with a few f-m servicing precautions and short cuts, you should be able to locate and replace bad parts in these sets just as readily as in other sets.

In an f-m radio the signal picked up by the antenna is converted to the intermediate frequency of 10.7 mc by the converter stage. After amplification in the i-f amplifier, the 10.7-mc signal is converted to an audio signal by the f-m discriminator or ratio-detector stage. The remainder of the receiver is essentially the same as the audio section of an a-m radio or a television set.

Up to the discriminator, an f-m radio has much higher frequencies than you work with in a-m radios. Coils and capacitors will therefore have

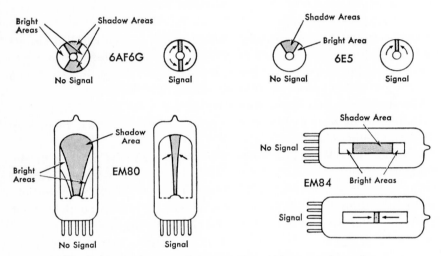

Fig. 16. Examples of patterns seen on cathode-ray tuning indicators in f-m radios

smaller values, making stray capacitances between leads and the chassis extremely important. Do not change the positions of leads or parts when hunting for trouble. When replacing a bad part, clip its leads to the same length as for the old part, so the new part will be in the same position.

Many f-m radios have tuning indicators. These help the customer to tune the receiver accurately to a station frequency. It is difficult to tune by ear alone because program material is continually changing. Distortion due to mistuning is not so noticeable on some types of music as on others.

Examples of cathode-ray tuning-indicator presentations for correct and incorrect tuning are shown in Fig. 16. A maximum bright-green area (minimum dark area) indicates correct tuning for each one. On strong signals the dark area may disappear completely and the bright areas may overlap.

A good antenna is needed for distortion-free f-m reception. If the customer is using only the antenna built into the set, and the tuning eye does not completely close on strong stations, try the set on a television antenna before removing the chassis. A good antenna may be all that it needs.

If the customer does not want to put up a separate outdoor antenna for f-m, install a two-set coupler. This permits operating both the television and f-m sets at the same time off the same antenna.

F-M Radio Troubleshooting. When an f-m radio goes bad, check tubes first just as with a-m radios. An ordinary tube tester will detect serious tube trouble, but substitution of new tubes one by one is the best way of tackling elusive troubles.

After clearing tubes of suspicion, tube voltages can be checked against those given in the service manual. When the trouble is isolated to a particular stage by such voltage measurements, pull out the line-cord plug and start checking parts in that stage with an ohmmeter.

When an f-m radio has several tubes of the same type, always return each tube to its own socket after a test. Interchanging such tubes may affect the tuning circuits.

Distortion in F-M Radios. Changes in tuning circuits are one of the commonest causes of distortion in an f-m radio. Even normal drifting during warmup can detune the set enough to cause noticeable distortion. This is why so many f-m radios have to be retuned more accurately about 5 minutes after they are turned on. To eliminate this nuisance, some of the more expensive f-m sets have automatic frequency-control circuits.

If an f-m receiver requires frequent retuning, it is likely that some part in the oscillator stage is defective. Try a new oscillator tube first. The special negative-temperature-coefficient ceramic capacitors in the oscillator stage are next in line as suspects. There is no practical way to check the effect of these capacitors on tuning, so the next step is replacing them all at once or one at a time. Remember to keep lead lengths and positions exactly the same as for the old part. Resistors and coils are the least likely to be the cause of the frequency drift.

When all tubes and parts check out good but careful tuning does not clear out distortion, it will usually be necessary to realign the set. This calls for special test instruments. It will therefore be best to pass up f-m receiver alignment jobs or turn them over to another serviceman until you have learned alignment procedures and obtained the necessary equipment.

Far too often the distortion heard in an f-m radio is due to poor program material or a defect at the transmitter. If there is one station in your locality that consistently has high-quality recorded programs or live music, always tune to it and listen for a few minutes before blaming the receiver for distortion.

If the receiver has a phonograph or phono input jack, listen to it while playing a record that is known to be free of distortion. If the record then sounds good, you know that the audio section of the receiver is good. The trouble is then either in the transmitter or ahead of the volume control in the receiver. If distortion occurs on all f-m stations, you can then concentrate on the receiver sections between the antenna and the volume control.

21

Repairing and Replacing Loudspeakers

Getting Acquainted with Speakers. The part that converts amplified audio-frequency signals back into the original sounds is the loudspeaker, more often called simply the speaker. No matter how perfectly the tube circuits in a receiver are operating, the sounds heard will not be right unless the loudspeaker also is in good operating condition.

The construction of a typical speaker in its simplest form is shown in Fig. 1. The large paper part that sets air in motion is called the cone or sometimes the diaphragm. Attached to the center of the cone is the

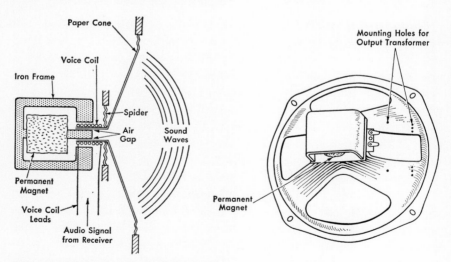

Fig. 1. Construction of a p-m dynamic speaker

voice coil that moves the cone back and forth in response to audio signals from the receiver output circuit.

The current-carrying voice coil is positioned in an air gap through which passes a strong magnetic field. In the commonest type of speaker, called a permanent-magnet dynamic speaker, or simply a p-m speaker, this magnetic field is produced by a permanent magnet.

In older electrodynamic speakers, the magnetic field is produced by a large field coil that carries direct current, as in Fig. 2. Here the field coil

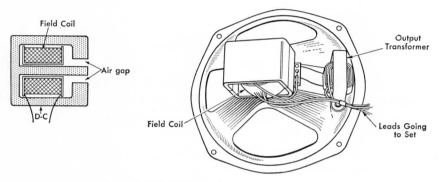

Fig. 2. Construction of an electrodynamic speaker

of the speaker generally serves also as the filter choke in the power supply of the receiver.

How Speakers Work. When current is sent through the voice coil, this coil becomes an electromagnet that has a definite magnetic polarity depending on the direction of the current. The magnetic field of the speaker also has a definite polarity. When both the voice coil and the speaker field have the same polarity, a repelling action is obtained because like poles repel, and the voice coil moves outward, as in Fig. 3A.

When the audio signal reverses its direction of flow, the voice coil changes its magnetic polarity. Now there are unlike poles and they attract each other, so that the voice coil is pulled inward, as in Fig. 3B. The cone is rigidly attached to the voice coil and hence moves out and in also.

Since an audio signal is changing its direction of flow thousands of times per second, the voice coil and cone must vibrate back and forth at this same high rate. As a result, the speaker produces essentially the same sound waves that act on the microphone in a broadcast studio.

The greater the strength of the audio signal at a particular instant, the

500 Television and Radio Repairing

greater is the strength of the voice-coil electromagnet produced by this signal. A stronger magnet has greater attracting and repelling power, and therefore a stronger audio signal makes the cone vibrate farther in and out. The resulting sound waves are then stronger, and the sound is louder.

Summing up, the number of vibrations of the cone corresponds to the frequency of the audio signal. The distance or amplitude through which the cone vibrates corresponds to the loudness of the program.

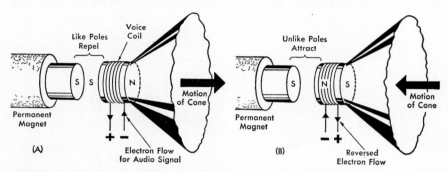

Fig. 3. The voice coil of a speaker reverses its direction of motion whenever the polarity of the audio signal reverses, thereby producing sound waves

Cone and Voice Coil. The complete cone assembly of a typical speaker is shown in Fig. 4, along with the speaker frame into which this assembly fits. The cone is made from stiff paper and is attached to the speaker frame only around its outer edge. The cone is always fastened at its rim with a special type of cement that can be loosened with acetone or a similar cement solvent. A paper or felt dust cap is cemented to the center of the cone to keep dust and iron particles out of the air gap.

The voice coil is wound on a thin paper, fiber, or aluminum cylindrical form that is just the right size to move in and out through the air gap without touching either the central pole piece or the outer pole pieces. In the assembled speaker, the voice coil is positioned accurately in the air gap by the paper cone and by a flexible corrugated paper disk called the spider. This spider is so constructed that it keeps the voice coil centered while permitting in-and-out movement of the voice coil. The spider is either cemented directly to a circular portion of the speaker frame or fastened to the frame with screws that can be loosened for easy recentering of the voice coil. The name spider is used because in old speakers this part had cutout slots somewhat resembling the legs of a spider.

Loudspeakers   501

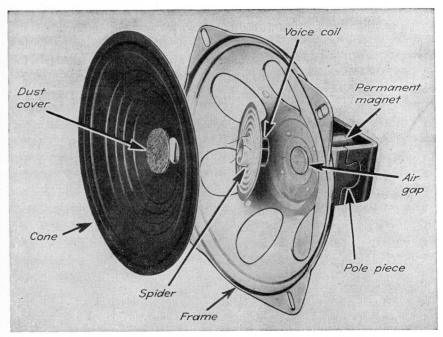

Fig. 4. How a typical modern speaker looks when taken apart for repair. The cone is glued to the front edge of the voice coil in an assembled speaker, and the dust cap or cover is glued over the hole in the cone. (Cletron photo)

Voice-coil Leads. The two wire leads of the voice coil are brought out through holes in the cone. Since the voice-coil wires are thin and extremely fragile, they are anchored to the surface of the cone with speaker cement on the inside. Stronger flexible wires are soldered to the voice-coil leads on the outside of the cone. Flexibility is highly important for these leads because they bring audio signals to a cone that is vibrating thousands of times per second for all the hours during which the receiver is in operation.

Speaker Troubles. The three commonest types of troubles you will encounter today in speakers are broken cones, off-center voice coils, and dirt in the air gap. Speaker repairs require considerable time, patience, and skill, however, along with special equipment in some instances. It will usually be less expensive and far better for you to install a new speaker.

If a replacement speaker is not available for any reason, have the repair work done by a shop that specializes in speaker work. Your jobber can usually give you the address of the nearest good speaker-repair service. If it is in another town, it is best to write first for shipping instructions and a price list. Some mail-order radio-supply firms handle speaker repairs also.

Although most speaker-service firms promise 48-hour repair service, shipping time adds considerably to this figure, and there is the nuisance of packing the speaker for shipment. Receiver manufacturers take still longer to make speaker repairs. Mail-order radio-supply firms may require 30 days or more for speaker-repair service, even though their rates are extremely reasonable. On most repair jobs the customer is not willing to wait this long. It then becomes justifiable and essential to make the repair yourself if possible, provided the repair will cost the customer appreciably less than a new speaker.

New permanent-magnet speakers in sizes up to 5 by 7 inches oval or 8 inches round, as used in table-model radios, most television sets, and many of the lower-priced home audio systems, can be obtained for only a few dollars each. This is often less than what you would pay to have a new cone put in. Here the new speaker is the best buy for both you and the customer, so get familiar with prices of replacement speakers at your jobbers.

Repairing Damaged Cones. Chattering or rattling heard along with a program can be caused by a rip or tear in the cone, a loose seam in the cone, a loose outer edge of the cone, a loose felt dust cap, a loose voice-coil joint, loose voice-coil turns or leads, or a loose spider. Normally it is better business to replace a speaker having such damage or have it fixed by a speaker-repair service.

When an emergency or inexpensive repair is desired by the customer, you can try cementing a rip or a loosened joint in a cone. Apply speaker cement to the loose parts or to the edges of a rip, then press the pieces together for a minute or two until the cement hardens. For long tears, a bit of fluffy absorbent cotton can be worked into the tear along with the cement.

Do not smear cement carelessly over the cone, because this stiffens the cone and may affect the tone quality. Do not use Scotch tape on a cone, because it is likely to work loose in damp weather.

Removing Dust Cap from Speaker. The felt dust cap at the center of the cone must be removed in order to recenter the voice coil or remove particles from the air gap, and must always be replaced after the work has been completed. Although it is possible to loosen the cap with cement solvent, a safer procedure is to pry and tear out the cap carefully with a small screwdriver. This eliminates the danger of surplus solvent loosening the cone-voice-coil joint. Any remaining pieces of the old cap can then be loosened safely with cement solvent and scraped off. A new cap is then put on when the repair job is done. Assortments of replacement dust caps can be obtained from most jobbers.

Loose Voice-coil Leads. The two very fine wires running from the voice coil to anchoring holes on the cone are cemented to the cone during manu-

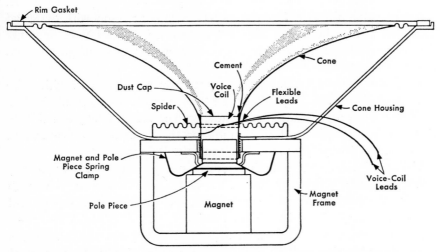

Fig. 5. The two voice-coil leads of a p-m dynamic speaker are cemented to the inside of the cone. Heavier flexible leads are similarly cemented to the outside of the cone and soldered to the inside leads

facture, as indicated in Fig. 5. If this cement gets loose because of heat or because of solvent used carelessly by a previous serviceman, the wires can vibrate enough to cause a faint sizzling sound. To repair, apply speaker cement to the leads and press them back against the cone.

Detecting Off-center Voice Coils. When a voice coil is sufficiently off-center so that it rubs against the pole pieces, a raspy scraping sound will be heard along with the program. There may also be noticeable distortion of loud bass notes. To test for proper centering, the front of the speaker must be accessible. With the set turned off, press evenly but gently on the outside edges of the cone with your fingers so as to move the voice coil in and out. Listen carefully for a scraping sound as you do this.

Sometimes you can even feel the vibration as the voice coil rubs against a pole piece. On a good speaker, nothing will be felt or heard when moving the cone.

Particles of dirt or metal in the air gap of a speaker can give the same rubbing sound. Clean the air gap before deciding whether or not the voice coil needs recentering.

How to Clean Air Gaps. Ordinary nonmagnetic dirt will usually work its way out of an air gap if you place the speaker face down on a clean part of your workbench and tune in a program with fairly loud volume. Gentle tapping of the back of the speaker magnet assembly with the wood handle of a screwdriver will help, but be sure it is gentle. Never hammer on a p-m speaker, because each shock causes a permanent magnet to lose a little of its strength.

Recentering Voice Coils. The corrugated paper spiders used with voice coils in modern p-m dynamic and electrodynamic speakers rarely give trouble when properly installed. These spiders are generally glued to a ring-shaped supporting surface mounted on the frame of the speaker, and the voice coil is rigidly glued to the center of the spider.

If a glued voice coil in an inexpensive speaker gets off-center for any reason, it is usually better to replace the entire speaker than to go through the time-consuming recentering procedure. Expensive speakers can be sent to shops that specialize in speaker repairs.

A few speakers are made with adjustable corrugated spiders. Here a metal clamping ring and two bolts hold the outer rim of the corrugated spider securely against the speaker frame, serving in place of cement. This permits easy adjustment if the voice coil requires recentering. All you need for this job are recentering shims and the appropriate size of screwdriver or wrench for loosening and tightening the spider adjusting bolts.

Choosing Shims. Nonmagnetic metal shims or plastic shims were once obtainable in sets containing four each of several different thicknesses. These may now be difficult to obtain, since very few servicemen do voice-coil recentering work.

You can cut suitable shims yourself from stiff paper for temporary use. You will have to try different thicknesses of paper, starting with business calling cards, until you find the thickness that just allows four equal-thickness strips to go snugly inside the voice coil without forcing. Even better is one long strip that goes all around without overlapping and fits snugly.

Recentering Procedure. The procedure for recentering the voice coil of a speaker that has bolts on the spider is: (1) remove the dust cap if present; (2) loosen the bolts that hold the spider, using a thin flat wrench or a miniature socket wrench; (3) insert speaker shims to center the voice coil with respect to the center pole piece; (4) tighten the spider mounting bolts firmly; (5) remove the shims; (6) test for rubbing by moving the cone carefully in and out with your fingers.

If the voice coil still rubs, repeat the procedure, being sure to get one of the shims in at the exact point where the rubbing is worst. Try using the next thicker size of shims if you can get them in without forcing.

If the rubbing sounds continue after recentering of the voice coil, they may be due to magnetized particles in the air gap inside or outside the voice coil or to loose turns of wire on the voice coil. Both these situations call for specialized speaker-repair services, to put in a new cone and voice coil after cleaning out the air gap.

In an emergency, an off-center speaker having a cemented spider can sometimes be fixed by inserting shims, then dampening the outer edge of the cone all around. The cone will usually recenter itself as it dries.

Replacing Speaker Cones. Replacement of a speaker cone is always considered to mean replacement of the entire cone-voice-coil assembly. The first decision to be made is whether to have the cone replaced or replace the entire speaker. An entire new speaker in the popular 5-inch size can actually be cheaper than the charge for repairing the cone. Even where a new speaker costs several dollars more than the repair charge, it may still be better business for you to put in the new one. You save all the time and bother of packing the speaker for shipment or taking it to a local repair service. You also save the expense of shipping charges and the nuisance of writing letters.

Ordering New Speakers. Since speakers vary greatly in size, in locations of mounting holes, and in electrical characteristics, time will be saved if you can get an exact-duplicate speaker from the manufacturer of the set (ordering through his nearest distributor) or from a jobber. When ordering, give the make and model number of the receiver and give the greatest outside diameter of the speaker in inches. If you have the service manual for the receiver, look up the manufacturer's part number for the speaker and give that also. It may pay to ask whether there is a credit allowance for turning in the old speaker, because some manufacturers have their own speaker-rebuilding service.

If the speaker is oval, measure two ways, to give the maximum and minimum diameters of the speaker frame. Common sizes of oval speakers are 4 by 6, 5 by 7, and 6 by 9 inches.

If the output transformer is mounted on the old speaker, it is desirable to have corresponding mounting holes on the new speaker. Replacement speakers generally have one or more sets of holes drilled for output transformers to simplify replacement problems.

Many servicemen simply take the old speaker in to a jobber and ask to have it duplicated. Catalogs do not usually give sufficient information on mounting details.

If you are replacing an old electrodynamic speaker, the new unit should have the same field-coil resistance as before. You can measure the resistance of the old coil with an ohmmeter if the coil is undamaged, or obtain the field resistance from the service manual for the set.

Output-transformer Considerations. If the old output transformer is going to be used with the new speaker, the voice-coil impedance rating must also be looked up in the service manual. This impedance cannot be measured with an ohmmeter because it represents the opposition that the voice coil offers at audio frequencies. Fortunately most manufacturers have standardized on 3.2 ohms as the voice-coil impedance for small speakers in single-speaker radio and television sets. Catalogs sometimes abbreviate voice coil as V.C.

Larger speakers in older sets used voice-coil impedances of 6 to 8 ohms. Using a 3.2-ohm speaker in place of 6 to 8 ohms would make a difference in tone and volume because of improper matching, but cause no other damage. Chances are that the customer will not even notice the change in tone unless he is extremely music-minded, or he may even prefer the new tone. When two or more speakers are used in a set, other voice-coil impedances are generally used. Here it is the combined impedance of all the speakers that must match the impedance of the output transformer.

Wattage Ratings of Speakers. The wattage rating of a speaker refers to the power-handling ability of the voice coil. The higher the wattage rating, the greater the volume at which the loudspeaker can be safely operated without risk of burning out the voice coil or tearing loose its turns.

The speaker wattage rating is generally the same as the receiver power-output rating in watts, so this receiver rating can be your guide for ordering the new loudspeaker. The higher the wattage rating of the speaker, the more it will cost. A typical audio wattage rating for speakers in table-model sets is 3.5 watts.

Mounting the New Speaker. New speakers come with slotted holes that allow considerable leeway as to positions of the mounting bolts.

The new speaker should be bolted tightly to the cabinet or chassis. There should be a lock washer under each nut to prevent loosening by vibration. If some lock washers and nuts are missing, replace them from your own stock. You can get assortments of commonly used sizes of nuts, bolts, and washers from your jobber to take care of requirements like this.

New speakers usually have terminal lugs for connections. If the old speaker has a socket or plug mounted on it for disconnecting purposes, as is often the case in large sets that have separately mounted speakers, transfer this socket or plug to the new speaker. If there is no convenient way of mounting this connecting unit directly on the new speaker, let it hang loose, using long enough leads from it to the speaker so the plug and socket do not bump against anything. Solder all new connections.

Tape exposed terminals if there is a possibility that the terminals or their bare wires can touch each other. Tape is essential if the speaker has a field coil, because the terminals of the field coil are often at a high voltage with respect to the chassis.

Phasing of Voice Coils. With sets having a single p-m dynamic speaker it does not matter how you make voice-coil connections. In sets having two or more p-m dynamic speakers, however, voice-coil connections should be such that both speakers push and pull the air in unison (in phase). If one cone pushes outward while the other is moving inward, they will cancel each other's work to a certain extent and will not sound right. Reversing the connections to one of the voice coils will cure this out-of-phase condition.

If you make a careful diagram of connections before removing the old speaker and connect the new speaker the same way, the two speakers should be in phase if using an exact-duplicate speaker. If in doubt, make temporary connections and try reversing the voice-coil leads of one speaker a few times until you find the connection that gives the most pleasing medium-range notes and bass notes.

Here is a simple phasing test. Disconnect the speaker leads from the output transformer and touch them momentarily to the terminals of a flashlight cell while watching the cones. Both cones should move in the same direction when contact is made to the battery, if the voice coils are connected in phase.

Examples of connections for two or more speakers are shown in Fig. 6. The large black dots identify identical voice-coil terminals on the various speakers. When the marked terminals are connected as indicated on the circuit diagram, correct phasing is obtained.

Speaker Jack Connections. Many radios and some television sets have a jack connected to the speaker, essentially as shown in Fig. 6. When the plug for an earphone or pillow speaker is inserted in this jack, the loudspeaker is automatically disconnected. This permits listening in private, without disturbing others. Dirt on the jack contacts can break the speaker circuit even when no plug is inserted, so check the jack before blaming the speaker or output transformer when the set is dead.

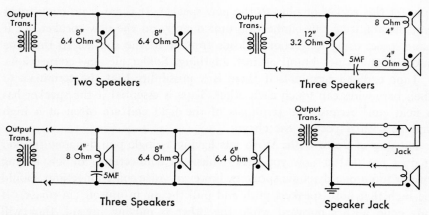

Fig. 6. Examples of connections for two and three speakers, and method of connecting jack in speaker circuit to permit use of earphone or pillow speaker

Magnet Weight. Practically all modern p-m dynamic speakers have permanent magnets made from Alnico V, a hard crystalline metal that provides a strong magnetic field. The greater the amount of Alnico V used, the stronger is the magnetic field and the greater is the volume obtained with a given set. This is why catalogs sometimes give magnet weight as one of the speaker specifications. Speakers for ordinary home radio receivers seldom have magnets weighing more than 2 ounces, but large high-fidelity speakers for good f-m receivers and custom-built installations can have magnets weighing several pounds. A heavier magnet permits more volume without distortion, but costs more.

Magnet Troubles in P-M Speakers. Loss of magnetism with age or through mishandling is about the only trouble that can occur in the magnetic-field structure of a p-m dynamic speaker. Distortion and loss of volume are symptoms of a weak magnet. Many other receiver defects can cause the same symptoms, however, so check everything else first. Magnets rarely go bad.

If no other cause of distortion can be found, hook up a spare speaker temporarily and see if the trouble clears up. If it does, install a new speaker, as repairs are not practical.

Field-coil Troubles. The commonest field-coil trouble in an electrodynamic speaker is an open coil, due to breaking or burning up of the wire with which the coil is wound. A break often occurs at the joints between the ends of the coil wire and the insulated leads that come out of the coil.

A field coil can be tested with an ohmmeter. A reading below 5,000 ohms means the field coil is not open. An infinity reading means the field coil is open. Disconnect one lead of the field coil before making an ohmmeter test.

Shorted turns make a field coil get so hot that the insulation becomes charred. A shorted paper or electrolytic capacitor can also make a field coil get hot, so do not blame the field coil too soon. Shorted turns cannot ordinarily be detected with an ohmmeter.

If the field coil is bad in a small speaker, the entire speaker should be replaced. For large and expensive speakers it may pay to have the field coil replaced by a speaker-repair service. Give the make and model number of the receiver when ordering. It may be a good idea to have the cone and voice coil replaced at the same time, especially in an old set that the customer is keeping for sentimental reasons.

Electrostatic Speakers. A few receivers, chiefly in the high-fidelity class, use a special electrostatic speaker for handling the higher audio frequencies. It consists of a stretched aluminum-coated plastic diaphragm spaced a small distance away from an aluminum back plate. This is equal to a two-plate air capacitor. When a signal voltage is applied between the diaphragm and the back plate, the diaphragm is alternately attracted and repelled.

The signal voltage is obtained directly from the output tube rather than from the output transformer, usually through resistors, capacitors, and coils that serve as a filter network.

There are ordinarily no adjustments or repairs that can be made on an electrostatic speaker. An exact-duplicate replacement is the only answer when one goes bad.

Test Speaker. An extra speaker can be mounted in a hole cut in the back of your service bench for test purposes. Use long leads, for convenience in connecting it to a chassis when you leave the speaker mounted in the console. An inexpensive 6- by 9-inch speaker will serve nicely, but a better check of tone quality can be obtained with special test speakers that are sold by some jobbers. These have jacks or a switch that provides the correct impedance for matching the output transformer of the receiver being checked.

Loudspeakers in Transistor Sets. When a loudspeaker is driven by a push-pull transistor output stage, there may not be an output transformer. Instead, the voice coil will have a center tap that goes to the battery or on-off switch. The outer two voice-coil leads go directly to the collector terminals of the output transistors.

A glance at the circuit diagram will show how to check the continuity of the voice coil.

If no thump is heard when a transistor auto radio is turned on, do not blame the output stage until you have checked the speaker circuit. This can be done quickly with an ohmmeter. Turn off the radio, set the ohmmeter to its $R \times 1$ range, connect one ohmmeter lead to the chassis or ground and touch the other lead to the case of the power transistor in the output stage. The case connects to the collector. A pop should be heard each time this is done. If no pop is heard, look for a defective speaker, a break somewhere in the speaker circuit, or a shorted collector bypass capacitor in the output stage.

22

Repairing Cabinets

Why Cabinet Repairing Is Important. A new scratch on the cabinet of a television or radio set that you are returning will be spotted faster than anything else you do to that set. Such damage is heartbreaking to the housewife, and yet accidents are bound to happen no matter how careful you are in handling sets and in working on them in your shop. By learning how to repair accidental damage to cabinets, you can avoid this embarrassing situation.

Since prevention is always better than cure, this chapter starts with suggestions for protecting and handling cabinets while in your care. These suggestions are important to you financially, because the repair of cabinets is often a slow and time-consuming process for which you cannot charge if you did the damage yourself.

It takes only a little extra time and effort to polish the cabinet and clean the chassis after you have repaired a set. As a serviceman, you are expected to make the set perform as well as when new. When you also make it look like new, you have done the unexpected, and the housewife will undoubtedly mention this to her friends. This can lead to more repair jobs for you.

Polishing a cabinet sometimes makes scratches, dents, and other defects stand out. If the customer had not noticed them before, you may get blamed for them. Furthermore, there are always a few unfair customers who deliberately use some minor defect as their excuse for not paying the bill. Because of this, always examine a cabinet carefully before taking it out of a home, and point out each defect to the customer. You can do this tactfully by asking if she would like to have the defects repaired.

Making a Profit on Cabinet Repairs. When cabinet work involves more than routine polishing, you are entitled to charge for the work and make a profit on it. Point out the defects tactfully to the customer and ask if she would like to have them fixed. If the answer is yes, give an estimate of the charges and get her approval, just as you do for other types of repair work.

If the damages are extensive, it may be better for you to take the cabinet to a professional furniture-repair shop. You can make arrangements with one shop to do such work for you at wholesale rates. Bill your customer enough more to cover your transportation costs on the cabinet and a nominal profit. The cabinet can thus be repaired while you work on the chassis in your shop.

Cabinet repairing requires a lot of practice. Before doing work on a customer's set, try out the repair on an old piece of your own furniture. Keep on making gouges, scratches, and cigarette burns on various types of cabinets and finishes until you are confident you can handle them all, if you want to add cabinet repairs to your services.

Avoiding Damage to Cabinets. The biggest risk of damage to table-model cabinets is right in your own shop. Never pile sets up on your bench because tools may slip, the sets may bump and scratch each other, and some may even fall onto the floor. Never put table-model sets on the floor, because in this location you are eventually bound to trip over one or drop something on it.

Shelves are the best place for table-model cabinets when not working on them. A plain bookcase-like arrangement will do if the back is covered with plywood to prevent knobs and small parts from dropping off behind. Oftentimes you will want to store a set without putting the knobs back on, while awaiting the required new parts.

Later, when the volume of your business justifies a bit more expense, you may wish to install a wall outlet at the back of each shelf compartment, for operating repaired sets at low volume a day or two to make sure that no new troubles develop.

Shelf-construction Hints. When building shelves, consider the possibility that you may have to move them some day to another location. Build the shelves in sections that will go through a standard door.

When carrying table-model sets in your car, always arrange them so they cannot bump against each other or fall down. Winding the line cord around the set horizontally is one way of obtaining a rubber cushion when it is necessary to put the sets close together in the car. Be sure the plug on the cord cannot gouge a hole in the cabinet. Newspapers or a blanket are safer for separating sets.

How to Move Consoles. With consoles it is usually best to leave the cabinet in the customer's home. On those occasional jobs where it becomes desirable to remove a console cabinet to your shop, such as for repairs to the cabinet itself, use standard furniture-moving techniques. This means wrapping a heavy blanket or commercial furniture-protecting pad around the cabinet and strapping it in place. Do all this before attempting to move the cabinet out of the room, because it is just as easy to damage the cabinet in the home as on the outside.

If the cabinet is heavy, do not try to lift it yourself. Get help. Do not try to move a large cabinet with an ordinary car; this is a job for a small truck or a station wagon. Even small consoles are difficult to get into an ordinary car without risking damage to the cabinet.

If the cabinet must be tied in place in the truck, use wide straps for this purpose. Wire or rope may cut through the pad and damage the cabinet. If using an open truck, be sure waterproof canvas is available to cover the set in case of rain or snow. Remove all loose objects that might damage the cabinet while the car or truck is in motion.

Cabinet-handling Mistakes. Common sense is just about all that is needed to prevent damage to cabinets, and yet cabinets are damaged by careless servicemen every day. They do this by placing the chassis carefully on top of a console cabinet and then accidentally bumping the chassis so its sharp edges gouge deeply into the finish. They damage cabinets by placing heavy or sharp tools on them. They damage cabinets by leaving a hot soldering iron near the cabinet on a crowded bench and later pushing the iron accidentally against the cabinet. Every one of these accidents can be avoided by using common sense.

A cabinet should never be placed near a hot radiator or other source of heat. Too much heat will damage the finish and may even dry out the glue at the joints of the cabinet. Excess moisture is equally damaging to the finish, and can swell the veneer as well as damage delicate coils in the chassis itself. Never leave sets in a damp location or near open windows.

Materials Needed for Cabinet Repairs. Inexpensive cabinet-repair kits developed especially for use by servicemen are available at jobbers. A kit usually contains furniture polish, scratch stick or scratch-removing polish, spirit stain (several shades), ivory enamel, alcohol, French polishing pad or felt, sandpaper or garnet paper, fine steel wool, and several small brushes.

Larger kits include an assortment of different shades of stick shellac, shellac rubbing fluid, an alcohol lamp, a spatula for applying the shellac, additional shades of spirit stain, touchup enamels and lacquers, a polishing cloth, and wood glue. Refills can be obtained for the kits, and the most-used items can be bought separately in larger quantities. It is suggested that you start with a small kit.

General Cabinet-renewal Procedure. Most cabinets can be made to look like new merely by cleaning and polishing. This takes only a few minutes when the cabinet is in your shop, yet makes a tremendous improvement in the appearance of the cabinet.

Cleaning a Wood Cabinet. Although wood cabinets are usually maintained in good condition by the housewife along with her other furniture, it is remarkable how the appearance of an older radio cabinet can be improved just by properly applying a special scratch-removing cream polish. This fills the multitude of minor scratches in the lacquer or varnish, removes any dirt that may have been overlooked by the housewife, and at the same time produces a high luster.

Always have two soft polishing cloths on hand, one for applying the polish and the other for rubbing the cabinet to a high gloss. Cheesecloth is ideal. Pour a teaspoonful or less of the polish on the cloth, fold the cloth into a flat pad, then rub the surface of the cabinet until all dirt and grease have been removed. All rubbing should be done in the direction of the grain of the wood. Do one surface at a time, allow about a minute for it to dry, then take the other cloth and rub with the grain until a high polish is obtained.

The cloth used for applying the polish can be used over and over again, because its cleaning and polishing ability increases with the number of times it is used. Wash the cloth occasionally in warm water to remove the dirt it has picked up. Keep this cloth in a paper bag when not in use, so it will stay clean.

The cream-base scratch-removing polish provides the final finish itself but is usually effective only on very shallow scratches. Being neutral in color, it can be used on any color of finish.

It is good business to carry a small bottle of cream-base polish in your tool kit along with a polishing cloth. Go over a wood cabinet lightly with this after you have finished putting the chassis back, to remove any finger marks, and see what a good impression it makes on the housewife.

Scratch-removing polishes that contain a colored stain are more effective on deeper scratches. These polishes come in a dark shade for walnut and mahogany and a light shade for maple and other light-colored finishes. Apply with a cloth just as for the cream polish, rub the surface thoroughly dry with the other clean cloth, then apply any ordinary good-quality wax furniture polish to produce the gloss of a new cabinet.

Cleaning Plastic Cabinets. When plastic cabinets are dirty, they can be cleaned by scrubbing with soap and water or with Bon Ami and water. Use a stiff brush if necessary, but work carefully at first until you are sure it will not scratch the surface. Rinse with clear water, and allow to dry. A coat of Simoniz wax can then be applied and polished with a chamois or any clean, soft cloth to restore the original luster. Plastic cabinets are easily scratched, so never use an ordinary rough cleansing powder on them.

Removing Beverage Stains. Milky areas or rings on a finish are generally produced by spilled beverages or wet glasses. Since this damage does not usually go through the finish, try rubbing it out first. For a dull finish, use oil and very fine steel wool. Pumice is best for a semigloss finish, and rottenstone for a high-gloss finish.

If rubbing does not remove the milky area, dampen a pad of cheesecloth in ammonia, wring out the pad tightly, then lightly and quickly brush the ammonia across the white area.

White spots on a lacquer finish can often be removed by similarly wiping lightly and quickly with lacquer thinner. Do not rub too hard or too long, or you will take off the finish along with the spot.

Use of a Scratch Stick. Another good item to carry in your toolbox is the scratch stick shown in Fig. 1. This contains solid stain at one end and a felt cork in a bottle of liquid at the other end. It is used for touching up minor scratches when you do not intend to do a complete repolishing job, and is also handy for making quick emergency repairs of unexpected scratches. To use, rub the solid stain lightly over the scratch until it is filled, then buff the area of the scratch with the felt on the other end. Remember that this is for emergencies and minor scratches only; it is not practical where there are a great many scratches or when the scratches go deep into the finish.

Repairing Cigarette Burns. A common type of cabinet damage is a burn made by a lighted cigarette that falls off an ash tray. The larger the area of the burn, the deeper it will be and the more difficult will be the repair job.

The first step in treating a burn is removing all the charred wood with a sharp knife. Next, smooth off the damaged area with very fine sandpaper, such as 2/0 or 4/0, then stain the exposed wood to match the surrounding finish.

A burn hole not more than $1/16$ inch deep can be filled in by a method known as French polishing. Make a small pad having about 20 layers of cheesecloth folded to a 1½-inch square. Dip the pad in wet shellac, squeeze out the excess shellac enough to stop dripping, then pull up the four corners in your fingers to form a rounded-bottom pad. Apply a few drops of linseed oil to the rounded portion, then rub the shellac into the depression with a brisk circular motion. Start with very little pressure, and

Fig. 1. Using a scratch stick to touch up scratches on a wood cabinet. (General Cement photo)

gradually apply more pressure as the shellac starts to harden. Continue working shellac into the burned area until it is even with the rest of the surface. If the finished repair is too glossy to match the surrounding finish, dull it down with the finest steel wool, 4/0 sandpaper, or pumice.

If the burned area is too deep for French polishing, fill it in with stick shellac just as for gouges.

Repairing Gouges and Other Deep Damage. Serious damage to a cabinet can be repaired by building up the wood with stick shellac. Scrape out the damaged or burned region first, then clean it with gasoline to remove dirt and old wax. Stain any bare wood to match the color of the finish.

To fill in a large gouge or hole with stick shellac, first select the proper shade of shellac. It is better to use a lighter shade than a darker one because the shellac can be darkened with stains if necessary. The shellac will also darken somewhat with application of heat.

Stick shellac is applied with a spatula that has been heated in the blue part of an alcohol flame. Larger furniture-repair kits contain an alcohol lamp for this purpose. Alcohol is required because it deposits the least amount of carbon on the blade.

Fig. 2. Applying stick shellac to blade of spatula after heating the blade in the flame of an alcohol lamp, for use in filling deep gouges in a cabinet. (General Cement photo)

After the spatula blade is hot enough to melt the shellac, quickly wipe it clean with a soft cloth and touch the shellac stick to the blade as in Fig. 2, so that some of it melts. If the shellac bubbles as this is done, the blade is too hot; wait until it stops bubbling. Now allow the hot shellac to run off the knife into the depression in the wood, as in Fig. 3. If the shellac will not flow freely, reheat the spatula and repeat the process.

When the scratch is completely filled, smooth the surface and lift off surplus shellac with the hot blade. If pinholes appear in the shellac, reheat with the blade to melt what is in the hole and allow it to flow together.

After the shellac has hardened, moisten a piece of felt with shellac rubbing fluid and polish the repaired region, as in Fig. 4.

On a large patch, use 7/0 garnet finishing paper or 400-grit sandpaper and a sanding block for preliminary leveling, with a few drops of fine oil or furniture polish on the paper. Work carefully here because this paper will also cut down the surrounding finish. Polish as before with rubbing felt.

One requirement for a successful repair with stick shellac is a close color match with the original finish. If the correct shade of stick shellac is not at hand, two colors may be mixed together on the hot knife to obtain the perfect match. To complete the stick-shellac repair after leveling the patch, clean the surface thoroughly and apply a wax polish.

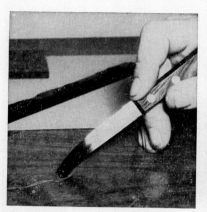 

Fig. 3. Spreading molten shellac into deep scratch with heated metal spatula. (General Cement photo)

Fig. 4. Rubbing shellac patch with felt pad moistened with shellac rubbing fluid. (General Cement photo)

Do not give your customers the idea that you can repair anything. A badly crushed corner or edge of a cabinet can be improved somewhat in appearance with stick shellac, but cannot be built out to its original shape because shellac is quite fragile. Such damage requires the services of an expert furniture repairman, who will set in a replacement piece of wood and refinish the entire area of the damage.

Synthetic Finishes. Some sets have a printed imitation wood grain pattern on the cabinet. When this is damaged, check with the manufacturer's distributor to see if special lacquers and inks are available for repairing cabinet damage and restoring the grain pattern.

Refinishing Enameled Cabinets. Some metal, wood, and plastic cabinets have a lacquer or enamel finish. Chips and scratches here leave an unsightly spot. This damage can be easily repaired by using a touchup lacquer enamel after smoothing the injury with fine sandpaper or steel wool. Two different shades of enamel can be mixed on a piece of glass or in a saucer to get a good color match.

After the patch has dried thoroughly, rub lightly with fine steel wool to blend the new and old finishes together, then polish the entire cabinet with a cream-base polish.

Repairing Cracked Plastic Cabinets. Plastic cabinets that have cracked but are not completely shattered can occasionally be repaired with the special plastic cements that are available for this purpose. Work the plastic cement into both surfaces of the crack, allow to dry, apply another heavy coat of cement to one of the surfaces, then clamp the two surfaces together or hold them together with large rubber bands until the joint has set and is thoroughly dry. Wipe excess glue off the surface of the cabinet before it hardens. Excess shellac or hardened glue should be removed with a sharp razor blade. The cabinet can then be polished with cream-base polish.

Table 1 lists the cements that will work on various types of plastic cabinets. When you do not know the exact type of plastic used in a cab-

Table 1. Cements for Repairing Broken Plastic Cabinets

Type of Plastic	Recommended Cement
Lucite or Plexiglas	Acrylic cement
Fiberglas and other laminated plastics	Epoxy cement
Celluloid	Nitrocellulose
Phenolic plastics (commonly called Bakelite)	Phenolic cement
Polyethylene coverings	Polyethylene cement
Polystyrene plastics	Styrene cement
Vinyl coverings	Vinyl cement

inet and do not need a full-strength bond, a general-purpose cement such as Duco or Pliobond may prove satisfactory.

A plastic can often be identified by touching a hot soldering iron to a part that does not show. A phenolic gives the same pungent smell as an overheated carbon resistor. Epoxies will be very little affected by the soldering iron. Other plastics have their own characteristic odors.

Badly cracked cabinets should be replaced rather than repaired. New cabinets can be obtained for this purpose from some jobbers and distributors. The new cabinet need not be the same color and shape, but must fit the chassis and have holes for controls in the right positions.

Repairing Loose or Blistered Veneer. The surface layer of veneer on a cabinet sometimes gets loose because of dampness or water damage. It is often possible to make a repair by forcing glue into the opening, spreading it with a thin, long knife, then applying pressure to the veneer until the glue sets. Do not get too much glue in, because it will squirt right back out when you apply pressure. A good wood glue for this purpose is Elmer, which is available from most hardware stores in small squeeze bottles.

Replacing Speaker Grille Cloth. Grille cloth serves the dual purpose of improving the appearance of the set and protecting the speaker. When this cloth is damaged to the extent that it is in need of replacement, remove the old cloth and scrape clean the surface to which it had been glued. Order a piece of new cloth from a jobber, cut it down to the correct size, then coat the inside surface of the cabinet with grille cloth fabric cement and apply the cloth. A small tack or even a thumbtack in each corner of the cloth will help to hold it in position until the cement has set. Watch the cloth for the first few minutes to make sure it does not loosen and wrinkle.

The pattern of cloth you get does not usually matter, because material sold for replacement purposes will harmonize with almost any cabinet.

Cabinet Hardware Trouble. One of the most common troubles with cabinet hardware is screws that have stripped their threads. Fill the screw holes with plastic wood and put the original screws back. Stop turning the screws as soon as the screw heads are flush, because continued turning will pull the plastic wood right out of the hole. If necessary, use clamps to hold the hardware in position for a few hours until the wood has set.

23

Installing and Repairing Television and Radio Antennas

Choosing a TV Antenna. The picture on a television receiver screen can be clear only if a strong signal gets to the receiver. In the early days of television, a roof-top antenna system was needed for almost every installation to give sufficient signal strength. Receiver designs have improved and television stations have boosted their power so much that indoor antennas give good results at many television receiver locations today. At the same time, good outdoor antennas give the best possible pictures at medium distances and make reception possible in fringe areas 50 to 100 miles away from the transmitters.

The type of antenna used most often in a given location is generally your best guide for choosing a new antenna. If few roof-top antennas can be seen, an indoor antenna will probably be adequate. If most outdoor antennas have rotators, your customer probably expects and needs one too. If elaborate arrays on high masts are common, you will probably need one too. This is why you need to get acquainted with the various types of television antennas being used today.

Built-in Antennas. Most television sets have wire or foil antennas attached to the back cover or inside the cabinet. These may give acceptable results if close to the stations, for temporary or even permanent use. Try the built-in antenna first; even if it does not work well, it serves to prove that a better antenna is needed.

Television antenna substitutes that plug into a wall outlet are generally even worse than built-in antennas, despite advertising claims. Some of these plug-in devices are actual frauds, with nothing inside the housing.

522   Television and Radio Repairing

Indoor Antennas. Indoor antennas usually have individually adjustable arms, as in Fig. 1A. Servicemen often call these antennas rabbit-ears. The antenna is usually set on top of the television set. The arms are set at a V angle, as a compromise between height and spread. The entire antenna can be rotated for maximum pickup in a particular direction. The tele-

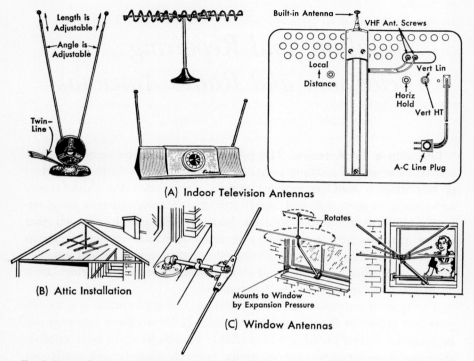

Fig. 1. Types of antennas that can be used where outdoor antennas are not permitted

scoping arms can be extended or shortened to give maximum pickup at a particular frequency. The arms should be longest for channel 2 and shortest for channel 13. Most users find a compromise of arm length and direction that gives acceptable results for all stations.

Some portable television receivers have a single telescoping rod at the rear that serves in conjunction with the metal cabinet as an indoor antenna. One conductor of the conventional twin-line goes to the telescoping mast, while the other goes to a cabinet screw. The cabinet connection must be removed before taking off the back cover of the set. Be sure to replace this connection after the cover is put back on. An example of this type of indoor antenna installed on a set is shown in Fig. 1A.

Some indoor antennas have a tuning capacitor or an adjustable coil in the base. The adjustment serves to provide the best possible match between the indoor antenna and the twin-line for a particular channel. The adjustment should therefore be made last after extending or shortening the antenna arms and changing their angles for best pickup. Adjust for the best signal on the channel giving the poorest signal, if the customer does not want to readjust each time he changes stations.

It is difficult to eliminate ghosts with an indoor antenna because it is not very directional. If ghosts are seen in the picture, try moving the antenna and receiver to other places in the room or to other rooms.

Most indoor antennas come with about 6 feet of twin-line. This should not ordinarily be lengthened, because long horizontal runs of twin-line may pick up more signal than the antenna itself. When this occurs, moving the indoor antenna will have little effect. The line can be run vertically without pickup troubles, but in a home there is seldom an opportunity to place an indoor antenna appreciably higher than the receiver. The best placement must be determined by trial.

In general, indoor antennas work well for close-in locations in cities having only one television station or perhaps two stations situated near each other on adjacent or nearly adjacent channels in the same band. Always warn the customer not to expect too much from an indoor antenna.

Portable television sets sometimes have permanently attached indoor antennas. The arms telescope out of sight when not in use.

One or more indoor antennas are handy on top of the service bench. They bring in sufficient signal for troubleshooting or for running a set after repair to see if new troubles develop. Every shop should also have a good outdoor television antenna, so that final tests of a receiver can be made with an antenna approximating that used by the customer.

Attic Antennas. In homes having sufficiently large unused space in the attic, good results can often be obtained with an ordinary outdoor antenna mounted upside down from the peak of the roof as in Fig. 1B. A floor-mounted mast is equally good. The antenna is then fully protected from the weather and should last much longer than if outdoors. If there is no convenient way to run the transmission line down between the walls, run it outside the house from the attic, then back in at the receiver location.

Be sure to point out to the customer that the signal may be weaker when the roof is wet or covered with snow. If within about 20 miles of the television transmitters, however, the reduction due to rain or snow on the roof will not normally be noticeable on the picture. Do not try to use an attic antenna if the roof is metal or slate, or if the roof has metal-foil insulation. Large metal objects and house electrical wiring should be kept at least a few feet away from an attic antenna.

Window Antennas. Because of restrictions on antenna installations on some apartment buildings, special television antennas have been designed for use on window sills. These are certainly worth trying in strong-signal areas, particularly for windows that face in the direction of the television stations.

Examples of window-mounted antennas are shown in Fig. 1C. Most of these have a horizontal telescoping bar which is extended to press against the sides of the window sill, thus mounting without bolts. Most window antennas have rotating or swivel joints to permit orienting for best reception.

Simple Dipole Antennas. The simplest television receiving antenna elements are the plain dipole and the folded dipole. Although these are seldom if ever used by themselves as outdoor television antennas, their characteristics determine the performance of all television antennas installed on roof tops, on masts, or in the attics of homes.

A plain dipole antenna has two rods placed end to end, separated at the center by an insulator as in Fig. 2. A transmission line connects to the rods at the center and brings the signals down to the receiver. This dipole is the basic antenna used in many television antenna systems today.

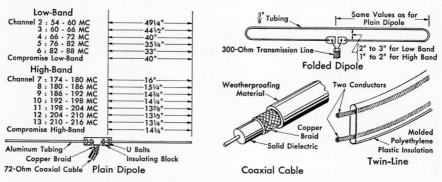

Fig. 2. Types of dipole antennas and transmission lines that form the basis for practically all vhf television receiving antenna installations. Both dipoles receive best broadside, which is at right angles to their sides. Note that coaxial cable has a solid-wire conductor centered inside a sleeve of copper braid that serves as the other conductor

A folded dipole is a single rod bent back on itself at each end as in Fig. 2. The twin-line connects to the ends of the rod. This dipole is easy to mount rigidly, uses a low-cost twin-line, and works well over a wide range of frequencies. For these reasons the folded dipole is widely used in television antenna systems.

Dipole Lengths. Both types of dipoles have one best length for each television channel, as indicated in Fig. 2. Channel 2 requires the longest dipole arms, and channel 13 the shortest.

In the fringe area just beyond normal reception range from a television station, it is generally necessary to use a separate dipole antenna array for each station. The dipole would then be cut to exactly the correct length for the station frequency.

In good-reception areas, a compromise length of about 40 inches for each arm of a dipole brings in all stations fairly well. When separate antennas are used for high and low bands to obtain better pickup of high-band stations, two compromises are used, one for the low-band stations (channels 2 to 6) and another for the high-band stations. The high-band antenna is generally placed above the low-band antenna on the same mast. The two antennas may have individual transmission lines or may be connected together with a critical definite length of line for use with a common transmission line.

Directional Qualities of Dipoles. All dipole antennas are directional. This means that they respond best to signals arriving from certain directions. A dipole having the right length for a signal receives that signal best at right angles to the dipole. Signals of other frequencies may come in best slightly off from right angles to the sides of the dipole. This is why rotation or orientation of a television antenna for best reception is so important.

When receiving direct signals from a nearby television station or when receiving a signal that is bent downward after passing over a hill, the antenna is aimed at the station. A compass and map are needed for rough orientation when the transmitting antenna cannot be seen.

When a reflected signal is the best that can be obtained, aim the receiving antenna at the hill serving as reflector.

Transmission Lines. The impedance of a plain dipole is about 72 ohms, while that of a folded dipole is about 300 ohms. The transmission line should have approximately the same impedance as the antenna. Maximum signal power is transferred from the antenna to the line when the impedances are equal or matched. Similarly, the transmission-line impedance should match that of the television-receiver input.

Modern television receivers have 300-ohm inputs. The lowest-cost transmission line is twin-line, which also has a 300-ohm impedance. This is why twin-line is used in the majority of television antenna installations today.

Coaxial Cable. When impedance values do not match, it is possible to use an impedance-matching arrangement between the different impedances and get good results. Inexpensive devices made for this purpose can be obtained from jobbers. An example would be a location requiring type RG-59/U 72-ohm shielded coaxial cable to minimize pickup of noise interference. With this cable, commonly called coax, a matching device is needed if the receiver has only 300-ohm input. No matching device is needed at the antenna end of coax if a plain dipole is used.

Coax is much more expensive than twin-line, hence is used only when absolutely necessary. Coax is required when the transmission line must be bunched with other lines or wires as in apartment buildings, when the line must run inside metal ducts, and when the line must run against metal of any kind. In general, coax is unsatisfactory for uhf reception, however.

Transmission Lines for UHF. Where fairly strong uhf signals are available, ordinary twin-line can be used and installed just as for vhf reception. The losses of any line are greater at higher frequencies, however, and these losses rise rapidly when the line is wet.

A tubular 300-ohm line is often used for uhf antenna installations where ordinary twin-line would reduce the signal strength too much on rainy days. Other low-loss lines are also available at most jobbers for use in uhf fringe areas, but are more costly and usually also more difficult to install. One type is perforated twin-line, where most of the insulating web between the parallel wires is punched out. The best type, but the most difficult to handle, is open-wire line. This has two parallel exposed wires that are held apart at regular intervals by low-loss plastic or ceramic spacing insulators.

Use of Reflectors and Directors. A metal rod placed a definite distance back of a folded dipole as in Fig. 3A acts as a reflector. The rod absorbs signals that get past the dipole and reradiates them back to the dipole. When the spacing between reflector and dipole is correct, the reflected signal adds to that arriving directly, and a stronger signal goes down the transmission line. A reflector also serves to suppress undesired signals coming from the back.

A metal rod placed ahead of a dipole as in Fig. 3B is called a director. Its length and position are such that it reradiates the desired signal to give addition of signals at the dipole.

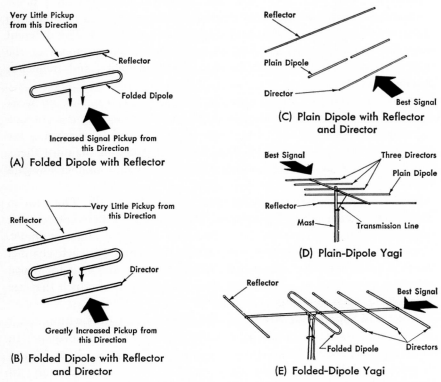

Fig. 3. Use of reflectors and directors with dipoles to improve signal pickup in one direction and reduce it in the opposite direction

Directors and reflectors may be used with plain dipoles if the lengths and spacings are as in Fig. 3C. A reflector is always slightly longer than the dipole, and a director is always slightly shorter than the dipole.

Yagi Antennas. Still greater pickup in the desired direction can be obtained by using additional directors with a plain or folded dipole, as in Figs. 3D and 3E. Any antenna having a reflector and one or more directors is commonly called a Yagi antenna.

The addition of dipoles and reflectors increases signal pickup on the channel for which the antenna is designed, but reduces pickup on all other channels. For this reason, a Yagi antenna is generally used as a single-channel antenna. A Yagi is much more directional than ordinary multichannel antennas, and must therefore be aimed more accurately.

Antenna Gain. Although at first glance it might seem that every home has a different kind of television antenna, all are basically a combination of the dipole, reflector, and director elements just described. The different shapes and arrangements are used to give different compromises of the three important antenna characteristics: gain, directivity, and frequency response.

The higher the gain of an antenna, the better it will pick up weak signals in fringe areas. Highest gain is obtained by designing the antenna to receive only one channel, rather than all twelve, and by narrowing the angle over which signals are received best. A high-gain antenna must therefore be aimed or oriented much more accurately than a simpler low-gain antenna.

Antenna gain is expressed in decibels, abbreviated as db. The gain in db tells how much better the antenna is than a plain dipole cut for that channel. A gain of 0 db means the antenna has the same gain as a dipole. The gain of the average roof-top antenna is different from each channel and may range from 0 to about 6 db. More elaborate antennas for fringe areas may have gains of 10 db or higher at the channel for which they are designed, but very low gains on other channels.

Fringe and Superfringe Areas. In city and suburban areas within about 25 miles of a station, antenna gains under 5 db are usually adequate. In the fringe area, normally thought of as the region between about 25 and 50 miles away from the station, gains up to 10 db become necessary. In the superfringe area beyond this, gains up to 15 db or higher become necessary just to receive one station. With such high-gain antennas, good reception is being obtained even out as far as 150 miles from a station.

Common Types of VHF Antennas. The folded-dipole and reflector arrangement of Fig. 4A is highly popular for close-in locations, being low in cost and easy to erect. It is usually cut for best reception in the low vhf band, hence may not work well enough on some of the high-band channels.

Adding short batwing loops at an angle as in Fig. 4B improves high-band performance of the folded dipole. Still better is use of a separate high-band folded dipole ahead of the low-band unit as in Fig. 4C. The two dipoles must be connected together by a definite length of twin-line, as specified in installation instructions that come with the antenna kit, to ensure best results. On low-band stations the signal is received by the longer folded dipole working with the reflector. On high-band stations the signal is received by the shorter folded dipole, with the longer folded dipole then serving as reflector.

Arranging dipole rods as if they were along the surfaces of two cones gives the highly popular conical antenna of Fig. 4D, also used with twin-

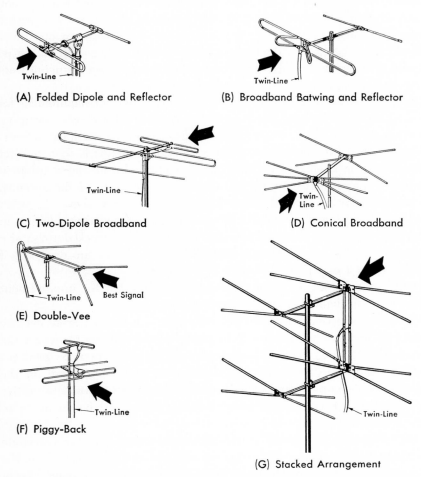

Fig. 4. Examples of commercially available vhf television antennas for normal service areas. Use high-quality antennas and hardware, because low-priced parts may rust quickly and stain white siding

line. This may have two, three, or even more rods on each side, and may have a single solid-rod reflector in place of the two-rod conical reflector shown. The reflector can even have three rods on each side. Conical antennas give fairly good results on all bands, as also does the double-V arrangement of Fig. 4E.

When high-band stations are a different direction than low-band stations, two antennas that can be aimed in different directions as in Fig. 4F are usually used. The small high-band antenna is usually placed above the other on the mast, hence the name *piggy-back*. If connected together correctly according to instructions, the two antennas work fairly well with a common transmission line. Better results are obtained with separate transmission lines going to a double-pole double-throw slide switch at the receiver. The switch must be operated by the listener when changing from low-band to high-band stations or vice versa.

In fringe areas, high gain is usually the most important requirement for a vhf antenna. In areas having good signals along with ghosts due to reflections from buildings or nearby mountains, a highly directional antenna having a good front-to-back ratio is needed. This antenna is able to reject signals coming from the rear, and may therefore also be the solution in locations where interfering signals arrive from the back of the antenna.

Stacked Arrays. Two identical antennas stacked one above the other and connected together by carefully cut lengths of transmission line as in Fig. 4G are called a *stacked array*. In superfringe areas, stacked arrays give surprisingly good results on the channel for which they are designed, if mounted on a sufficiently tall mast or tower.

Three or more identical antennas can be stacked, but the improvement over two is so slight that larger arrays are rarely used.

The vertical spacing between the antennas of a stacked array may be either $\frac{1}{4}$ or $\frac{1}{2}$ wavelength at the frequency of the channel for which the antenna is designed. The $\frac{1}{2}$-wavelength spacing gives about 33 per cent better pickup than the smaller spacing but requires more room on the mast. Only experiment and experience can tell whether a more sensitive stacked array will be more economical than greater mast height in a particular fringe location.

UHF Antennas. Reception of uhf television stations calls for a separate uhf tuner in the receiver and a separate uhf antenna system, because the channel frequencies are so much higher than for the vhf band. A vhf antenna rarely works satisfactorily even for a nearby uhf station.

When only uhf reception is required, you simply order and install the particular type of uhf antenna recommended by your jobber for the location in question. Examples of a few types currently being used are shown in Fig. 5. All are much smaller than vhf antennas, because of the higher frequencies involved.

Practically all uhf antennas have some form of sheet reflector. This may be perforated sheet metal, wire netting, or parallel rods. The reflector acts as a mirror for uhf radio waves, while the openings or spaces serve to reduce wind resistance.

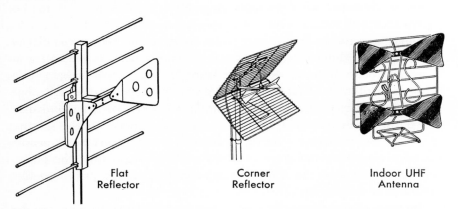

Flat Reflector   Corner Reflector   Indoor UHF Antenna

Fig. 5. Examples of uhf antennas

The uhf antenna itself is often made from punched sheet metal instead of rods. It can have a variety of unusual geometric patterns, to give essentially uniform performance on all uhf channels.

Antenna height is critical for uhf reception. The greatest height does not always give the strongest signal. After orienting a uhf antenna, slide it up and down on the mast a few times, or slide the entire mast up and down in its mounting bracket. There may be a lower position at which a stronger signal is obtained.

In uhf fringe areas, stacked uhf antennas and even more elaborate arrays are required. These must be oriented and adjusted even more carefully than corresponding vhf arrays. In some locations they may even have to be tilted up or down.

Color Television Antennas. In localities having good signal strength, practically all standard vhf antennas are suitable for color reception. However, when signal strength gets down near the minimum needed for good black-and-white reception, a better antenna will usually be required for satisfactory color reception. The reason for this is that color reception requires a reasonably flat frequency response over a wider frequency range than is normally used for black-and-white signals. Here also, your jobber can give you practical advice on selecting antennas for color reception in a specific locality.

Choosing a Television Antenna. Although dozens of different types of television antennas are on the market, the variations are due more to the large number of antenna manufacturers than to the needs of users. Your jobber will generally have several types of good antennas for the combination of stations and signal strengths encountered in the area. Ask for his advice whenever you are in doubt as to what a particular location needs.

Television Antenna Prices. When two makes of antennas are similar in appearance but differ greatly in price, the more expensive one will generally have better materials and better workmanship. It will probably also be easier to assemble. Performance will probably be the same initially, but the cheaper antenna may not withstand strong winds or icing, and may corrode in a year or two. Here your choice will depend on the severity of weather in your locality and the desires of your customers. Some people want the cheapest possible job, especially if they are renting the home, while others will gladly pay a few dollars more to get a high-quality trouble-free antenna installation.

Assembly of Television Antennas. A snap-into-shape antenna can be taken out of its shipping carton and set up almost as easily as opening an umbrella. Other types require assembly and tightening of bolts, but this rarely takes more than 5 minutes once you have done it a few times for a particular antenna design. The choice as regards method of assembly is thus entirely personal preference.

Types of Masts. Masts which do not require guy wires are called self-supporting. This type of mast is usually limited to about 15 feet in height, unless you use a heavy, rigid pole or fabricated steel tower. Self-supporting masts are widely used because they give a neat and strong installation with a minimum of labor, and can be used where guy wires are not allowed.

Guyed masts are useful when extra height is needed. With extra mast sections and guys, you can obtain heights up to 30 feet or even more. Since guyed masts are light in weight, they can be erected on roofs. The height of the building is then added to that of the mast.

The strain or tension on guy wires should be kept at a minimum. If the wires are correctly located, there is no reason to make them so tight that they sing like violin strings. Make them only tight enough to provide adequate sideways support for the mast.

Antenna Rotators. In locations where stations are in different directions from the antenna and no single antenna position brings them all in satisfactorily, an antenna rotator may be the best solution. This is a reversible geared-down (slow-speed) electric motor that mounts on the mast and is controlled remotely from the receiver to turn the entire antenna.

A rotator is useful in fringe areas even when all stations are in the same direction. With most antennas the exact direction of maximum pickup changes a little for each channel. Thus, an antenna aimed for channel 2 may have to be rotated as much as 30 degrees to get maximum pickup on channel 11.

Two types of rotator systems are available. The simpler and less expensive type has just a start-stop-reverse switch on the control unit. There is no indicator to tell which way the antenna is pointing. The user must flip the control switch back and forth while watching the screen, until the best possible picture is obtained for the desired channel. Few people have the patience required to make this adjustment each time they change stations, so it is best not to recommend this type of rotator.

More satisfactory are rotators having a control unit that tells the direction in which the antenna is pointing. This eliminates searching for the best position.. Once the correct setting of the control is found for each channel, changing the antenna direction is just as easy as tuning the receiver to a different channel.

Installing Rotators. Installation instructions come with each rotator. In general, the motor unit is clamped to the top of the antenna mast and has an upward-projecting shaft on which the antenna itself is mounted. Special flat insulated cable made for rotators, having three or more wires, is run from the motor to the control unit near or on the television set. Order the type of cable specified in the installation instructions, estimating the length needed just as when installing transmission line. This cable is supported by standard twin-line standoff insulators and is kept at least 6 inches away from the transmission line. The control unit has a line cord that is plugged into any a-c wall outlet to get power for the motor.

With some types of rotators, enough extra transmission line is needed around the motor to permit rotating the antenna a full 360 degrees. These rotator motors have stops that permit only one complete rotation, as otherwise the transmission line would get wrapped around the mast too much. Other motors have slip rings for the antenna connections; here the antennas can be turned around and around as much as desired without tangling the transmission line.

Two-set Installations. More and more homes are keeping their old television receiver when buying a new set. To use two sets with the same antenna, a coupler unit is needed. This is mounted at the back of one of the sets or in the basement and connected to the existing antenna. Transmission lines to the two receivers are then run from the two pairs of terminals at the bottom of the coupler. The coupler prevents the two receivers from interfering with each other and maintains correct impedance match. There is some loss in signal strength, however, when using two receivers on one antenna. Separate antennas for each give the best results.

A coupler may also be used for connecting two or more antennas to a single television receiver. This situation occurs when separate antennas are needed for each channel being received, or when separate vhf and uhf antennas are used. Many types and makes of couplers are available, to meet the various reception conditions encountered, so be sure to ask your jobber for advice when ordering a coupler. Some couplers are better than others for preventing receivers from interfering with each other when connected to the same antenna.

When older receivers are connected to an antenna through a coupler and tuned to certain combinations of stations, one receiver may interfere with the other. This situation can occur when the sets are tuned to channels 2 and 5, to 3 and 6, to 7 and 11, to 8 and 12, or to 9 and 13. These pairs of channels have a frequency differential equal to the 24-mc i-f value of old receivers. A herringbone type of interference is then seen on the screen. The trouble can be eliminated by using a coupler that provides greater isolation between the receivers, though with some sacrifice in signal strength.

Never connect two sets directly to the same transmission line without checking operation of both sets thoroughly on all combinations of all stations. Technically, it is possible for such a direct connection to cause interference patterns on the picture when both sets are in use and tuned to different channels.

Antenna-installation Requirements. Installation of outdoor antennas, particularly for f-m and television, involves construction work rather than actual servicing. It generally requires a ladder, special tools for drilling holes in brick and cement where necessary, and a certain amount of ability to work safely and without fear on roof tops. Some men like this outdoor work better than others.

There is nothing wrong with saying to a customer that you are not equipped to do antenna work. You can then recommend someone else

Television and Radio Antennas   535

who you know is reliable. Get acquainted with other servicemen in your vicinity as soon as possible. It pays, because if you pass on jobs to them, they will try to help you in other ways.

Equipment Needed. For antenna work, two ladders are usually needed. One should be a two-section aluminum or magnesium extension ladder

Fig. 6. Use of double-extension ladder and single ladder with roof hooks to get up to the chimney of a house safely for installing a television antenna. (RCA Victor photo)

having an over-all length of at least 24 feet. The second should be a single-section ladder 12 or 14 feet long, provided with roof hooks. These grip the top of steep-peaked roofs, so the ladder can be used for climbing up the roof to the peak, as in Fig. 6.

All ladders should be equipped with pads to prevent slipping and to protect roofing material. Aluminum or magnesium ladders are ideal for antenna work because they are light in weight, easy to handle, withstand abuse, and are long-lasting. A roof rack on your car or truck is all that you need for carrying the ladders to jobs. The rack is also used for transporting long masts and large antenna arrays.

A 100-foot extension line cord will be needed for using your electric drill and soldering gun outside. Masonry drills are needed for installing antenna hardware. A few electrician's long wood drills for making holes through walls and floors should be carried also, along with your regular tools.

Supplies Needed. At the beginning of your service career, you can get along without special supplies if you make an extra trip to size up the installation. You can then order only the supplies actually needed for the job. Later, when you are doing enough antenna jobs to justify the investment, you can keep in your car or truck a few standard antennas and enough hardware and transmission line to take care of practically any job.

The list of supplies is best determined by actual experience in your own locality. It will usually include several types of antennas, several different lengths of masts, guy wire, grounding wire, several types of mast brackets and chimney straps, lightning arresters, ground rods, an assortment of standoff insulators and other hardware for both wood and masonry, a can of weatherproofing plastic spray for antenna connections, a can of roofing cement for use under roof-top brackets, and caulking compound for sealing lead-in holes. Order transmission line in 1,000-foot lengths, to minimize waste and avoid splices.

Locating the Television Set. When you arrive at a home to install a television antenna, the set will usually be unpacked and already in the customer's preferred position. Make tactful suggestions for a change of location only if the choice is obviously bad. Bright glare from a window behind or alongside the set would be one reason for suggesting a change. A location at which sunlight hits the screen at some time during the day should likewise be avoided. Do not allow the set to be placed in front of a radiator, because this would overheat the set. Never place the set near a heating-system thermostat, because normal heat from the set would make the thermostat click off prematurely. In a fringe area, try to avoid locations that require long runs of transmission line.

Point out that the antenna installation must be planned for a specific location. Moving the set at a later date usually means rerouting the transmission line or perhaps even moving the antenna.

Plug in the set, turn it on, and try it out. Without an antenna you should at least get a raster of white lines on each channel. Better yet, connect an indoor antenna temporarily for a more thorough preliminary check of the set. If the set does not work, fix it or get it fixed before starting on the antenna. Bad tubes and loose tubes are the commonest troubles in new sets and sets that have been moved to a new home.

There is no point in installing the antenna at a fringe location if the set needs to be taken back to the shop for repairs, as you would have to put the ladders up again later to orient the antenna.

Once the set is working, leave it on while installing the antenna. This may show up defects that can be fixed right away.

Locating the Antenna. With the receiver location decided on, the antenna location and the transmission-line route come next. For most jobs it is best to bring the line up behind the set from the basement. This is the easiest route from an installation standpoint and is preferred by the customer because no line is exposed in the living room. Check the basement layout, choose the point of entry for the line from the outside, and estimate how much line will be needed from that point to the set.

As the next step, study the ends of the house for antenna mounting possibilities. Wall brackets for masts are inexpensive and easy to install, with no danger of roof leaks. A chimney mount can be your second choice, but use a roof mount only as a last resort.

It is not always necessary to install the antenna on the highest peak of the roof, but it should be put as high as convenient. If the roof is metal or slate or if a neighboring roof is slate, however, it may be necessary to get the antenna well above the roof level. Extra height is of no particular value as long as a good picture is obtained, free of flashing white dots called snow.

Though power lines do not ordinarily cause interference, try to keep the antenna well away from power lines because of the possible shock hazard during installation or if the antenna falls during a storm.

Do not mount a television antenna mast on a soil pipe that projects above the roof, no matter how convenient this may be. Vibration of the mast in heavy wind would be transferred to the pipe and would eventually cause the roof to leak near the pipe.

Apartment-house Installations. The best place to locate an antenna at an apartment house is usually the parapet, since the transmission line must be run down the outside wall. There are certain fire regulations which must be followed on all installations where the roof may be used for fire fighting. Check with your local fire department to determine the local requirements.

In general, an apartment-house antenna must be located at least 8 feet above the roof to avoid possible injury to a person. No guy wires should be attached to the roof or other structure in the area that may be used by firemen, unless the wires are at least 8 feet above the roof at all points.

Transmission line should be supported by an insulator at least once every 10 feet or at each floor level. Where line is run on the roof surface, it should be protected against mechanical injury. All transmission line and ground wires carried above a flat roof should be at least 8 feet above the roof surface. No wires should cross a public street.

Observe Safety Precautions. Never underestimate the importance of observing routine safety precautions to protect your own life as well as others. Most accidents are the result of carelessness or overconfidence.

Climbing and roof work are dangerous. Make sure your ladder is on secure footing before going up. Avoid precarious positions, especially when working near the edge of a roof. See that tools and materials are placed where they will not endanger yourself or others.

Wear sneakers, rubbers, or other soft shoes that will give good traction when climbing steep roofs. Walk on the roof as little as possible. Any damage that you may accidentally do can be charged to you.

Be extremely cautious if the roof is wet or icy. Always keep one hand free for climbing the ladder. Always face the ladder when going up or down, as in Fig. 7. When mounting tall ladders, keep your eyes on the ladder rails and rungs instead of looking into the sky, to prevent losing balance. Lash metal ladders to the house with rope whenever possible, because even a light gust of wind can blow down aluminum or magnesium ladders.

Never ask for or accept the help of bystanders or members of the customer's family, because you may then be liable for their safety and for damage they do.

Be extra careful when handling metal ladders, masts, and antennas in the vicinity of power lines and other overhead wires. Keep as far away from wires as possible.

If the job calls for additional help, hire it. No job is worth a broken arm or leg. No customer appreciates having his living room turned into an emergency ward.

If on a metal ladder while drilling holes, drill slowly and no deeper than necessary. If applying too much pressure at the instant when your drill goes through the wood, the drill may also go through electrical BX cable that happens to be just at that point. This means sudden death if the drill contacts the hot wire inside the cable, because the metal ladder completes the path from the drill through your hands and heart to ground.

There is no danger if you watch what you are doing, because it is easy to feel the change in pressure as the drill breaks through the wood. A hand drill is better here, because an electric drill can bite through the BX before a careless user realizes that the hole is drilled. Several fatal elec-

Fig. 7. Correct method of carrying an assembled antenna up a ladder. Face the ladder and use one hand for gripping the ladder. (RCA Victor photo)

trical accidents have occurred when electric drills were used on metal ladders.

Protecting Your Customer's Property. Always remember that you are liable for any damage to a customer's house, garden, or lawn. Always tell the customer beforehand how you plan to make the installation, and ask permission to drill the required holes. Be particularly careful if you have to rest your ladder against brittle asbestos-shingle siding. Use pads on your ladders to distribute the pressure over a wider area where they touch the house. Wear rubber-soled shoes or sneakers when you have to walk on the roof. Try to avoid setting up a ladder in a flower bed. Do not drag wires carelessly over flowers.

It is best to notify the owners before you go up on any roof. If the building is rented or is an apartment, see the landlord or building superintendent. Get permission in writing if possible. This may save you embarrassment and money later.

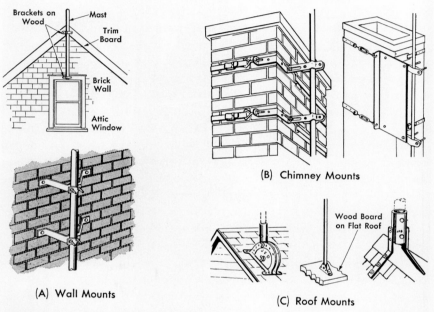

Fig. 8. Examples of the three commonest types of antenna mounts. Many other types of brackets and antenna hardware are also available from jobbers

Wipe your feet and hands each time you come down from a roof, particularly after installing an antenna on a sooty chimney. When the job is done, clean up all packing cartons, scraps of wire, and other refuse.

Installing Wall Mounts. The wall mount shown in Fig. 8A is the most popular mount for home use. It holds the bottom of the mast far enough away from a vertical wall so the mast clears the projecting peak or the eaves of the home and projects up above the roof. This mount works equally well on other vertical surfaces, such as on the inside surface of the parapet that runs around the edges of apartment-building roofs.

Large rustproof wood screws or square-headed lag screws are used to fasten wall-mount brackets to wood. On walls having wood or asbestos shingles, drill holes first to avoid cracking the shingles. Try to find solid mounting spots well away from joints in shingles.

It is a good idea to seal the screw leads and the edges of the brackets where they touch the wall, to prevent leaks and minimize corrosion of the brackets and screws. Use white caulking compound or black roofing cement, choosing whichever is less conspicuous against the color of the wall.

On brick or masonry homes, it is often possible to mount the brackets on the wood roof trim and on window frames, to avoid having to put screws into the bricks. On some jobs, however, it will be necessary to mount one or both brackets on a brick wall. A hole must be drilled in the masonry for each mounting screw and a special soft lead anchor sleeve inserted. Various types of anchors are available at jobbers and hardware stores for this purpose, and are easily installed once the holes are made.

The simplest yet hardest way of making a hole in brick is with a star drill and hammer. The star drill is turned slowly with one hand while pounding on the drill with a hammer. For brackets, always drill in the middle of a brick, not in the mortar between bricks. Screws anchored in mortar may work loose because water seeps in around the drilled hole. Holes can be made safely in mortar only for standoff insulators, as there is no strain on them.

A slow-speed electric drill and a carbide-tipped masonry drill greatly speed up drilling holes in brick walls. A 600-rpm drill is best. At least 100 feet of heavy extension cord will be needed with the drill, to reach an a-c wall outlet.

Installing Chimney Mounts. Chimney mounts are popular with many servicemen because they require no drilling. Installation involves placing one strap near the top of the chimney and the other 12 to 18 inches below. Adjust each strap carefully so that it is level all around the chimney, as illustrated in Fig. 8B. The straps should be over the bricks, not over the mortar between bricks. Straps generally are longer than necessary, so surplus length should be cut off after the straps are secured in their buckles. Tightening the bolts or turnbuckles on the straps makes the mount ready for the antenna mast.

Installing Roof Mounts. Examples of roof mounts are shown in Fig. 8C. If it is necessary to mount the antenna on the roof, try to locate it on the overhang so as to avoid damage from leaks. Whenever the antenna is mounted on the roof, a layer of roofing compound should be applied

542 *Television and Radio Repairing*

to the roof under the mounting bracket and the mounting bolts should be covered with tar after they are tightened. Most roof-mount installations will require guy wires. These should be run in three equally spaced directions.

Assembling the Antenna. Television antennas come disassembled, but assembly is easy. Put the antenna together on the ground and clamp it

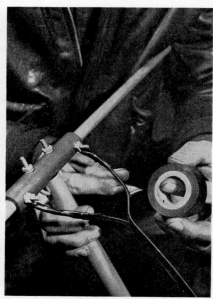

Fig. 9. Taping antenna connections with No. 33 Scotch electrical tape to give added strength against breakage by the wind. Note use of leather tool holder on belt for carrying tools safely while working on roofs. (Minnesota Mining & Mfg. Co. photos)

tightly to the top of the mast. Where spacings are critical in the more complicated arrays, instructions are included and should be followed carefully. Next, connect the transmission line to the antenna.

To prevent the transmission-line wires from bending back and forth in the wind and eventually breaking at the terminals, tape the connections with No. 33 Scotch electrical tape as in Fig. 9. Even better in cold climates is No. 88 Scotch electrical tape, because this tape remains flexible even at temperatures down to 0°F.

For extra protection, you can spray the connections and assembly screws with one of the insulating sprays made for television work. This is especially desirable near the ocean or where corrosive gases are present, as on chimney installations.

Attach several line standoff insulators to the mast and insert the line in them on the ground, since this will be difficult to do once the antenna is up.

Installing the Antenna. Carry the assembled antenna up to the mounting site. Be careful not to trip over the attached transmission line. Insert the mast in the brackets and tighten only partially, to hold up the mast yet permit turning later during orientation.

If in a strong-signal area, try aiming the antenna the same as others in the neighborhood, and tightening the mounting bolts permanently. If the set then works satisfactorily, you will not have to make another trip up to the roof.

Installing the Transmission Line. The transmission line should take the most direct route consistent with good appearance, because extra line costs money and increases line loss. For vhf antennas, use standoff insulators that give a clearance of 2 to 4 inches. For uhf antennas the clearance should be 4 to 6 inches.

Transmission line should not be run diagonally down the side of a house because of appearance. Whenever possible, the line should be run straight down so that rain, sleet, and snow will have less tendency to cling to it. Avoid sharp right-angle bends. If a horizontal run is necessary, it should be made under an eave or other protection. Suggestions for locating transmission lines are given in Fig. 10.

Try to avoid running a line down the front of a building, since it will detract from the appearance of the building and the installation. In most homes the line can be run down one side to the basement. Once in the basement, the line can easily be run up to a receiver in any downstairs room.

The line can enter the house through a basement or first-floor window sill or go right through the house wall into the basement as in Fig. 11. Where the line enters the building, leave a loop to allow water to drip off.

The entrance hole for transmission line can be sealed with caulking compound, putty, or ordinary household paraffin wax to keep out insects. If no cement is available, a small cork can be cut in half lengthwise and the halves pushed in the hole on each side of the twin-line.

Twin-line can be brought in over a wood window sill where necessary, as in apartments or rented rooms, provided it will not be crushed and

544 *Television and Radio Repairing*

broken as the window is opened and closed. Generally, it is better to pass the transmission line through a hole in the window frame.

In drilling entrance holes, use a ½-inch bit for twin-line and a 5⁄16-inch bit for RG-59/U coaxial cable. The hole should be made through the

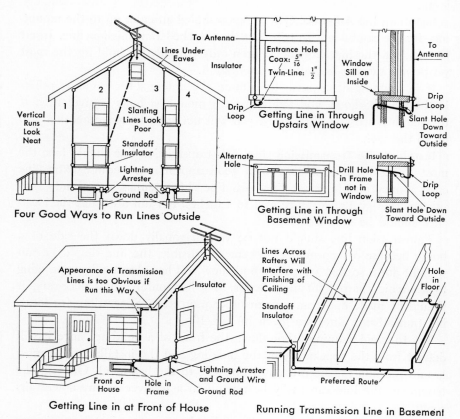

Fig. 10. Preferred routes for transmission line outside a house and in the basement

window sill or sash in as inconspicuous a spot as possible. Slant the hole slightly downward toward the outside, so that any moisture will run out of the house. Where a hole is drilled through the floor, it should be as close to the wall as possible, as in Fig. 12.

Do not hang the transmission line on water pipes or other metal fixtures. Use standoff insulators at least every 4 feet inside the house. Keep the line at least 4 inches away from other wiring. Run the line on the sides of floor joists wherever possible, to allow the basement ceiling to be

Television and Radio Antennas 545

finished later if desired. Make cross runs at right angles to the joists along the wall of the building. On finished ceilings, use standoff insulators.

At the set, leave at least 3 feet of extra line to permit moving the set away from the wall for cleaning or for servicing. Arrange the extra line carefully in U-shaped bends back of the set. Bunching or coiling of twin-line here can cut down signal strength, particularly on uhf channels.

Twin-line should be twisted about once in every 2 feet of line outside the house, to minimize flapping in the wind. Twisting also reduces undesired pickup of signals by the line itself.

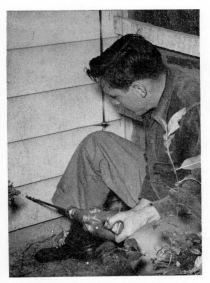

Fig. 11. Drilling a hole through the wall of a house into the basement for bringing in the television transmission line. (RCA Victor photo)

Fig. 12. Drilling a hole in the floor close to the wall behind the selected receiver location, for bringing up the transmission line from the basement. (RCA Victor photo)

Where twin-line goes over the edge of a building, use long standoff insulators to keep the line from rubbing. If there is no way to mount an insulator, protect the line by wrapping with electrical Scotch tape, so the insulation on the line will not wear through at the point where it touches the building edge.

When using tubular twin-line, as for uhf installations, cut or drill a drain hole in the bottom of the drain loop. This prevents condensation from accumulating inside the line.

The antenna end of tubular line should be sealed to keep out rain. This is done by heating the insulation with a cigarette lighter or matches until it is soft, then squeezing it together with pliers to form the seal shown in Fig. 13. As an extra precaution, run the line upward from the antenna for a few inches to form a rain loop, so rain runs off the line rather than into any small opening left in the seal.

Transmission line should not be painted. Most paints contain lead or other metallic pigments that can reduce signal strength. In addition, the oils in paints can damage the insulation.

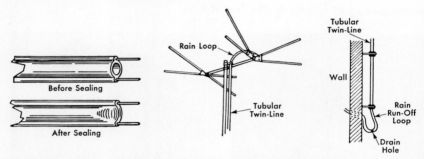

Fig. 13. Installation procedures for tubular twin-line, used chiefly for uhf antenna installations

Supports. Outside the house, twin-line should be supported about every 8 feet by standoff insulators. Some examples of the available types are shown in Fig. 14. On horizontal runs the standoffs should be about 4 feet apart. After the line is inserted in the slot of the standoff insulator, turn the slotted plastic disk so the line cannot slip out. The metal loop around the insulator is then squeezed inward with combination pliers, to make the plastic insulator grip the line and prevent it from sliding.

Be sure to leave a little slack in the twin-line between the antenna connections and the first standoff insulator, so there is no strain on the connections. Squeeze the first standoff enough so the line cannot slip down later.

Drive-in standoffs work well only on certain types of masonry mortar. Some mortar is so hard that several standoffs are bent for each one that goes in properly. Other mortar is so soft that it is impossible to get a rigid support.

On masonry a more reliable job can be obtained by drilling a small hole in the mortar, inserting a standard fiber sleeve such as a Rawl plug, then using a standoff having an ordinary wood-screw thread. The hole should be about $\frac{1}{8}$ inch in diameter and 1 inch deep. It can be made with a

Rawl drill that is hammered in much like a star drill, or with a ⅛-inch carbide-tipped bit in an electric drill.

Short runs of line inside the living room can be made by stapling the line to the molding. Staples or nails should be carefully placed so they do not short or damage the two conductors.

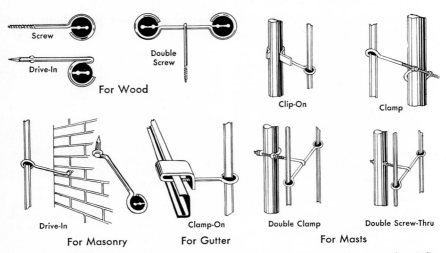

Fig. 14. Types of standoff insulators that are available in various lengths for use with twin-line and coax. Double types are used when high-band and low-band antennas have separate lead-ins or when a power line for an antenna rotator must be brought down along with transmission line. For vhf, use insulators that give 2 to 4 inches clearance. For uhf, the clearance should be 4 to 6 inches

Coaxial cable may be allowed to touch a building, since the outer conductor is a shield which is connected to ground. It may even be taped directly to the mast, saving the cost of insulators. Standoff insulators should be used only to provide proper support and a neat appearance. Twin-line standoffs can be used.

After holes are drilled for twin-line or coax, pull the line through to the set from the outside of the house, then estimate the amount required to reach the antenna. This minimizes waste of line because only about half the total length has to be estimated. Do not fasten any line inside the house until the antenna is up, however.

Splicing of twin-line or coax should be avoided because every splice is a possible point of trouble. When necessary, however, use the procedures shown in Fig. 15.

548 *Television and Radio Repairing*

Protecting against Lightning. An outdoor television antenna system must be grounded for protection against lightning. A properly installed lightning arrester protects both the home and the television receiver. Many installations are made without this protection, chiefly to save a few dollars, but more and more communities are passing local ordinances requiring that all outdoor antennas have lightning protection.

Where the antenna and mast are insulated from each other, both must be protected from lightning. This is done by connecting the mast directly

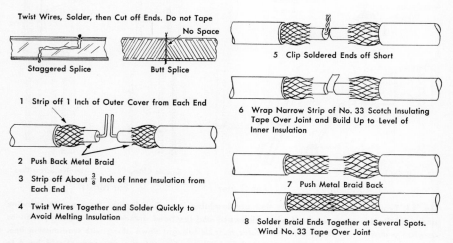

Fig. 15. Two methods of splicing twin-line, and eight steps in splicing coaxial cable

to ground and using a lightning arrester on the twin-line. Examples of lightning arresters for twin-line are shown in Fig. 16.

It is permissible to ground the antenna mast to a soil pipe only if you are certain it gives a continuous path to earth. Use a metal grounding strap around the pipe to ensure positive contact.

Use No. 6 or No. 4 bare wire for grounding a mast. Loop the wire under the head of one of the mast bracket bolts to get a good connection. Run the wire in as straight a line as practicable to the ground, on the outside of the building. Bare solid aluminum wire is widely used for grounding purposes because it is easy to handle. It can be nailed or stapled directly to the house, without insulators.

A ground rod is usually required, as outside water pipes are rarely in the right location. The rod will have a screw or clamp for making the ground wire connection. The ground rod should be at least a foot away from the house foundation, because the soil next to the foundation is often too dry to provide a good ground.

The lightning arrester for a television antenna should be as close as possible to the point at which the twin-line enters the house. The lightning arrester will have provisions for making connections to both conductors of the twin-line, plus a ground terminal. The ground wire can be the same as that used for the mast, or can be a smaller bare wire such as No. 10.

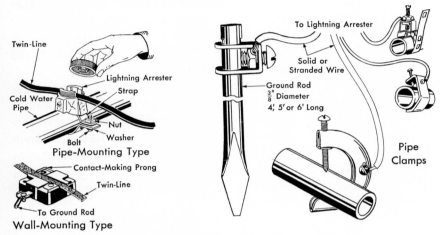

Fig. 16. Examples of lightning arresters, ground rods, and pipe clamps

The ground for the lightning arrester can be the same as for the mast, if convenient.

If there is a water pipe close to the point at which the twin-line enters the house, a lightning arrester can be clamped on the pipe to give automatic grounding. This type of arrester is available at jobbers and is the easiest of all to install.

When an antenna is used with coaxial cable, connect the cable shield to the mast at its upper end, and ground the cable shield somewhere near its lower end. This will ground the entire antenna system, and a lightning arrester will not be needed.

Orienting the Antenna. In cities having good television service, aiming the antenna in the general direction of the stations is generally adequate for orientation. Antennas on neighboring houses can be used as guides. A street map and compass can also be used for rough orientation.

Only in fringe areas is accurate orientation needed. Here the antenna must be adjusted for the best possible picture. This can be done by two men without special equipment if set and antenna are within shouting distance of each other.

Portable sound-powered telephone systems are made especially for servicemen, to give a more refined means of communication. Each man wears a telephone-operator type of headphone and mouthpiece. Up to 200 feet of two-conductor wire can be used between the antenna and the set. No batteries are needed.

When two men are orienting an antenna, they should talk to each other continuously. This allows the man on the roof to know from moment to moment whether reception is getting better or worse. Talking also lets the man at the receiver know the approximate direction in which the antenna is pointed, so he can identify the best position.

Temporary use of an antenna rotator makes it possible for one man to adjust an antenna accurately. The rotator is installed at the bottom of the mast, and its cable is run through an open window to an a-c wall outlet and to the control box in front of the set. By rotating the control knob back and forth while watching the picture, the best antenna orientation can quickly be found. You then go up on the roof and mark the antenna mast position precisely, raise the mast a few inches to remove the rotator, and install the mast permanently.

When two or more stations are to be received from different directions with a single fixed antenna, orientation must be a compromise. This means that reception must be checked on each station after each change in antenna position. If such a compromise is not satisfactory, you can recommend a permanently installed rotator or a separate antenna for each direction of reception.

Customer Instructions. While showing a customer how to adjust his newly installed television receiver, tune to all the stations one by one and point out any ghosts or interference patterns that cannot be improved. Explain how ghosts are produced by reflected signals arriving a bit later than direct signals. Often the reflection occurs off buildings near the transmitter, and nothing can be done about it. If the antenna has been oriented to favor one station at the expense of another, this, too, should be pointed out.

Repairing Television Antennas. When antenna trouble is suspected, make sure the transmission line is not loose or shorted across the terminals at the back of the set. Check also the soldered connections for the short piece of transmission line running from these terminals to the tuning unit inside the set. Jiggle the transmission line to check for an intermittent connection.

If transmission-line connections at the set are good, disconnect the line and try an indoor antenna. This can be a plain or folded dipole made from twin-line for test purposes, or one of the commercial V-shaped indoor antennas. If better reception is obtained with the temporary indoor antenna, the regular antenna system is definitely at fault. Another way of checking a suspected antenna system is to connect it to a spare or test receiver that is known to be good.

If the antenna system is found to be defective, check the lightning arrester by disconnecting it from the circuit. If this clears the reception, clean the arrester with a damp rag and recheck, or better yet, replace it.

Connections at the antenna itself come next. This usually means ladder work. Check line connections to the antenna for breaks, shorts, or corroded joints. Check the orientation of the antenna by comparison with other antennas in the neighborhood, to see if it is way out of line. Wipe off soot and dirt from all insulating blocks on the antenna.

If the trouble still exists, it is probably in the transmission line. The quickest check for this is to connect in a new length of transmission line temporarily. It is better to replace a bad line than to repair it. There is very little difference between final costs of repair and replacement, and a new line will give much longer service.

Importance of Radio Antennas. A defect in the antenna system of a radio set can make just as much trouble as a break in the wiring of the set itself. The customer will complain that the set does not work right. Your job will be to find and fix the trouble. If the defect turns out to be somewhere in the antenna system, you charge just as much for making the repair as for locating and fixing a defect in the set itself.

Built-in Radio Antennas. Most of the radios made today have built-in ferrite-rod antennas or loop antennas. These sets are ready to play just as soon as the line cord is plugged into a wall outlet. On a good set, these built-in antennas bring in just as many stations as did a high outdoor antenna on a receiver 20 years ago. A better antenna is not required unless the listener has special preferences or is located in the country beyond the service range of the stations he wants to hear.

Loop Antennas. In table-model radio sets a flat pancake-type loop like that in Fig. 17 was widely used a few years ago. This loop is usually on the back cover of the set, and must therefore be removed to get at the tubes and the chassis. The loop may be made from insulated wire, spaced bare wire, or metal foil stamped into the back cover.

A loop antenna replaces one coil of the first r-f transformer in the radio set, or sometimes both coils. A loop may thus have two, three, or more leads going to the chassis. These leads take a lot of punishment during normal servicing of a set, and often break. You will be able to repair many a set by resoldering a broken loop-antenna lead, so always inspect these leads carefully after you remove the back cover.

Fig. 17. Loop antenna glued to plastic back cover of three-way portable radio. If the glue weakens, the turns of wire in the loop get loose and may break. (Howard W. Sams photo)

On some older console radios the loop antenna is supported by pivots so it can be rotated to get maximum pickup from a particular desired station.

Loop Antennas Favor Two Directions. A loop favors stations located in line with the turns at the bottom of the loop. You can demonstrate this yourself by tuning in a weak station and rotating the loop or the entire table-model set to see how it affects volume.

A loop receives equally well from two directions. You cannot tell whether it is pointing directly toward or directly away from the station being received, but for radio reception no one cares anyway.

Replacing Loops. Since a loop antenna is actually a part of the receiver circuit, a damaged loop should be replaced with an identical unit to ensure best results. This means ordering from the manufacturer of the set. Fortunately, loops do not often need replacement, even though they often need repair.

Several types of replacement loops are usually available from jobbers.

Adjustable ferrite-rod antennas for use in place of loops are also available. Mount these as high as possible inside the cabinet.

In small table-model sets the loop is necessarily quite close to the metal chassis. The large amount of metal in the chassis often neutralizes most of the directional effect on the loop, so do not expect smaller sets to show much change in volume as you rotate the set. The closeness of the loop to the chassis is taken into account when designing a set, so do not change the method of mounting the loop.

Loop Connections. Loop antennas in console sets and in many table-model sets have leads that go to screw terminals on the chassis, for convenience in disconnecting the loop and removing the chassis. Make a habit of examining these screw connections whenever working on a set. Tighten the screws if necessary. Where plugs and jacks are used in place of screws, examine the leads carefully at this point for breakage. Be sure each plug is tight in its jack. When disconnecting the antenna, pry out each plug with a small screwdriver instead of pulling on the wires, to avoid breakage.

Some loop antennas are cemented to the inside of plastic radio cabinets. These loops frequently get loose, as also do pancake loops on back covers. Use service cement for remounting loose loops.

Loops for Portables. On older battery portable receivers, loop antennas are often concealed under the fabric covering on the hinged cover of the set. The covers are designed to stay upright while the set is in use, to get the loop away from the metal chassis and obtain maximum signal pickup. The directional characteristics of a loop will be still more noticeable now.

Some of the older camera-size portables have the loop antenna wires concealed in the shoulder strap used for carrying the set. Others use the earphone wires as an antenna.

Effect of Metal Objects on Loops. Radio waves are blocked by large areas of metal. For this reason, sets with built-in antennas will not generally work well in steel-framed skyscrapers, in rooms using metal lath as a backing for plaster, in metal-covered house trailers, or in railroad cars. Radio waves will pass through glass, however. A set can therefore be placed directly in front of a window in a metal building to obtain satisfactory reception of at least a few stations. Since windows face in various directions, some windows will give better results than others, depending on the location of the station that is tuned in.

Ferrite-rod Antennas. Today practically all a-m radios use a built-in antenna consisting simply of wire wound around a pencil-size core of powdered-iron magnetic material. One name for this material is ferrite, so this antenna is often called a ferrite-rod antenna. It is also known as a loopstick antenna. The iron core serves to multiply the sensitivity of the loop many times, making it just as effective as the larger pancake-type loops.

Ferrite-rod antennas have about the same directional characteristics as loops. Servicing problems are likewise the same. An example of a ferrite-rod antenna installation is shown in Fig. 18.

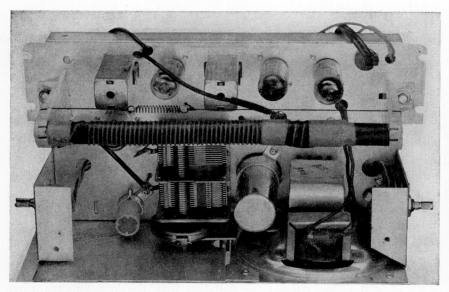

Fig. 18. Ferrite-rod antenna in three-way portable radio. (Howard W. Sams photo)

Terminal-shorting Bars. Radio sets having more than two antenna terminals will sometimes have a metal shorting link or bar that can be used to connect together two adjacent terminals when using one type of antenna. Loosening the terminal screws permits swinging the bar up out of the way when individual terminals are required for another type of antenna system. Instructions for using the bar are usually printed on the chassis or back panel of the set.

Built-in Hank Antenna Wire. Many old table-model radio sets had a permanently attached hank of insulated wire. When this wire was unwound and run under the rug or around a window frame, good reception was obtained in most locations. The higher the wire, the better the results.

A hank antenna wire should not be connected to any metal object. Keep it away from radiator pipes or other grounds, because some troubles that develop in universal sets can make the antenna electrically hot. A ground connection would then burn something out in the set or even blow the house fuse.

When the insulation on a hank antenna wire gets worn or damaged, replace the entire wire with a 25-foot length of flexible plastic-insulated

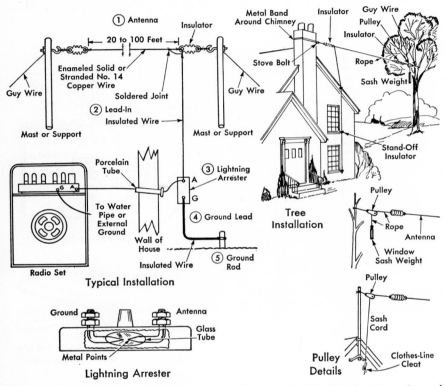

Fig. 19. Practical ideas for outdoor antenna installations used with old radio sets or in rural locations. The antenna wire should be soldered to the lead-in wire beforehand. Either a porcelain tube or a window strip may be used to bring the lead-in through the house wall

wire. Do not splice the new wire to the old one at the back of the set. Instead, remove the chassis from the cabinet, unsolder the old wire, and solder the new wire in its place to give a professional repair job.

Simple Outdoor Antenna. An outdoor antenna is today used only with very old a-m radio sets or in locations that are hundreds of miles from radio stations.

The simplest and most-used outdoor antenna for homes is the inverted-L antenna shown in Fig. 19. It has a long horizontal antenna wire mounted

556   Television and Radio Repairing

high in the air, sometimes called the flat top. The insulated lead-in wire runs more or less vertically downward from one end of the flat top to the antenna terminal of the radio set. For broadcast-band reception the direction of the antenna is unimportant.

A lightning arrester should be used with an outdoor radio antenna. It is mounted on the outside of the house, at the point where the lead-in

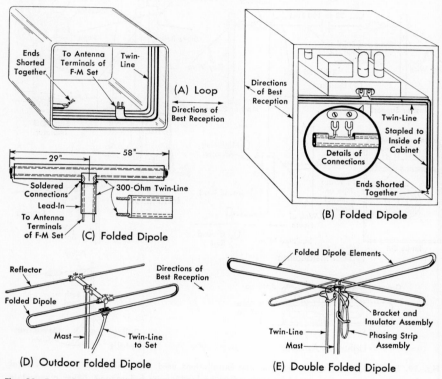

Fig. 20. Examples of antennas used with f-m radio receivers

wire enters through a wall or window. One terminal of the lightning arrester is connected to the lead-in, and the other is grounded to a nearby water pipe or to a rod driven into the earth below. When lightning strikes the antenna, it jumps the gap between the metal points in the lightning arrester and travels harmlessly to the ground.

F-M Antenna Problems. Most frequency-modulation radio receivers are provided with a built-in antenna and provisions for an external antenna. The loop is smaller or has fewer turns than a broadcast-band loop, because the 88–108-mc f-m band is so much higher in frequency than the broadcast band. There may even be a long, narrow loop made by shorting one end of a length of 300-ohm television twin-line and connecting the other end to the two antenna terminals of the set, as in Fig. 20A.

Another type of built-in f-m antenna made from twin-line is the folded

Fig. 21. Taping f-m antenna connections with No. 88 Scotch electrical tape. This tape remains pliable at temperatures down to 0°F, whereas No. 33 tape becomes brittle at freezing temperatures. Either tape may be used during warm summer months. (Minnesota Mining & Mfg. Co. photo)

dipole, shown in Fig. 20B. This is strung out around the cabinet and tacked or stapled to the cabinet.

Built-in f-m antennas rarely give the best possible reception, even for local f-m stations. Both height and length are needed, along with correct orientation as to direction.

A folded dipole strung between rafters in the attic usually provides good pickup of f-m signals. Such a dipole, easily assembled from twin-line, is shown in Fig. 20C. This antenna also works well when spread across the top of two adjoining windows, with the transmission line or lead-in coming down between the windows.

Running twin-line down to the set from the attic involves the same techniques as for television antennas. The solution is different for every house. Often the only way is to bring the line out through an attic window so it can run down outside the house.

If the received f-m signal is weak, orientation of the f-m antenna will help to give maximum pickup of the signal. Tune the set to the desired station and rotate the antenna until background noise heard with the program is a minimum.

Roof-top F-M Antennas. Outdoor antennas for f-m radio receivers are very similar to those used for television, since the f-m band is between television channels 6 and 7. That in Fig. 20D is widely used when all the stations to be received are in approximately the same direction from the receiver. That in Fig. 20E receives equally well in all directions. Installation and repair procedures for outdoor f-m antennas are the same as for television antennas.

Twin-line connections to an outdoor f-m antenna should be covered with electrical insulating tape as in Fig. 21, for protection against vibration and breakage in strong winds, just as for outdoor television antennas.

Index

A battery, 326
AB battery pack, 325
A-c outlets, bench, 36
A-c voltages, 118
 measuring, 132
Accuracy, voltage readings, 131
Adapter, picture-tube, 181
A-f signals, 45
Agc control, 82
Agc switch, 82
Air-core coils, 440
Air gap, speaker, 504
Air trimmers, 480
Alternating current, nature of, 117
Alternating voltage, 118
Aluminum-backed screens, 175
A-m broadcasting system, 47
A-m radio broadcast band, 46
A-m superheterodyne receiver, 49
Ammeter, 111
Amperes, 110
Antenna rotators, 533
Antenna-system trouble, 89
Antennas, built-in television, 521
 color television, 531
 dipole, 524–530
 ferrite-rod, 553
 f-m, 556, 558
 gain, 528
 hank, 554
 indoor television, 522
 loop, 86, 87, 551–553
 outdoor radio, 555
 radio, 551–558
 television, 521–551
 vhf television, 528–530
 workbench, 41
 Yagi, 527
Aquadag coating, 172
Arcing, 190

Attic antennas, television, 523
Audio amplifier, 48, 50
Audio frequencies, 45
Audio voltage measurement, 134
Auto ignition interference, 91
Auto-radio fuses, 322
Auto-radio spark plates, 323
Auto-radio transistors, 293
Auto-radio tubes, 324
Auto radios, 53, 322–325
 hybrid, 294, 324
 signal-seeking, 485
Automatic tuning, 482
Automobile for servicing, 21

B battery, 326
Ballast resistors, 371
 testing, 236
Bases, tube, 144
Batteries, checking, 330
 mercury, 297
 nickel-cadmium, 298
 rechargeable, 298, 330
 shelf life, 332
 silver-cadmium, 298
 transistor radio, 297, 328
Battery charger, 299
Battery pack, **325**
Battery rejuvenator, 332
Battery troubles, 327
Bayonet sockets, pilot lamps, **167**
Bead chains, 479
Bench design, 34
Bench lighting, 39
Breast drill, 32
Brightener, picture-tube, 28
Brightness control, 82
Brilliance control, 82
Broadcast band, 46
Broadcasting systems, **46**

560 *Television and Radio Repairing*

Built-in antennas, 521
Business card, 19
Buying tubes, 228

Cabinet repairing, 511–520
  materials, 513
Cable-type remote controls, 493
Caddy, tube, 27, 225
Candohms, 365
Capacitance, 401
Capacitors, 68
  action, 396
  ceramic, 69, 398, 429–434
  discharging, 436
  electrolytic, 69, 398, 413–429
  mica, 434–436
  paper, 68, 397, 404–413
  parallel, 404, 436
  ratings, 401
  symbols, 403
Capacity (*see* Capacitance)
Carbon resistors, 66, 333–357
  appearance, 333
  color code, 344
  failure, 339
  high-voltage, 350
  installing, 349
  ordering, 343
  power rating, 336
  printed, 352
  removing, 348
  stock, 351
  testing, 342
  tolerance, 346
Carbon tetrachloride, 41
Card, business, 19
Cardboard electrolytics, 415
Carrier signal, 48
Carrier-type remote controls, 493
Cartridge fuses, 311
Catalog, file, 17
Cathode, 147
  indirectly heated, 147
Cathode-ray tuning indicators, 150, 495
Cements for plastic cabinets, 519
Center punch, 33
Centering magnets, 177
Ceramic capacitors, 69, 398, 429–434
  color code, 432
  disk, 431
  flat-plate, 431
  high-voltage, 432
  replacing, 434
  testing, 432
  tubular, 429
Ceramic trimmers, 480

Channel-selector switch, 82
Charge accounts, 8
Charges, cost-multiplying table, 12
  electrical, 107
  parts, 11
Chart, filament-continuity checking, 213
  filament pin numbers, 211
  mica capacitor color code, 435
  pilot lamps and sockets, 167
  pin numbers of sockets, 153
  resistor color code, 344
  soldered joints for wires, 258
  soldering-lug joints, 256
  television interference, 91
  television picture defects, 81
  television tube troubleshooting, 219
  transistor radio troubles, 300
  transistor symbols, 285
  tube symbols, 150
  wirewound resistor color code, 361
Chassis, cleaning, 102
  removal, 98
  removing bolts, 97
Cheater cord, 28
Checking account, 11
Chimney mounts, 541
Chisel, 34
Choosing, multimeter, 123
  soldering gun, 248
  soldering iron, 264
  tools, 23
  tube testers, 233
Circuit, nature of, 112
  reading, 355
Circuit breakers, 80, 311
Circuit diagrams, getting, 13
Clamp, tube, 157
Cleaning, chassis, 102
  picture-tube face, 103
  safety window, 103
  television glass surfaces, 103
Clock motors, replacing, 395
Clock-radio timers, 393
Clock radios, 52
Clock shafts, replacing, 395
Clocks, transistor radio, 392
Coaxial cable, 526
  splicing, 548
Coils, 71, 437–448
  air-core, 440
  construction, 439
  failure, 443
  impedance, 439
  inductance, 440
  iron-core, 441
  ordering, 450

Index

Coils, parallel, 441
  powdered-iron-core, 441
  repairing, 444
  replacing, 450
  series, 441
  shorted turns, 445
  television tuner, 455
  testing with ohmmeter, 443
  video peaking, 446
Cold chisel, 34
Collecting money, 104
Color code, ceramic capacitors, 432
  crystal diodes, 305
  flexible resistors, 361
  i-f transformers, 454
  line-cord resistors, 375
  mica capacitors, 435
  paper capacitors, 411
  resistors, 344
  tubular paper capacitors, 412
  wirewound resistors, 360
Color television, antennas, 531
  reception, 62
  servicing, 63
Combination pliers, 24
Component packs, 354
Concentric potentiometers, 387
Condensers (see Capacitors)
Conductor, 111
Cone, speaker, 500
Connector, high-voltage, 172
Console-cabinet problems, 99
Consoles, 63
Continuity check, filament, 212
Contrast control, 82
Control knobs, removing, 93
Controls, operation, 377–380
  replacing, 382–388
  television, 82–85
Converter, 50, 51
Cord, cheater, 28
Corona, 190
Credit for old picture tubes, 201
Crystal diode testers, 306
Crystal diodes, 65
  color code, 305
  replacing, 306
  testing, 306
  use, 305
Current, nature of, 110
Current flow, 113
Cycles, 44

D-c power lines, 321
D-c voltages, measuring, 127

Dealer service, 18
Decimal multipliers, 360
Deflection-yoke thermistors, 460
Deflection yokes, 175, 458
Diagonal cutters, 24
Diagrams, socket layout, 156
  tube base, 153
Dial-cord tuning systems, 473
Dial cords, replacing, 475
  tightening, 474
Dial pointer, removing, 95
Diodes, crystal, 65, 305–306
  tube, 146
Dipole antennas, 524–530
Direct current, nature of, 117
Directors, 526
Discharging capacitors, 436
Disk ceramic capacitors, 431
Distortion, f-m radios, 496
  pincushion, 176
Distributors, receiver, 10
Dressing a joint, 257
Drilling out rivets, 33
Drills, 31
Dry electrolytics, 413
Dual controls, 377
Dual potentiometers, 380
Dust cap, speaker, 502

Effective value, 118
EIA color code, i-f transformers, 454
  line-cord resistors, 375
  mica capacitors, 435
  resistors, 344
EIA tube-numbering system, 151
Electric drill, 32
Electrical pushbutton tuning, 484
Electricity, nature of, 108
Electrodynamic speaker, 499
Electrolytic capacitors, 69, 398, 413–429
  can-type, 427
  checking by substitution, 420
  leakage, 417
  ordering, 422
  polarity, 415
  reactivating, 419
  removing, 425
  shelf life, 418
  stock, 428
  surge voltage rating, 423
  tubular, 427
  use, 413
Electromagnets, 120
Electron flow, direction of, 109
Electron-ray tube, 150, 495
Electronic multimeters, 126

Electrons, nature of, 106
Electrostatic focus, 179
Electrostatic speakers, 509
Enameled wire, cleaning, 253
Envelope, tube, 143

Ferrite-rod antennas, 553
Field coils, 499
    troubles, 509
Filament-continuity checking, with a-c voltmeter, 213
    with neon-type tester, 214
    with ohmmeter, 212
Filament warmup, 208
Filaments, 147
    intermittent, 215
    series-string, 209
Files, 30
Filter chokes, 468
Fine tuning, preset television, 481
Fine-tuning control, 82
Fixing tough receivers, 18
Flashlight, 26
Flexible resistors, 361, 370
Flux, soldering, 250
F-m antennas, 556–558
F-m broadcast band, 47
F-m broadcasting system, 55
F-m radio troubleshooting, 496
F-m receiver operation, 56
F-m tuning systems, 494
Focus, picture-tube, 179
Focus control, 83
Folded dipole, 525
Folders, Photofact, 15
Foreign tubes, 165
French polishing, 516
Frequencies, audio, 44
    signal, 45
Fringe antennas, 528
Frozen yoke, 195
Fuse resistors, 363
Fuse troubles, 309
Fuses, 73, 80
    auto radio, 322
    checking in television set, 207
    receiver, 311
    slow-acting, 315
Fusible resistors, 313
Fusible surge limiter, 315

Gang tuning capacitors, 70, 470
Gas pliers, 25
Gas test, tube, 240
Gassy tubes, 163
Germanium crystal diodes, 305

Germanium rectifiers, 320
Getter, 143
Ghost images, 91
Gimmicks, 450
Grille cloth, speaker, 520
Ground rods, 549
Guarantees, tube, 229

Hacksaw, 30
Halo light, 203
Hammer, 33
Hank antennas, 554
Hearing limits, 45
Heat sink, 391
Heater, 147
Height control, 83
Hex nut drivers, 25
High-fidelity home audio systems, 63
High vhf band, 46
High-voltage ceramic capacitors, 432
High-voltage connector, 172
High-voltage power supply, 221
High-voltage precautions, 223
High-voltage resistors, 350
Home audio systems, 63
Home radios, tube stock, 227
Home servicing calls, 75
Horizontal afc control, 83
Horizontal centering control, 83
Horizontal drive control, 83
Horizontal frequency control, 84
Horizontal hold control, 84
Horizontal linearity control, 84
Horizontal output transformers, 456
Horizontal stages, 221
Horizontal sweep circuits, 60
Horizontal sync control, 84
Hot-chassis sets, 38, 87
House-fuse troubles, 309
Hybrid auto radios, 294, 324

I-f amplifier, 50
I-f transformers, action, 543
    color code, 454
    identifying, 452
    ordering, 455
    television, 455
    troubles, 454
I-f value, 50
Impedance, 400, 439
Implosions, picture-tube, 192
Indirectly heated cathodes, 147
Indoor antennas, television, 522
Inductance, 440
Infinity, 136
Insulating wire joints, 259

Index **563**

Insulation, removing from wire, 252
Insulators, 111
Intercarrier sound system, 60
Interchangeable tubes, 164
Intermittent filaments, 215
Inventory guide, tube, 224
Ion-spot damage, 173
Ion trap, adjusting, 174
Ion-trap magnets, 174
Iron-core coils, 441
Isolation transformer, 38

Jobbers, parts, 7
Joints, insulating, 259
  for soldering lugs, 255
  for wires, 258

K, 137, 335
Kilocycles, 46
Kilovolts, 110
Kilowatthours, 115
Kilowatts, 114
Knife, 25
Knobs, removing, 93
  tuning, 472

Ladders, television antenna, 22, 535
Lamps, pilot, 166
Lap joint, 257
Leakage current, electrolytics, 417
Leakage resistance, capacitors, 408
Letter sample, service data, 14
Letters in tube type numbers, table, 151
Lighting, workbench, 39
Lightning arresters, 548
Line-cord plugs, polarized, 321
Line-cord resistors, 373
  color code, 375
Line cords, repairing, 310
Linearity coils, 448
Lines of force, 120
Local-fringe switch, 84
Local-oscillator interference table, 90
Loktal bases, 146
Long-nose pliers, 24
Loop antennas, 86, 551
Loudspeakers (*see* Speakers)
Low vhf band, 46
Low-voltage power supply, 221
Lubriplate, 27

M, 137, 336
Magazines, servicing, 18
Magic-eye tube, 150, 495
Magnetic fields, 120
Magnetic lines of force, 120

Magnetism, 118
Mail-order houses, 11
Manuals, Rider's, 15
  service data, 15
  tube, 17
Manufacturers, receiver, 10
Manufacturers' service data, 13
Masts, antenna, 532
Measurement, a-c voltages, 132
  audio voltages, 134
  d-c voltages, 127
  direct current, 139
  high d-c voltages, 131
  resistance, 134
Mechanical pushbutton tuning, 482
Megacycles, 46
Megohms, 112, 335
Mercury batteries, 297, 331
Metallic rectifiers, 72, 315
Meter scale divisions, 129
Meter scales, reading, 128
Meter sensitivity, 124
Mica capacitors, 69
  color code, 435
  silvered, 434
  testing, 434
Mica trimmers, 480
Microammeter, 111
Microamperes, 111
Microfarads, 401
Micromicrofarads, 401
Microphone, 48
Microvolts, 110
Milliammeter, 111
Milliamperes, 110
Millivolts, 109–110
Mirrors, bench, 37
Mixer-first detector, 51
Modulated carrier signal, 49
Modules, 71, 278
Molded paper capacitors, 411
Multimeter, choosing, 123
  electronic, 126
  kits, 126
  repairs, 141
  replacing batteries, 140
  sensitivity, 124
  testing, 139
  troubles, 141
Multiple-section electrolytics, 421
Multipliers, 360
Multiplying factor, ohmmeter, 125, 137
Multirange tester, 122
Multiterminal parts, unsoldering, 277
Multitester, 122
Multivalue resistors, 363

Negative charges, 107
Nickel-cadmium batteries, 298
Noise test, tube, 240
Noisy tubes, 216
North pole, magnet, 119

Octal bases, 145
Ohmmeter, checking filament continuity, 212
  multiplying factor, 125, 137
  precautions in transistor circuits, 138, 304
  reading scale, 136
  replacing batteries, 140
  testing transistors, 288
Ohm's law, 114
Ohms (units), 112, 334
Omega, 136
On-off switches, 84, 390
Orientation, television antenna, 549
Oscillator interference table, 90
Outdoor antennas, radio, 555
Output transformers, action, 461
  ordering, 462
  ratings, 461
  testing, 462
Outside foil, 411

Paper capacitors, 68, 397, 404–413
  checking, 407–408
  failures, 404
  leakage resistance, 408
  ordering, 410
  outside foil, 411
  replacing, 412
Parallel resistors, 356
Parts jobbers, 7
Pencil-type soldering iron, 264
Pentodes, 149
Permanent magnets, 120
Permanent soldered joint, 256
Permeability-tuned trimmers, 481
Permeability tuning, 479
Phasing of voice coils, 507
Phillips screwdriver, 23, 25
Photoelectric remote controls, 493
Photofact folders, 15
Picture-centering magnets, 177
Picture controls, 84
Picture defects, 81
Picture tubes, brighteners, 28, 188
  built-in safety glass, 201
  cleaning, 103, 201
  color television, 203
  corona and arcing, 190
  credit for old, 201
  destroying, 202

Picture tubes, discharging second anode, 193
  8-inch check, 182
  extension cables, 197
  focus, 179
  gassy, 186
  halo light, 203
  handling, 96, 191
  installing, 197
  intermittent, 187
  ion-spot damage, 173
  loosening frozen yoke, 195
  measuring voltages, 189
  operation, 170
  rebuilt, 201
  removing, 193, 195
  resoldering pins, 185
  shorted, 186
  spot of light after turning off, 187
  stocking, 204
  testing, 181
  troubleshooting table, 183
  type numbers, 171
  warranties, 202
  weak, 186
Piggy-back television antennas, 530
Pilot lamps, 72, 166
  connections, 97
  testing, 236
Pin numbers, tube, 152
Pin straightener, tube, 162
Pincushion correction magnets, 176
Pins, resoldering picture-tube, 185
Pipe clamps, 549
Plastic cabinets, cleaning, 515
Plastic wire, 253
Pliers, combination, 24
  long-nose, 24
  slip-joint, 25
Plugs, polarized, 321
P-m dynamic speaker, 498
Pneumatic remote controls, 494
Pointer, dial, 95
Polarity, in auto radios, 325
  crystal diodes, 305
  electrolytic capacitors, 415
  selenium diodes, 307
  voltages, 116
Polarized line-cord plugs, 321
Poles, magnet, 119
Polishing cabinets, 514
Portable radios, 325–332
  tube stock, 226
  tube-type, 325
Positive charges, 107
Potentiometers, 67, 377
  action, 378

Potentiometers, dual, 380
  noisy, 381
  ordering, 382
  repair, 381
  shafts, 383
  stock, 388
  symbols, 378
  taper, 385
  testing, 382
Powdered-iron-core coils, 441
Power, 114
Power lines, a-c, 117
  d-c, 117, 321
  voltages, 308
Power ratings, resistor, 336
Power-supply trouble, 89
Power transformers, action, 464
  ordering, 466
  testing, 465
  troubles, 465
Power transistors, 286
  replacing, 291
Preset fine tuning, 481
Primary coil, 442
Printed circuits, construction, 262–264
  locating bad joints, 271
  locating breaks, 272
  measuring, 274
  removing, 270
  repairing cracked board, 282
  repairing wiring, 273
  soldering iron, 264
  troubleshooting, 271–275
  unsoldering parts, 275
Printed resistors, 71
  construction, 352
  replacing, 354
  testing, 353
Protons, 107
Punch, 33
Push-on connections, 261
Push-pull on-off switch, 383
Pushbutton tuning, 482

Q of coils, 449

Radio antennas, 551–558
Radio frequencies, 45
Radio set, trying out, 78
Radio-type remote controls, 493
Radio waves, 49
Radios, auto, 53, 294, 322–325
  clock, 52, 392–395
  consoles, 63
  f-m, 56, 494, 496
  portable, 325–332

Radios, transistor, 52, 292–294, 328
Range, television signal, 59
Ranges, multimeter, 125–137
Ratings, capacitor, 401
"RCA Receiving Tube Manual," 17
Reading meter scales, 128–136
Rebuilt picture tubes, 201
Receiver manufacturers, 10
Receiver service data, 13
Recentering voice coils, 504
Rechargeable batteries, 298, 330
Record-changer leads, 96
Record changers, removing, 100
Rectifiers, metallic, 72, 315
Reflectors, 526
Rejuvenators, battery, 332
Remote controls, television, 61, 488–494
Rescaps, 70, 353
Resistance, 111, 334
Resistance measurement, 134
Resistor-capacitor units, 70, 353
Resistors, 66
  action, 334
  ballast, 371
  color code, 344
  fuse, 363
  fusible, 313
  line-cord, 373
  multivalue, 363
  parallel, 356
  series, 356
  surge, 363
  tapped, 364
  wirewound, 358–376
Resonant capacitors, 413
R-f chokes, 446
R-f interference, 91
R-f oscillator, 51
R-f signals, 45
R-f transformers, 449
Rheostats, 67, 377
Rider's manuals, 15
Rivets, removing, 33
Roof mounts, 541
Rosin-core solder, 41, 250
Rotators, antenna, 533
Rotors, tuning capacitor, 470

Safety check, receiver, 133
Safety window, cleaning, 103
  removing, 201
Sample service data letter, 14
Sams Photofact folders, 15
Scale divisions, meter, 129
Scales, meter, 125
  reading, 128

Scratch-removing polish, 514
Scratch stick, 515
Screwdriver controls, 80
Screwdrivers, 25
Second-anode connector, picture-tube, 193
Secondary coil, 442
Selenium diodes, 307
Selenium rectifiers, construction, 315
   replacing, 319
   testing, 317
Selling tubes, 228
Sensitivity, multimeter, 124
Separate sound channel, 61
Series resistors, 356
Series-string filaments, 209, 314
Series-string tube substitution, 223
Service bench, 34
Service cement, 41
Service data, getting, 13
Service data letter, 14
Service data subscriptions, 17
Service manuals, 15
Service wholesale dealer, 18
Servicemen, organizations, 18
Servicing vehicle, 21
Set tester, 122
Shafts, control, 383
Shelf life, batteries, 332
   electrolytics, 418
Shelves, tube, 227
Shield, tube, 156
Shims, 504
Shock, normal receiver, 320
Short-wave region, 46
Shorted tubes, 163
Shorts test, tube, 239
Signal, carrier, 48
Signal frequencies, 45
Signal-seeking auto radios, 485
Silicon rectifiers, construction, 319
   replacing, 320
Silver-cadmium batteries, 298
Silvered mica capacitors, 434
Slide-rule dial-cord drive, 475
Slip-joint pliers, 25
Slo-Blo fuse, 312
Slotted plates, tuning capacitors, 471
Slow-acting fuses, 314
Slug tuning, 479
Socket wrenches, 25
Sockets, layout diagram, 156
   replacing, 155
   troubles, 154
   tube, 151
Solar cells, 331
Solder, 41
   choosing, 250

Solder pot, 267
Soldered joints, 255
Soldering gun, 24
   operation, 248
   use, 248–261
Soldering iron, pencil-type, 264
Soldering rules, 260
Solderless connections, 261
Solid wire, tinning, 254
Sound, 221
Sound system, intercarrier, 60
   separate-channel, 61
Sound waves, 45
Source, voltage, 108
South pole, magnet, 119
Spaghetti, 350
Spark plates, auto radio, 323
Speakers, 50, 498–510
   action, 499
   disconnecting, 95
   electrodynamic, 499
   electrostatic, 509
   grille cloth, 520
   ordering, 505
   p-m dynamic, 498
   replacing cones, 505
   test, 509
   transistor radio, 510
   troubles, 501
   wattage ratings, 506
Spin-type socket wrenches, 25
Splicing, coaxial cable, 548
   twin-line, 548
Stacked arrays, 530
Stage names, television receiver, 221
Station frequencies, 46
Station-selector switch, 84
Stators, tuning capacitor, 470
Stereo, home audio systems, 64
   records, 64
   tapes, 64
Stick shellac, 517
Stock, parts, 6
   resistors, 351
   tubes, 224
Storage shelves, tube, 227
Stranded wire, tinning, 254
Substitution test, television tube, 218
Superfringe antennas, 528
Superheterodyne receiver, 49
Surge resistors, 363
Surge voltage rating, electrolytics, 423
Surgistor, 313
Sweep circuits, 60
Switch, bench, 38
Switches, 68, 389–391
   clock radio, 391

Index

Switches, on controls, 389
  thermal-delay, 312
Symbols, capacitor, 403
  tube, 150
Sync control, 84
Sync stages, 221

Table, cements for plastic cabinets, 519
  electrolytic capacitor stock, 429
  letters in tube type numbers, 151
  picture-tube troubleshooting, 183
  pilot lamp replacement, 167
  pin numbers of sockets, 153
  radio troubles caused by tubes, 217
  recommended charges for parts, 12
  soldering rules, 260
  television channels, 47
  television controls, 82–85
  television interference, 90
  television receiver stage names, 221
  television troubles caused by tubes, 219
  transistor radio troubles, 300
Taper, potentiometers, 385
Tapped resistors, 364
Telephone calls, 74
Telephone ordering, 8
Television antennas, 59, 521–551
  choosing, 532
  installation, 534
  orientation, 549
  repairing, 550
Television broadcasting system, 57
Television camera, 58
Television channels, table, 47
Television controls, 81–85
Television i-f transformers, 455
Television interference, 90
Television oscillator interference table, 90
Television receiver operation, 59
Television receivers, stage names, 221
Television receiving antennas, 59
Television remote controls, 488
Television set, trying out, 79
Television tube troubleshooting chart, 219
Television tuners, 486
Temporary soldered joint, 256
Test leads, 126
Test speakers, 509
Testers, transistor, 287
Testing, multimeter, 139
  tubes free, 228
Tetrodes, 149
Thermal-delay switches, 312
Thermistors, 292, 314, 355
  deflection-yoke, 460

Three-way portables, operation, 326
  troubleshooting, 327
Timers, clock-radio, 393
Tinning soldering lugs, 255
Tinning wire, 254
Tolerances, resistor, 346
Tone control, 84
Toolbox, 26
Tools needed for servicing, 23
Top-cap connectors, 157
Tough receivers, fixing, 18
Transformer, isolation, 38
Transformerless television sets, 223
Transformers, 71, 449–458, 461–468
  operation, 442
  r-f, 449
Transistor batteries, 297
Transistor electrodes, 284
Transistor radios, 52, 292–294
  replacing batteries, 328
Transistor radio troubleshooting table, 300
Transistor receiver troubleshooting, 296
Transistor testers, 287
Transistors, 65
  auto radio, 293
  home radio, 295
  interchangeability, 286
  Japanese, 286
  ohmmeter precautions, 138
  operation, 284
  power, 286
  in remote controls, 294
  symbols, 285
  in television sets, 295
  testing with ohmmeter, 288
  type numbers, 286
  unsoldering, 280, 290
Transmission lines, 525
  installing, 543
Trimmer capacitors, 70, 480
Trimmers, 479
Triodes, 148
Troubleshooting, with ohmmeter, radio, 138
  three-way portables, 327
Troubleshooting chart, radio, 217
  television, 219
  transistor radio, 300
Truck, servicing, 22
Tube-base diagrams, 153
Tube bases, 144
Tube caddy, 27, 225
Tube clamps, 157
Tube manuals, 17
Tube pin straightener, 162
Tube-pulling tools, 158
Tube shields, 156

568   Television and Radio Repairing

Tube sockets, 151
  repairing, 280
  replacing, 281
Tube stock, 224
Tube storage shelves, 227
Tube tester, accuracy, 244
  choosing, 233
  gas test, 240
  kits, 246
  manufacturers, 245
  noise test, 240
  operation, 230
  roll chart, 234
  secondhand, 246
  shorts test, 239
Tube testing by substitution, 218
Tube troubles, 89
Tubes, 65
  buying, 228
  checking filaments, 210–215
  emergency repairs, 165
  faint markings, 166
  foreign, 165
  guarantees, 229
  identifying sockets, 211
  interchangeable, 164
  noisy, 216
  ordering, 164
  picture (see Picture tubes)
  replacing, 159
  selling, 228
  types, 226
Tubular ceramic capacitors, 429
Tubular paper capacitors, 397
  color code, 412
Tuner coils, 455
Tuners, 221
  exchange service, 45
  switch troubles, 487
  uhf, 488
  vhf, 486
Tuning action, 51
Tuning capacitors, 51, 70, 470
Tuning devices, 469–488
Tuning knobs, 472
Tv-phono switch, 84
Twin-line, installing, 543
  splicing, 548
Type numbers, tube, 150

Uhf antennas, 530
Uhf band, 46
Uhf tuners, 488
Ultrasonic remote controls, 489–493
Unsoldering, 251
  multiterminal parts, 277
  transistors, 290

Variable capacitor, 51
Vertical centering control, 85
Vertical hold control, 85
Vertical linearity control, 85
Vertical output transformer, 463
Vertical size control, 85
Vertical stages, 221
Vertical sweep circuit, 60
Vhf television antennas, 528–530
Vibrators, replacing, 322
Video amplifier, 60, 221
Video detector, 60
Video i-f, 221
Video peaking coils, 446
Vise, 34
Voice coils, 500
  off-center, 503
  phasing, 507
  recentering, 504
Volt-ohm-milliammeter, 122
Volt-ohmmeter, 122
Voltage drop, nature of, 115
Voltage sources, 108
Voltage values, typical, 109
Volts, 109
Volume control, 85
VOM, 122

Warmup, filament, 208
Warranties, picture tube, 202
  receiving tube, 229
Watts, 114
Wave traps, 447
Waves, radio, 49
Weller soldering gun, 23
Wholesale dealer service, 18
Width coils, 448
Width control, 85
Width-control sleeve, 460
Window antennas, television, 524
Wire-wrap connections, 260
Wirewound potentiometers, 380
Wirewound resistors, 67, 358–376
  color codes, 360
  construction, 358
  installing, 362
  power ratings, 362
Wood cabinets, cleaning, 514
Workbench antennas, 41
Workbench design, 34
Workbench lighting, 39
Workbench tools, 30
Wrenches, socket, 25

Yagi antennas, 527
Yoke, frozen, 195

Zero adjustment, meter, 140